Energy Conservation

Energy Conservation

S. C. Bhatia

BE (Chemical), MBA

Edited by

Sarvesh Devraj

B.Tech (Mechanical), UPTU
M.Tech (Renewable Energy Engineering and Management), TERI University,
(Research Associate – TERI, New Delhi)

Published by Woodhead Publishing India Pvt. Ltd.
Woodhead Publishing India Pvt. Ltd.,
303, Vardaan House, 7/28, Ansari Road,
Daryaganj, New Delhi - 110002, India
www.woodheadpublishingindia.com

First published 2016, Woodhead Publishing India Pvt. Ltd.
© Woodhead Publishing India Pvt. Ltd., 2016

Woodhead Publishing India Pvt. Ltd. ISBN: 978-93-85059-23-0

Woodhead Publishing India Pvt. Ltd. e-ISBN: 978-93-85059-73-5

Typeset by Asian Enterprises, New Delhi
Printed and bound by Replika Press

Contents

Preface

Energy is an essential basic need for not only human beings, but also for national economic and social development. Energy conservation promises to fill the gap between supply and demand. Several measures for conservation of energy are very important for consideration. There is a good scope of energy conservation in various sectors, viz., industry, agriculture, transport and domestic. The energy audit can unearth huge profits to the industry. Saving of usable energy, which is otherwise wasted, has direct impact on economy, environment and long term availability of non-renewable energy resources. Energy conservation implies reduction in energy consumption by reducing losses and wastage by employing energy efficient means of generation and utilisation of energy.

This book on Energy Conservation summarises various aspects of energy consumption and conservation and is divided into 21 chapters.

Chapter 1 deals with energy conservation: A review. Chapter 2 is devoted to energy savings in industry as energy plays a key role in achieving the desired economic growth.

Chapter 3 focuses on energy conservation in agricultural sector, which is the major consumption of energy such as-farm equipments, irrigation pumps and motors, etc. Chapter 4 concentrates on energy conservation in food industry, various energy consuming and energy saving aspects in baking, biscuits and brewery are discussed.

Chapter 5 focuses on energy conservation in dairy industry. Approximately eighty per cent of the energy requirements are for thermal uses to generate hot water and produce steam for process applications (e.g. pasteurisation, evaporation, and milk drying) and cleaning purposes. Chapter 6 deals with energy conservation in chemical process industries. Chemical industry is the fountain head for the manufacture of a variety of chemical products beginning with products like soda ash, paper, plastics, petrochemicals, petroleum products, organic and fine chemicals, fertilisers and pharmaceuticals.

Chapter 7 concentrates on energy conservation in pharmaceutical industries. Variety of opportunities exist within pharmaceutical laboratories, manufacturing facilities, and other buildings to reduce energy consumption while maintaining or enhancing productivity. Chapter 8 focuses on energy conservation in textile industry. Textile industry uses large quantities of electricity and fuels. There are significant losses of energy within various operations of textile plants such as spinning, weaving and dyeing. Various energy conservation/saving aspects of textile processing are also discussed. Chapter 9 deals with energy conservation in cement, ceramic and glass. Various energy saving aspects in kilns, spray dryers, boilers are discussed in detail.

Chapter 10 is devoted to energy conservation in electrical system. Various energy efficient techniques in transformers, electronic ballast, adjustable speed drive (ASD) and DG sets are discussed. Chapter 11 concentrates on energy conservation in thermal power plants. There is tremendous scope for energy potential in various areas of thermal power plants including waste heat recovery for power. Chapter 12 focuses on energy efficiency in motors, fans and compressors. When considering energy-efficiency improvements to motor systems, a systems approach incorporating pumps, compressors, and fans must be used in order to attain optimal savings and performance.

Chapter 13 deals with energy efficiency in boilers. Various energy efficiency opportunities in boiler system can be related to combustion, heat transfer, avoidable losses, high auxiliary power consumption, water quality and blow down. Chapter 14 is devoted to heating, ventilation and air conditioning systems. Replacing energy inefficient systems with energy-efficient alternatives and using HVACs sparingly offer huge opportunities for companies to save electricity in many instances.

Chapter 15 concentrates on energy conservation in pumps. Pumping systems account for a significant percentage of energy consumption of the total industrial energy usage. Various energy saving options in pumping systems are discussed. Chapter 16 focuses on energy conservation in iron and steel industry. Various energy conservation technologies are discussed related to arc, melting and reheating furnaces along with various heat recovery operations.

Chapter 17 is devoted to energy conservation in furnaces. Energy consuming and saving aspects in induction and melting furnaces along with waste heat recovery methods are discussed. Chapter 18 deals with energy conservation and nanotechnology. Nanotechnology provide the potential to enhance energy efficiency across all branches of industry and to economically leverage renewable energy production through new technological solutions and optimised production technologies. Chapter 19 is devoted to industrial waste heat recovery. Chapter 20 concentrates on energy audit. Energy audit is the key to a systematic approach for decision making in the area of energy management. It attempts to balance the total energy inputs with its use and serves to identify all the energy streams in a facility. Chapter 21 focuses on case studies. Three case studies related to: Honeywell Farms Dairy Milk—(USA), Tata Power Company Ltd.—Mumbai and Hindustan Petroleum Corporation Ltd. (HPCL)—Mumbai are discussed.

Such wide coverage makes this book a treatise on the subject. Diagrams, figures, tables and index supplement the text. All topics have been covered in a cogent and lucid style to help the reader grasp the information quickly and easily.

This book could not have been completed without the help of Mr Aman Bhatia (my nephew) who worked hard in locating and organising the material and spent many hours checking the manuscript. Appreciations are also extended to Mr Harinder Singh, Senior DTP operator, who drew and labelled the flow diagrams and worked long hours to bring the book on time. I am thankful to Mr Sarvesh Devraj (Research associate – TERI, New Delhi) who helped me in editing the book. I am also thankful to the editorial team of Woodhead Publishing India Pvt. for their wholehearted cooperation in bringing out the book in time.

It may not be wrong to hold that this book on *Energy Conservation* is essential reading for professionals and students pursuing engineering courses. Besides students, this book will prove useful to industrialists and consultants in the respective fields.

It has been prepared with meticulous care, aiming at making the book error-free. Constructive suggestions are always welcome from users of this book.

S. C. Bhatia

1

Energy conservation: A review

1.1   Introduction

'Energy conservation' and 'energy efficiency' are often used interchangeably, but there are few differences. At the most basic level, energy conservation means using less energy and is usually a behavioural change, like turning lights off or setting thermostat lower. Energy efficiency, however, means using energy more effectively, and is often a technological change. Energy efficiency measures the difference between how much energy is used to provide the same level of comfort, performance or convenience by the same type of product, building or vehicle.

Conservation certainly reduces energy use, but it's not always the best solution because it may impact comfort or safety as well. Efficiency, on the other hand, maintains the same level of output (e.g. light level, temperature) but uses less energy to achieve it. A combination of both energy conservation and energy efficiency measures yields an ideal solution.

1.2   Industrial sector energy efficiency

The industrial sector represents more than one third of both global primary energy use and energy-related carbon dioxide emissions. In developing countries, the portion of the energy supply consumed by the industrial sector is frequently in excess of 50% and can create tension between economic development goals and a constrained energy supply. Further, countries with an emerging and rapidly expanding industrial infrastructure have a particular opportunity to increase their competitiveness by applying energy-efficient best practices from the outset in new industrial facilities.

Integrating energy efficiency into the initial design or substantial redesign is generally less expensive and allows for better overall results than retrofitting existing industrial facilities, as is typically required in more developed countries. Conversely, failure to integrate energy efficiency in new industrial facility design in developing countries represents a large and permanent loss in climate change mitigation potential that will persist for decades until these facilities are scheduled for major renovation.

Despite the potential, policymakers frequently overlook the opportunities presented by industrial energy efficiency to have a significant impact on climate change mitigation, security of energy supply, and sustainability. The common perception holds that energy efficiency of the industrial sector is too complex to be addressed through public policy and, further, that industrial facilities will achieve energy efficiency through the competitive pressures of the marketplace alone. Neither premise is supported by the evidence from countries that have implemented industrial energy efficiency programmes. The opportunities for improving the efficiency of industrial facilities are substantial, on the order of 20–30% even in markets with mature industries that are relatively open to competition. The principal business of an industrial facility is production, not energy efficiency. This is the underlying reason why market forces alone will not achieve industrial energy efficiency on a global basis, 'price signals' notwithstanding. High energy prices or constrained energy supply will motivate industrial facilities to try to secure the amount of energy required for operations at the lowest possible price. But price alone will not build awareness within the corporate culture of the industrial firm of the potential for the energy savings, maintenance savings and production benefits that can be realised from the systematic pursuit of industrial energy efficiency. It is this lack of awareness and the corresponding failure to manage energy use with the same attention that is routinely afforded production quality, waste reduction and labour costs that is at the root of the opportunity.

Industrial energy efficiency is dependent on operational practices, which change in response to variations in production volumes and product types. Due to this dependence, industrial energy efficiency cannot be fully realised through policies and programmes that focus solely on equipment components or specific technologies. Companies that actively manage their energy use seek out opportunities to upgrade the efficiency of equipment and processes because they have an organisational context that supports doing this wherever cost effective, while companies without energy management policies do not. Providing technology-based financial incentives in the absence of energy management will not result in significant market shifts because there is no organisational context to respond to and integrate the opportunity into ongoing business practice. A portfolio of industrial policies is needed that is designed to assist companies in developing this supporting context, while also providing consistency, transparency, engagement of industry in programme design and implementation, and, most importantly, allowance for flexibility of industry response. When these criteria are met, industry has shown that it can exceed expectations as a source of reductions in energy use and corresponding greenhouse gas (GHG) emissions, while continuing to prosper and grow.

1.2.1 Industrial sector trends

The industrial sector uses 160 exajoules (EJ) of global primary energy, which is about 37% of total global energy use. Primary energy includes upstream energy losses from electricity, heat, petroleum and coal products production. The industrial sector is extremely diverse and includes a wide range of activities. This sector is particularly energy intensive, as it requires energy to extract natural resources, convert them into raw materials, and manufacture finished products. The industrial sector can be broadly defined as consisting of energy-intensive industries (e.g. iron and steel, chemicals, petroleum refining, cement, aluminium, pulp and paper) and light industries (e.g. food processing, textiles, wood products, printing and publishing, metal processing). The aggregate energy use depends on technology and resource availability, but also on the structure of the industrial sector. The share of energy-intensive industry in the total output is a key determinant of the level of energy use.

Economic development trends

Historical trends show that the importance of industry within an economy varies by its stage of economic development. The structure of the industrial sector varies between countries and their level of development since the materials demanded by an economy differ through successive stages of development. Industrialisation drives an increase in materials demand for construction of basic infrastructure needs such as roads, railways, buildings, power grids, etc. As countries develop the demand for energy conservation increases.

Even though these general trends can be observed, economic development trends vary by country and there is no standard development path. India, for example, has a very high share of the service sector, accounting for 51% of total value added; even so, the industrial sector continues to grow, particularly in material production.

1.3 Energy consumption and energy-related carbon dioxide emissions trends

Energy use in the industrial sector varies widely between countries and depends principally on the level of technology used, the maturity of plants, the sector concentration, the capacity utilisation and the structure of subsectors. A recent study compares regional levels of energy use intensities in 2011 and calculates that if all developing countries met the developed country average manufacturing energy use intensity, energy consumption could potentially be reduced by 70%.

The largest emissions are from industrial energy use in the Centrally Planned Asia region, with more than a third of global CO_2 emissions, due to increasing energy-intensive industrial production and the heavy use of coal in the industrial and power sectors. Developed countries account for 35% and transition countries for 11% of global CO_2 emissions from the industrial sector, while the remaining countries account for 54%.

1.4   Industrial energy efficiency

Industrial energy efficiency—or conversely, energy intensity, which is defined as the amount of energy used to produce one unit of a commodity—is determined by the type of processes used to produce the commodity, the vintage of the equipment used, and the efficiency of production, including operating conditions. Energy intensity varies between products, industrial facilities, and countries depending upon these factors.

Steel, for example, can be produced using either iron ore or scrap steel. Best practice primary energy intensity for producing thin slab cast steel from iron ore using a basic oxygen furnace is 16.3 gigajoules (GJ) per metric ton, while production of the same product using scrap steel only requires 6.0 GJ/T. The energy intensity of the Chinese steel industry dropped from 29 GJ/T steel in 2003 to 23 GJ/T steel in 2011 despite an increased share of primary steel production from 79% to 84%, indicating that the efficiency of steel production improved over this period as small, old inefficient facilities were closed or upgraded and newer facilities were constructed.

Within industry, systems that support industrial processes that can be found to varying degrees in virtually all industrial sectors, regardless of their energy intensity. These industrial systems, which include compressed air, pumping, and fan systems (referred to collectively as motor systems), steam systems, and process heating systems are integral to the operation of industrial facilities, providing essential conversion of energy into energised fluids or heat required for production processes.

Motor and steam systems account for 15% and 38%, respectively, of global final manufacturing energy use, or approximately 46 EJ/year.

Because these systems typically support industrial processes, they are engineered for reliability rather than energy efficiency. Industrial systems that are oversized in an effort to create greater reliability, a common practice, can result in energy lost to excessive equipment cycling, less efficient part load operation, and system throttling to manage excessive flow. Waste heat and premature equipment failure from excessive cycling and vibration are side effects of this approach that contribute to diminished, not enhanced, reliability.

More sophisticated strategies, made possible through the emergence of modern controls, create reliability through flexibility of response—and redundancy in the case of equipment failure—rather than by brute force. The energy savings can be substantial, with savings of 20% or more common for motor systems and 10% or more for steam and process heating systems.

1.4.1   Opportunities for industrial energy efficiency

Opportunities to improve industrial energy efficiency are found throughout the industrial sector. Assessments of cost-effective efficiency improvement opportunities in energy-intensive industries in the United States, such as steel, cement and paper manufacturing, found cost-effective savings of 16% to 18% even greater savings can often be realised in developing countries where old, inefficient technologies have continued to be used to meet growing material demands. An estimate of the 2010 global technical potential for energy efficiency improvement in the steel industry with existing technologies identified savings of 24% in 2010 and 29% in 2020 using advanced technologies such as smelt reduction and near net shape casting.

While the energy efficiency of individual system components, such as motors (85–96%) and boilers (80–85%) can be quite high, when viewed as an entire system, their overall efficiency is quite low. Motor systems lose approximately 55% of their input energy before reaching the process or end use work and steam systems lose 45%. Some of these losses are inherent in the energy conversion process; other losses are due to system inefficiencies that can be avoided through the application of commercially available technology combined with good engineering practices.

At present, most markets and policymakers tend to focus on individual system components (motors and drives, compressors, pumps, boilers) with an improvement potential of 2–5%—that can be seen, touched and rated (rather than systems).

While systems have impressive improvement potential—20% or more for motor systems and 10% or more for steam and process heating systems—achieving this potential requires engineering and measurement.

The presence of energy-efficient components, while important, provides no assurance that an industrial system will be energy efficient. System optimisation requires taking a step back to determine what work (process temperature maintained, production task performed, etc.) needs to be performed.

Improved energy system efficiency can also contribute to a company's bottom line by increased production through better utilisation of equipment assets, greater reliability and reduced maintenance costs. Payback periods for system optimisation projects are typically short—from a few months to three years—and involve commercially available products and accepted engineering practices.

1.4.2 Barriers to industrial energy efficiency improvement

The decision-making process regarding investments in energy-efficient technologies is shaped by firm rules, corporate culture and the company's perception of its level of energy efficiency. Researchers found that most firms view themselves as energy efficient even when profitable improvements are available. Lack of knowledge or the limited ability of industrial commodity producers to research and evaluate information on energy-efficient technologies and practices is another barrier. Uncertainties related to energy prices or capital availability can lead to the use of stringent investment criteria and high hurdle rates for energy efficiency investments that are higher than the cost of capital to the firm. Capital rationing is often used within firms as an allocation means for investments, especially for small investments such as many energy efficiency retrofits. The relatively slow rate which industrial capital stock turns over can prove to be a barrier to adoption of energy efficiency improvements since new stock is typically more energy-efficient than existing facilities. Another barrier is the perceived risk involved with adopting new technology since reliability and maintenance of product quality are extremely important to commodity producers.

Optimising industrial systems for energy efficiency is not taught to engineers and designers at university—it is learned through experience. Systems are designed to maintain reliability at the lowest first cost investment, despite the fact that operating costs are often 80% or more of the life cycle cost of the equipment. Facility plant engineers are typically evaluated on their ability to avoid disruptions and constraints in production processes, not energy-efficient operation. Equipment suppliers also have little incentive to promote more energy-efficient system operation, since commissions increase when equipment size is scaled upward and educating a customer to choose a more efficient approach requires extra time and skill.

Plant engineering and operations staff frequently experience difficulty in achieving management support. Industrial managers are rarely drawn from the ranks of facilities operation—they come from production and often have little understanding of supporting industrial systems. This situation is further exacerbated by the existence of a budgetary disconnect in industrial facility management between capital projects (including equipment purchases) and operating expenses. In addition, most optimised industrial systems lose their initial efficiency gains over time due to personnel and production changes. Detailed operating instructions are not integrated with quality control and production management systems. Without well documented maintenance procedures, the energy efficiency advantages of high efficiency components can be negated by clogged filters, failed traps and malfunctioning valves.

1.4.3   Energy efficiency

Typically, the process for setting energy efficiency or GHG emission reduction targets requires a preliminary assessment of the energy efficiency or GHG mitigation potential of each industrial facility, which includes an inventory of economically viable measures that could be implemented.

Identification of energy-saving technologies and measures

Countries with strong industrial energy efficiency programmes, whether or not they are associated with agreement programmes, provide information on energy efficiency opportunities through a variety of technical information sources including fact sheets, brochures, guidebooks, technical publications, energy efficiency databases, software tools and industry- or technology-specific energy efficiency reports.

Benchmarking

Benchmarking provides a means to compare the energy use within one company or plant to that of other similar facilities producing similar products or to national or international best practice energy use levels. Benchmarking can compare plants, processes or systems.

Energy efficiency audits or assessments

Energy efficiency audits or assessments involve collecting data on all of the major energy-consuming processes and equipment in a plant, documenting specific technologies used in the production process and identifying opportunities for energy efficiency improvement throughout the plant. An audit is an essential first step in identifying opportunities that can contribute to an organisation's energy efficiency targets.

Energy saving action plans

An energy action plan outlines a company's plan for improving energy efficiency during the period covered by energy efficiency targets and is a required component of compliance with an energy management standard. The energy action plan provides primary guidance for the internal implementation of the activities that will be undertaken to reach the energy-saving target. It also serves as a reference to evaluate progress on an annual basis. The plan, which is typically reviewed by an independent third party and updated as needed in response to changes over time, includes a description of the facility's energy uses, a description of the energy efficiency measures considered, a description of the planned energy efficiency measures, a time-frame for implementation of the energy efficiency measures, and expected results in terms of energy efficiency.

Monitoring

Monitoring guidelines for energy efficiency and GHG mitigation projects have been developed by numerous entities in order to understand the progress and results of specific energy efficiency projects.

1.5   Energy management standards

The purpose of an energy management standard is to provide guidance for industrial facilities to integrate energy efficiency into their management practices, including fine tuning production processes and improving the energy efficiency of industrial systems.

An energy management standard requires a facility to develop an energy management plan. In companies without a plan in place, opportunities for improvement may be known but may not be promoted or implemented because of organisational barriers. These barriers may include a lack of communication among plants, a poor understanding of how to create support for an energy efficiency project, limited finances, poor accountability for measures.

Typical features of an energy management standard include:

1. A strategic plan that requires measurement, management and documentation for continuous improvement for energy efficiency.
2. A cross-divisional management team led by an energy coordinator who reports directly to management and is responsible for overseeing the implementation of the strategic plan.
3. Policies and procedures to address all aspects of energy purchase, use and disposal.
4. Projects to demonstrate continuous improvement in energy efficiency.
5. Creation of an Energy manual, a living document that evolves over time as additional energy saving projects and policies are undertaken and documented.
6. Identification of key performance indicators, unique to the company, that are tracked to measure progress.
7. Periodic reporting of progress to management based on these measurements.

1.5.1   System optimisation and capacity building

System optimisation seeks to design and operate industrial systems (i.e. motor/drive, pumping, compressed air, fan, and steam systems) to provide excellent support to production processes using the least amount of energy that can be cost-effectively achieved.

The process of optimising existing systems includes:

1. Evaluating work requirements.
2. Matching system supply to these requirements.
3. Eliminating or reconfiguring inefficient uses and practices (throttling, open blowing, etc.).

4. Changing out or supplementing existing equipment (motors, fans, pumps, compressors) to better match work requirements and increase operating efficiency.
5. Applying sophisticated control strategies and variable speed drives that allow greater flexibility to match supply with demand.
6. Identifying and correcting maintenance problems.
7. Upgrading ongoing maintenance practices.

A system that is optimised to both energy efficiency and cost effectiveness may not use the absolute least amount of energy that is technically possible. The focus is on achieving a balance between cost and use that applies energy resources as efficiently as possible.

Building technical capacity: The goal of capacity-building is to create a cadre of highly skilled system optimisation experts.

Documenting for sustainability: With the renewed interest in energy efficiency worldwide and the emergence of carbon trading there is a need to introduce greater transparency into the way that industrial facilities identify, develop and document energy efficiency projects. In order to ensure persistence for energy efficiency savings from system optimisation projects, a method of verifying the ongoing energy savings under a variety of operating conditions is required.

Tax and fiscal policies

Tax and fiscal policies for encouraging investment in energy-efficient industrial equipment and processes operate either through increasing the costs associated with energy use to stimulate energy efficiency or by reducing the costs associated with energy efficiency investments. Various forms of these instruments have been tried in numerous countries over the past three decades, including energy or CO_2 taxes; grants and subsidies such as energy efficiency loans and innovative funding mechanisms; and tax relief such as accelerated depreciation; tax rebates, deductions, exemptions. In addition, integrated policies that combine a variety of financial incentives in a national-level energy or GHG emissions mitigation programme are also found in a number of countries. Such integrated policies are often national-level energy or GHG programmes that combine a number of tax and fiscal policies along with other energy efficiency mechanisms such as voluntary agreements.

Energy or energy-related carbon dioxide (CO_2) taxes have been used in a number of countries to provide an incentive to industry to improve the energy management at their facilities through both behavioural changes and investments in energy-efficient equipment. Taxes on energy or energy-related CO_2 emissions are now found in Austria, the Czech Republic, Denmark, Estonia,

Finland, Germany, Italy, the Netherlands, Norway, Sweden, Switzerland and the United Kingdom.

1.6 Capacity-building through training experts and suppliers

A comprehensive training programme is typically required to create a cadre of system optimisation experts who are prepared to identify energy efficiency measurements and to develop efficiency improvement projects. For maximum effectiveness, the training should be targeted to plant and consulting engineers, as well as equipment suppliers.

1.6.1 Experts training

The purpose of this training is to prepare a group of experts who will be expected to: (i) Provide awareness training to encourage plants to undertake system optimisation improvements, (ii) conduct plant assessments to identify system optimisation opportunities, (iii) Work with plants to finance and develop projects based on these findings and (iv) Prepare case studies of successful projects. A one-to-one, one-to-many, training and implementation scheme has been tested and proven effective.

In this approach, international experts are engaged in the initial capacity-building to create a core of highly-skilled experts who will become a resource to their country and the region for years to come. To ensure success of the training, selection of the individuals to be trained must be rigorous and based on technical and training capabilities.

Successfully negotiating this selection process will require the international team and the country coordinators to develop a shared vision of the project goals, which will vary somewhat from country to country in response to cultural, organisational, and social requirements. This cadre of experts will form the nucleus for future training of additional experts as well as conduct awareness training for factory personnel.

1.6.2 Suppliers training

Concurrent with experts training, training should be conducted to introduce equipment suppliers, manufacturers' representatives and vendors to system optimisation techniques. The purpose of this training is to prepare manufacturers, suppliers, and vendors to: (i) participate in reinforcing the system optimisation message with their customers and (ii) assist them in identifying what will be required to reshape their market offerings to reflect a system services approach. Combining the expert training and vendor training is not recommended, as their needs are different.

1.7 Building industrial awareness

A core element of any industrial energy efficiency programme is an information campaign. This campaign is designed to introduce industry to the basic concepts of energy management and industrial system optimisation. The message needs to be appropriate to plant managers and needs to make a direct link between industrial energy efficiency and cost savings, improved reliability, and greater productivity. If international corporations have already established or plan to establish industrial facilities in the country, they may be important allies in this campaign. Once the in-country system optimisation experts have been trained, additional awareness messages will be needed to help them build the market for system optimisation services. It is important for the government to be active during this early stage of market transformation by hosting factory awareness training sessions as part of the programme response to the announcement of the energy management standard. A list of the trained experts can be kept and made available to companies seeking energy efficiency services.

1.7.1 Developing and enabling partnerships

For an industrial energy efficiency policy to become effective, government officials will need to form partnerships.

These enabling partnerships are needed to:

1. Build ownership in the proposed efforts to change existing practices and behaviours for greater energy efficiency.
2. Reach many industrial firms with the energy efficiency message through existing business relationships (such as with suppliers, trade associations, etc.).
3. Develop credibility within specialised industrial sectors.
4. Ensure that proposed policies are practical given the current situation of industry in the country.
5. Engage the financial community and assist them in understanding the financial benefits of industrial energy efficiency.
6. Recruit the best talent to become trained in system optimisation techniques.
7. Successfully launch an industrial energy efficiency programme.

The specific organisations that make effective partners will vary from country to country, but generally include: industrial trade associations, professional engineering societies or associations, equipment manufacturers and suppliers and their associations, leading and/or growing industrial companies, energy suppliers, technical universities and commercial lenders.

To sum up industrial energy efficiency is frequently overlooked by policymakers concerned about energy supply and use. Although designing an industrial energy efficiency programme takes time and must be undertaken with some care, the opportunities for improving the efficiency of industrial facilities are substantial, even in markets with mature industries that are relatively open to competition. Developing countries with an emerging and expanding industrial infrastructure have a particular opportunity to mitigate GHG emissions while increasing their competitiveness by applying energy-efficient best practices from the outset in new industrial facilities.

Evaluations of experience with target-setting agreements show that while results have been varied, the more successful programmes have seen significant energy savings even resulting in a 50% increase over historical autonomous energy efficiency improvement rates and they can be cost-effective.

These agreements have important longer-term impacts including changes of attitudes and awareness of managerial and technical staff regarding energy efficiency. Overall, international experience shows that target-setting agreements are an innovative and effective means to motivate industry to improve energy efficiency and reduce related emissions, if implemented within a comprehensive and transparent framework.

International experience with energy management standards in industry has been very positive. Because energy management standards have only been in force since 2000 or later, most programmes have not yet been subject to an independent evaluation. Their effectiveness can be inferred by the number of companies that seek affiliation with them, even when there is no penalty assessed for non-participation. Once a company meets the requirements of an energy management standard—establishing a cross divisional management team led by an energy coordinator who reports directly to management; establishing a strategic plan that requires measurement, management and documentation for continuous improvement for energy efficiency; developing policies and procedures to address energy purchase, use and disposal; initiating projects to reduce energy use on an ongoing basis; establishing key performance indicators to measure progress, and regularly documenting and reporting this progress—energy efficiency becomes part of organisational culture. This is the goal of the industrial standards framework.

System optimisation offers a way for companies to quickly realise cost, productivity and operational benefits that can provide the reinforcement needed for management to proceed with the organisational changes required to fully integrate energy efficiency into daily operational practices. Capacity building training creates a cadre of highly skilled system optimisation experts that can provide the necessary technical assistance for industrial facilities to identify and develop energy efficiency improvement projects.

The system optimisation library standardises and streamlines the process of developing and documenting energy efficiency improvement projects, while also increasing the likelihood that the resulting energy savings will be sustained over the project life (typically 10 years or more).

Evidence of sufficient documentation to support the persistence of energy savings is a critical prerequisite for considering energy efficiency projects for emerging financial instruments such as white certificates or carbon credits. Taken together, these elements comprise an effective industrial policy package that combines energy reduction targets, energy efficiency standards, system optimisation training, and documenting for sustainability. The industrial sector represents more than one third of global primary energy use and 36% of carbon dioxide emissions.

Energy savings in industry

2.1   Introduction

Industrial sector is the major consumer of energy in modern society, accounting for roughly 40% of final energy use. Coal or oil are heavily used, especially by primary industry, manufacturing and refining. Finally, energy efficiency is an important component of a company's environmental strategy. End-of-pipe solutions are often expensive and inefficient, while energy efficiency can often be the cheapest opportunity to reduce pollutant emissions. Gas is being used increasingly to replace coal because it is a cleaner fuel producing less impact on the environment. Electricity is only a minor component of industrial energy use, although its use in driving electric motors is very important.

The major sectors within industry can be categorised as follows:

Manufacturing: This includes the processing of primary resources into consumer products. Mineral refining, oil refining and chemical manufacturing are some areas of energy use where considerable savings could be made. Such activities often occur in the industrial zones of major cities.

Power generation: The power generation industry is a massive user of fossil fuels and accounts for more than 50% of international greenhouse gas emissions. Many power stations are very inefficient and there are strong economic and environmental incentives to save energy in the power supply industry. Most cities have major power stations and these are often a cause of air pollution as well.

Mining: This is a primary industry which generally occurs outside cities, often in remote parts of the country. Energy intensity is high in most mining operations but there is an incentive to save energy because energy wastage is reflected in the cost of the minerals.

Agriculture: Another major user of primary energy which takes place in rural areas.

Construction: Construction is a modest user of energy, particularly liquid fuels because this activity often takes place at sites where electric power is not readily available. Considerable savings are available in this sector because there is often a large amount of wastage in construction activities.

The main focus will therefore be on energy savings in manufacturing and power generation as these are the major users of industrial energy in cities.

2.2 Energy auditing in industry

Energy auditing in industry takes a similar approach to audits undertaken in the commercial sector:

1. An analysis of existing energy consumption records to determine where, how and how much energy is being used in the plant. It will also seek to identify trends in consumption data.

2. A walk through audit that documents where the main areas of energy consumption exist within the plant. This phase will identify any obvious areas of wastage together with the most promising areas for potential savings.

3. A detailed analysis phase which will take the data obtained in the previous two phases and prepare detailed plans for energy savings options. These plans will include details on the energy use and cost of each stage of the production process as well as costings and expected payback periods of the various energy saving options proposed.

In the case of the industrial sector, the main focus should be level 3 auditing (as mentioned above), where the individual processes are analysed, for example, the production of steam for use in commercial laundries. Although level one auditing, which focuses on the analysis of energy use through and investigation of the tariff structure of existing energy purchases, should be undertaken, the greatest potential for savings in the industrial sector will usually revolve around the selection, operation and maintenance of efficient equipment in the process.

2.2.1 Planning for energy efficiency in industrial processes

Once an initial energy audit has been undertaken, it will provide an important first step in monitoring and achieving the progress towards energy efficiency goals. This information is the baseline energy consumption, or the energy usage associated with current practices in the factory as well as existing equipment. Known as T_0, this is the energy consumption prior to any systematic energy efficiency measures being undertaken.

In conjunction with the result from the screening survey, the establishment of this baseline information allows energy managers to set targets for reduced energy consumption which can be achieved through changes in the management and operation of the industrial process as well as targets which would be possible through the implementation of energy efficient technologies.

Short term energy efficiency targets

Energy efficiency targets, which can be achieved in the short term, as a result of streamlined operation of the plant, are known as T_1, or housekeeping targets.

These energy savings will usually be the result of the efficient use of energy consuming equipment, a reduction in the amount of waste energy, timely maintenance of equipment and continual monitoring of the energy consumption of the industrial process.

Long term energy efficiency targets

Further reductions in energy consumption which can only be achieved through purchase with a high capital cost are known as T_2, or investment targets, and should ideally be based on the lowest energy consumption of best practice examples of similar industrial processes. As the purchase of expensive capital equipment is required to achieve these targets, careful modelling should be undertaken to ensure that the investment is sound, i.e. that the payback period of the equipment is not greater than the working life of the equipment.

The establishment of investment targets is a complex process, requiring a large amount of technical knowledge of similar industries, the options available for energy savings through the investment in new capital, as well as knowledge and skills in economic modelling. However, a large number of international best practice examples and case studies from industrialised nations, particularly in Europe and the United States of America, has increased the amount of information and data available on international best practice. Much of this information is available through CADDET–Energy Efficiency, the US Office of Industrial Technologies and the World Energy Efficiency Association.

Innovation energy efficiency targets

Energy efficiency is an area of increasing technological innovation and some consideration should also be given to setting T_3, or innovation targets. These targets are based on the energy consumption of state of the art technologies, which are still economically viable. Innovation targets, whilst not immediately achievable, may become achievable in the medium to long term as a result of changes in the economic environment (i.e. greatly increased profitability of the industry), the production environment (i.e. the need for a higher quality or specialised product for niche markets) or regulatory changes (i.e. the introduction of legislation governing pollution control, energy consumption or the Kyoto protocol).

2.3   Strategies for energy savings in industry

The strategies for achieving energy savings in industry are quite different to those for most other sectors. Industry is very diverse and is often controlled by very large multi-national corporations. In this context, the appropriate approach needs to be carefully considered. Industry is generally receptive to efforts to cut its energy costs, but it is less likely to be attracted to regulatory measures which increase its operating costs.

2.3.1 Technical options

The technical options available for energy savings in the industrial sector are as diverse as the industries themselves. However, they principally revolve around the saving of energy in areas such as:

1. Electric motors
2. Compressed air
3. Steam
4. Furnaces
5. Heat recovery

The production of onsite power and heat (or steam) through Cogeneration systems, or Combined Heat and Power (CHP) systems can also result in energy savings, through the utilisation of waste energy associated with the production of power. Opportunities for energy savings in relation to the operation and maintenance of industrial buildings also exist, although these are often similar to the commercial sector.

2.3.2 Energy savings and electric motors

Electric motors usually account for almost half of total industry energy consumption, and represent a significant opportunity for financial savings from energy consumption.

Four areas offer potential savings with regard to the selection and operation of electric motors:

1. Energy efficient motors (T_2).
2. Variable speed drives (T_2).
3. Correctly size motors (T_2).
4. Regular maintenance (T_1).

Energy efficient motors

By definition, one motor will be more efficient than another motor if it uses less energy to produce the same rated output. Most energy efficient motors are usually constructed with higher quality materials and advanced manufacturing techniques and result in less waste energy being produced through reduced vibration, noise and heat. Few countries have adopted minimum energy performance standards for new electric motors, many others have developed standards, which motors must meet in order to be sold as energy efficient motors. These regulatory measures offer the potential for long-term savings, although are unlikely to result in wide scale energy reductions in the short term as they are rarely retroactive, relating only to future purchases, which may be made five to ten years in the future.

Variable speed drives

Electric motors, which are able to operate at different speeds according to the amount of power supplied to the drive unit, are known by a variety of terms including, variable or adjustable speed drives and adjustable or variable frequency drives, as well as inverters (although not all inverters are variable speed drives). Variable speed drives are ideal for situations where a motor, or the device the motor drives, does not operate at full capacity during its entire operation, for example fans and pumps in HVAC systems and distribution systems in processes. In these situations, the variable flow rate of the fluid (i.e. air, water, acid, etc.) is often obtained by physically restricting the system to achieve the lower flow rate, or installing vanes and throttles. Variable speed drives allow the speed of the drive, and hence the flow rate of the fluid, to be reduced by decreasing the amount of power supplied through the use of power control units. Pump flow rate comparison between fixed and variable speed drives is shown in Fig. 2.1.

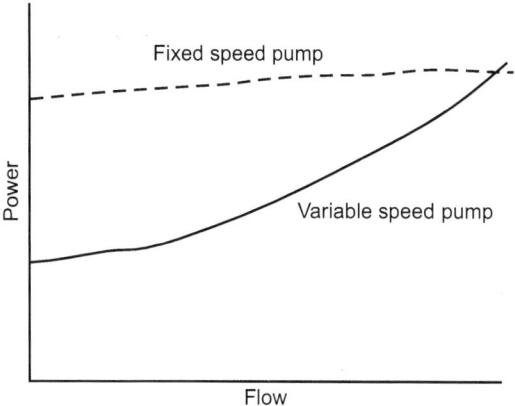

Figure 2.1: Pump flow rate comparison between fixed and variable speed drives.

The main advantage of these drives is when the speed of the fluid fluctuates between low and high flow rates. For example the flow rate of conditioned air in a temperature controlled building, a smaller amount of power can be used to drive the unit.

Correctly sized motors

In many applications, the speed of a device powered by an electric motor is relatively constant. In these situations, high efficiency single speed motors are ideal as they are usually more efficient near the rated load of a motor than variable speed drives. However, careful attention should be paid to ensure that the motor is not significantly oversized, given the usual load. As with applications where the load fluctuates, motors, which are operated at less than full load, are operating far less efficiently than those at or near the rated load.

Maintenance of electric motors

As with other pieces of capital, electric motors and the devices they drive should be regularly serviced and maintained to:

1. Ensure components are clean and free from dust and oil.
2. Operating at peak performance as compared to the manufacturers specifications.
3. Identify areas of wear or damage before the performance of the motor is degraded.
4. Increase the operating life of the motor.

Frequently when electric motors fail, it is due to a fault in the stator wire. In this situation, rewinding the wire usually repairs the motors. Whilst motors can be rewound to have about the same level of efficiency, some reduction in efficiency will usually occur. It is usually not practical, or cost effective, to have an electric motor of a lower efficiency rewound to a higher efficiency as the material used in the stator core and rotor will also influence the overall efficiency of the motor. However, a motor failure does represent an opportunity to upgrade to a higher efficiency motor.

2.3.3 Energy savings and compressed air systems

In many industrial processes, compressed air systems can consume a large component of energy use and hence offer the potential for large financial savings from reduced energy consumption. The largest component of a compressed air system is the compressor unit. Compressors can utilise a variety of fuel sources, including diesel, petrol and electricity. Whilst this discussion will focus on electric compressors, the principles apply across all fuels, although predicting financial savings may be difficult where the fuel price, especially for oil based fuels, fluctuates rapidly.

Energy savings from compressed air systems will usually result from savings in two areas:

1. Compressor unit (T_2).
2. Distribution system (T_1/T_2).

Compressor units

At the heart of all compressors, irrespective of fuel type, there are four areas that are essential for energy efficiency:

1. Compressor motor.
2. Compressor element (also known as the air end).

3. Compressor control system.

4. Hear recovery.

Compressor motor: Energy savings for motors consideration should be given where feasible to replace electric motors with diesel fuels and vice versa, if energy and hence financial savings are possible.

Compressor element: The air end of the compressor is the component that is responsible for compressing the air in the compressor unit. The performance of compressor elements will depend largely on the type of element in the unit (rotary screw, rotary vane, reciprocating or centrifugal) and may vary by as much as 20% between the styles. The size of the element is also an important factor as larger, underutilised or poorly functioning compressors will use more energy than smaller units operating at full capacity. Reducing the operating pressure of the larger compressor may also result in energy savings.

Control systems: The development of microprocessors has had an important effect on the efficiency of air compressors. Modern controls are able to match the air supply to the demand much more efficiently than is possible manually, and savings of up to 45% may be possible, through the installation of a number of smaller compressors which can be brought online automatically to match the demand for air. Aside from savings associated with energy consumption from the compressor components, financial savings will result from compressor units, which are appropriately located, correctly installed, maintained and serviced regularly as indicated in Table 2.1.

Table 2.1: Energy savings through compressor maintenance.

Component	*Potential saving through maintenance*
Oil filter	1%
Inlet filter	1%
Separator	2%
Cool running, synthetic oil	4%

Typically, compressors are located out of sight in the factory and consideration should be given to ensure that they are not subjected to extremes of temperature and are appropriately ventilated, to prevent overheating. Reducing the compressor inlet air temperature by shading the location or through increasing ventilation of this inlet can also offer significant savings. The correct installation and commissioning of new equipment as well as regular recommissioning of existing equipment after major services and repairs should ensure that the system is operating efficiently.

Heat recovery: One of the results of the process of compressing air is the generation of considerable amounts of heat. The use of a heat recovery system can increase the overall efficiency and cost-effectiveness of the operation. In many applications, the waste heat from compressed air systems, particularly from oil cooled rotary screw compressors, can be used in process and space heating applications. Water cooled oil coolers can supply water for process heating applications at between 50°C and 70°C. Space heating is available from systems where cooling air is reticulated through pipes for distribution through buildings.

The distribution system

After the compressor unit has compressed the air, it will need to be transported either through reticulated pipes or bottles to the location of the end use. Whilst some air leakage is almost unavoidable, care should be taken to eliminate audible air leaks, especially in reticulated compressed air systems. Audible leaks result in large amounts of energy being wasted, (between 25 and 35%), maintaining pressure to the compressed air tool or device. There by increasing the amount of air that must be compressed to complete the task. Condensation in the distribution system should be minimised by eliminating its presence in the inlet air or providing systems for removing it from the distribution network. In reticulated systems, condensate traps, which collect and remove moisture from the distribution system, also pose as potential areas of energy loss. Traps which are not functioning correctly, or are manually controlled are especially prone to wasting energy. Filters should be regularly checked and cleaned as blocked or partially blocked filters will increase the pressure, and hence the energy, required to operate the system.

2.3.4   Energy savings and steam generation

Steam is used for a multitude of purposes in industrial plant. It can provide heat for chemical processing, hot water for cleaning purposes, steam for input to turbines for producing power and so on. Steam is generally produced by boilers. Boilers typically operate well below their optimum efficiency and savings of approximately 15% should be readily achievable.

As with all examples of industrial energy efficiency, it is important to consider the whole steam system from generation to recovery. Heat (and thus energy) losses in the steam generation and distribution systems will result in poor heating at the location of the end use.

2.3.5   Boilers

Energy savings in the generation side of steam use are usually the result of efficiency improvements in the operation of the boiler. In maximising the efficiency of boilers, two key principles need to be addressed; first, the level of excess air (the extra air needed to ensure good combustion of the fuel in the boiler) and secondly the temperature of the flue gases needs to be kept as low as possible (otherwise a large part of the heat that was produced in the boiler will go up the chimney). Good monitoring can be used to assist in achieving these outcomes. In addition, to these, the utilisation of high quality water, free from contaminants, ensures that the minimum amount of heat is required to produce steam.

In boiler plants, there are typically four areas of potential savings:
1. Monitoring equipment (T_1/T_2).
2. Load management (T_1).
3. Condensate return (T_1/T_2).
4. Fuel selection (T_1).

Monitoring equipment

Boilers are a potential source of energy savings since they are frequently inadequately monitored, even at the simplest level, resulting in efficiency losses. Simple, but regular analysis of the flue gases, including chemical analysis of the gases and its temperature, will help determine if the boiler is operating efficiently. Care should be taken to ensure that the tests are conducted with load levels of at least 65–70% and that the load and gas (steam) pressures are constant. Once this level of analysis is well-established, additional monitoring equipment which can determine the gross thermal efficiency of the boiler, may be required.

Load management

As with electric motors and air compressors, boilers do not run efficiently when they are operating at less than their recommended operating pressures. Significant cost savings can result where load management strategies, such as only operating the number of boilers to produce the required amount of gas/steam and advance warning of changes in the gas/steam load are given to boiler plant staff are implemented.

Condensate return

Unfortunately, there will always be some efficiency losses in process heating due to boilers as a result of condensate. Boilers and reticulation systems which are fitted with condensate return systems are far more efficient than those where the condensate enters a waste stream. The efficiency gains are largely the result of chemical profile of the steam condensate, which is typically hot and free of oxygen. This liquid requires less energy to convert the already heated and deoxygenated liquid to gas (especially steam).

Fuel selection

There are seven common types of fuels available for boilers:

1. Coal
2. Natural gas
3. Liquefied petroleum gas (LPG)
4. Furnace oil
5. Diesel
6. Electricity
7. Wood or wood wastes (biomass)

In many parts of the world, coal is used as a boiler fuel as it is usually the cheapest industrial fuel source. However, many countries are looking towards natural gas and biomass as alternatives due to the increasing cost of the traditional fossil fuels, diesel, coal and electricity as well as regulatory changes.

When selecting or reviewing the fuel selection for boilers, careful consideration should be given to ensure that the full cost of the fuel, including transportation cost, is considered. For example, a boiler in a pulp and paper mill may be more cost effective if it utilised the wood waste from the pulp process than coal or natural gas, despite a supply of both nearby.

Cogeneration, the simultaneous production of heat and power is also a potential area of energy savings, through the onsite generation of heat/steam and electricity.

2.3.6 Energy savings in steam distribution systems

As with compressed air systems, the distribution of steam throughout an industrial facility is a potential area of energy loss, and hence increased operating costs. Steam traps are used in steam distribution systems to remove condensate as it forms. They often have the dual function of removing any entrapped air in the system.

The presence of air and condensate in steam systems reduces the effectiveness of heat transfer in these systems as they tend to form insulating layers on heat transfer surfaces.

This means that temperatures have to be higher in order to achieve the same rate of heat transfer. Also, the presence of air reduces the overall temperature of the system which is governed by the pressure of the steam. If part of the pressure of the system is caused by entrapped air, the net pressure of the steam is less than that read on the steam gauges and so the temperature will be lower than expected. Regular checking of steam traps and air vents is essential for the efficient operation of steam plant.

Energy is also wasted in steam distribution systems where heat is lost to the environment through inadequate insulation of the reticulation system. Care needs to be taken with valves and fittings that, if not properly insulated, lead to significant heat loss. Any heat lost in the distribution system means that additional fuel has to be consumed in the boiler to make up for this loss.

Condensate is an inevitable product of any steam system either as a result of heat loss or simply as a result of using the steam to transfer heat to a process. This condensate represents a source of hot, very pure water and so is an ideal feedstock for the boiler input. Assuming that the condensate has not been contaminated by the process, the use of condensate is a very effective example of using waste heat (or heat recovery).

2.3.7 Energy savings and furnaces

Furnaces are widely used in the manufacturing and mining industries. Although similar to boilers, they are usually used to melt metals for casting. Many of the potential areas for energy savings are the result of high capital cost, or require detailed changes in the current operation of the factory or smelter. These include rescheduling to reduce the occurrence of a furnace being heated with less than an optimum load, automatic control of furnaces, insulation of the furnace as well as modifications to the furnace. Although these items require large amounts of capital, consideration should be given to these issues, especially where the furnace is due to be replaced, or where a new furnace is to be purchased. Cost effective, simple strategies for reducing the energy consumption of furnaces are very similar to those discussed in the section on boilers, and includes fuel selection, monitoring equipment (to ensure there is not excessive air in the melt) and load management. Furnace systems often offer good potential for heat recovery systems where the very high temperatures in the exhaust air can be used to preheat the combustion air entering the system.

2.3.8 Energy savings through heat recovery

In many processes considerable amounts of waste heat are produced. Examples include the exhaust stacks of engines, boilers or furnaces, condensate in steam systems, and waste streams from washing, heating applications as well as compressed air systems.

Heat recovery involves the use of these waste heat streams to provide useful heat for another part of the plant. Heat exchangers are used to extract heat from the waste stream and transfer it to a second fluid flow. In many instances, the waste heat from one part of the process can be used to preheat a fluid for use in that same process. For example, the hot air in the exhaust of a furnace can be used to preheat the combustion air used in the same furnace. Waste heat is usually best identified as part of an overall energy audit of the industrial process or facility. The audit should identify the fluid type (liquid or gas), the amount of fluid generated, either as a volume or flow rate, temperature of the fluid, the time of production, its location as well as the location of heat using processes. From this information, the monetary value of the waste heat should also be determined. Once the size and location of the waste heat product is known, a detailed analysis of the energy saving potential as well as the process of selecting an appropriate heat exchanger can be undertaken. Heat exchangers are devices which recover the waste heat from one process for use in another process. There are a variety of heat exchangers available on the market, suitable for both batch and continuous feed operations. Recuperator type heat exchangers are able to work in continuous feed processes as the heat recovery from the fluid is steady. In regenerator type heat exchangers, heat recovery is delayed due to the storage period required for the release of heat from the fluid, and thus are best suited to batch processes. Careful consideration must also be given to the physical and mechanical performance of the proposed heat exchanger prior to purchase.

2.3.9   Energy savings through cogeneration

Cogeneration is best explained by its alternate name of Combined Heat and Power (CHP). As it suggests useful process heat is generated simultaneously with power. Usually, electrical energy is produced through the combustion of fuel, generating heat as a by-product. Excess heat is recovered and used for process heat applications, including steam. Aside from increases in energy efficiency, CHP has the added advantage of increased reliability of supply for industry, especially where electricity supply is irregular or unreliable. Fuel substitution provides greater flexibility for CHP, where cheaper or renewable fuels are available to replace conventional fuels. Cogeneration has been extensively used throughout the world in industry and non-industrial applications, such as district heating. However, regulated electricity industries pose significant institutional barriers for cogeneration projects.

2.3.10   Awareness raising and education

Awareness raising and education of the need and potential for energy savings is as important to the industrial sector as it is to the domestic and commercial sectors. Operator behaviour can have a significant effect on the effectiveness of energy efficiency measures in industrial plant. Simply turning off equipment when it is not being used can provide an easy way of saving both energy and money. Education programmes can provide considerable potential for energy savings.

2.3.11 Regulatory measures

There are several regulatory measures that can produce energy savings in the industrial sector. These include:

1. Minimum energy performance standards. Few countries have introduced regulations requiring office equipment to meet minimum standards of energy efficiency.
2. Deregulation of the electricity industry to encourage the development of cogeneration.
3. Building codes can be used to ensure that new buildings meet minimum standards for passive solar design in order to reduce energy use.
4. Changes in the regulatory environment as a result of the control of emissions, such as carbon dioxide and international obligations, such as United Nations protocols.

2.3.12 Economic measures

The economic measures available to encourage energy efficiency in the industrial sector are similar to those offered to the commercial and domestic sectors and include:

1. Time of use tariffs.
2. Subsidies for substitute fuels (including biofuels).
3. Subsidies for the use of renewable energy.
4. Tax relief for investment in energy efficient equipment.
5. Levies and penalties on energy use.

3

Energy conservation in agricultural sector

3.1 Introduction

The world relies on agriculture to feed humanity. We simply cannot survive without food, and therefore, without agriculture. Energy is an essential component of agricultural production. It fuels the equipment, irrigates the crops, fertilises the soil, sustains the livestock, transports the food, and processes the food into its final forms. As the population continues to grow, more agricultural production is required to support the increased food demand.

At the same time, energy and environmental constraints mandate that agricultural production be accomplished effectively with minimal energy consumption. It is necessary to increase agricultural yields per unit area of land, while preserving the soil integrity and environment. Efficient energy management practices will help achieve and maintain this delicate balance. This chapter presents a wide variety of energy-efficiency opportunities as they relate to sustainable agriculture.

Agriculture is a core sector of the global economy. As countries mature and grow, it stimulates the establishment of regional manufacturing and service sector spin-offs. The existence of many industries depends on agricultural activity. As technological sophistication in agriculture progresses, this sector will transform into a more autonomous economic entity. It will be able to control terms for the purchase of goods and services. It is likely and even probable that some manufacturing processes will shift to agricultural sites, as onsite quality control solutions are developed.

Agricultural production, however, is increasingly strained by limitations in energy and arable land. This problem is growing due to large part to population growth—more food production is necessary to meet the requirements of a growing population. With the increase in population, more land is required to produce the needed output, or higher yields are necessary per unit area of land. To intensify the challenge, a considerable quantity of arable land is lost each year to erosion, increased salinity and water logging. This loss equated to about 1.5 billion hectares of land. To alleviate the increased land requirements, deforestation and increased fertilisation practices are continuing, with obvious adverse environmental impacts. Future technological efforts in agriculture should focus on increased yields per unit area of cropland by means that minimise energy usage and promote sustainability.

3.2 Energy usage

Energy is required for many aspects of the agriculture industry and the industries that support agriculture. Agriculture can be defined in several ways. In its most general sense, agriculture includes livestock, food crops, energy crops, fibres, ornamental plants, forestry activities, hunting, fishing, agriculture, and aquaculture.

3.2.1 Energy consumption relative to total energy consumption

In the current treatment, the focus is on energy use in agriculture for food production. Specific attention is placed on crop and livestock production. Therefore this treatment excludes land activities such as hunting, forestry, fibre crops, and ornamental plants, and water activities such as fishing, aquaculture, and mariculture. Though emphasis is placed on food production, support industries that bring the food from the farm to the table are identified.

Figure 3.1 shows a simplified flow diagram of food from agriculture to the table. Support industries and activities include transportation, food processing, the wholesale and retail trade, and final preparation. Transportation can be in the form of food transport between farms, post-processing plants, wholesale and retail locations, and restaurants and homes.

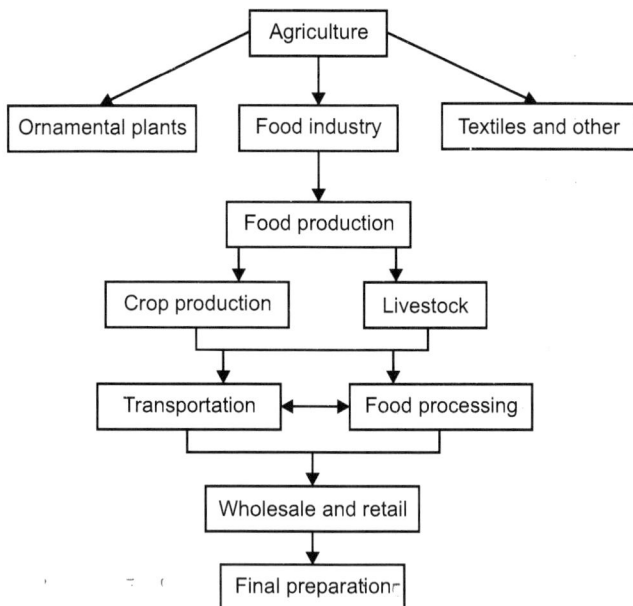

Figure 3.1: Simplified diagram of the flow of food from agriculture to final preparation.

Transportation is also required for delivering agricultural inputs, such as fertilisers, livestock, feeds, and seeds, to farms. Food processing requires a large amount of energy.

Wholesale and retail activities require energy for refrigeration and general storage. Final preparation entails energy for refrigeration, cooking, dish washing, and general storage. There are four main factors that contribute to food production energy consumption. Each factor is an input to the food production process. Figure 3.2 shows these energy consuming inputs: fertilisers and other chemicals, farm equipment, irrigation and pumping, transportation, and other miscellaneous inputs. The three most significant energy consumers are fertilisers, farm equipment, pumping and irrigation.

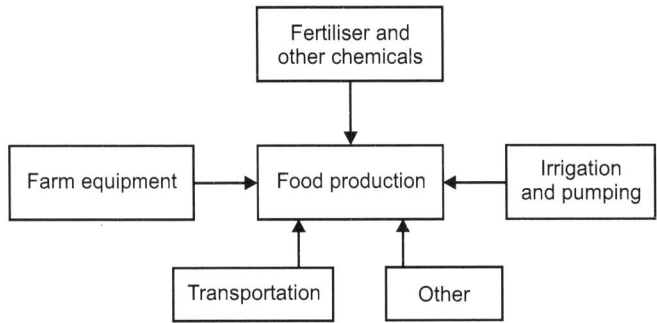

Figure 3.2: Energy-consuming inputs to food production.

The basic function of food manufacturing or processing is to convert raw agricultural outputs to food products. Specific activities within the food manufacturing sector include food preservation, packaging, and refrigeration and storage. Each of these main activities requires a significant quantity of energy. In addition, buildings and vehicles, as well as miscellaneous support functions associated with food processing activities, require substantial energy inputs. Direct process uses are process heating, process cooling and refrigeration, machine drive, and other miscellaneous process uses, and account for 45% of total fuel consumption. Boiler fuel is an indirect process use, and accounts for 46% of total fuel consumption. Direct nonprocess uses account for 9% of total energy use and include heating, ventilation, and air conditioning (HVAC), lighting, onsite transportation, and other nonprocess uses. Process heat is required for such processes as drying, cooking, pasteurisation, and sterilisation, and consumes a large share of energy. Process cooling and refrigeration includes activities such as simple food cooling, freezing, and storage, and utilises about 6% of total energy. Machine drive systems and other processes each represent about 11% of total energy use. Nonprocess uses, including lighting, HVAC, transportation, and onsite electricity generation, require only about 9% of total energy use. For canned food, the majority of energy is used in manufacturing the metal cans, but little is required for subsequent storage. The opposite is true for frozen foods, in which the packaging requires little energy compared with that required for frozen storage.

3.3 Energy efficiency in pumping and irrigation systems

A well designed and maintained irrigation system will keep the environmental impacts, as well as the operating costs, of irrigation low. Both the water and the energy consumption required to distribute the water can be minimised with careful planning and implementation of irrigation efficiency measures. In addition, the installation of high-efficiency pumping equipment, or improvements to an existing pumping plant, will help reduce unnecessary pumping losses. This section summarises efficiency measures for agricultural irrigation, and places particular emphasis on electric pumping plant efficiency, computerised irrigation scheduling, and irrigation load management.

Irrigation accounts for a substantial portion of agricultural energy consumption. In the United States, irrigation is the fourth largest end use of energy in food production, following chemicals (fertilisers and pesticides), agricultural equipment, and transportation. Improving an irrigation system's efficiency can save both water and energy. An optimised irrigation plan will use the minimum amount of irrigation required. As the quantity of irrigated water is reduced, the energy required to distribute the water will likewise be reduced. Another essential factor of energy-efficient irrigation is the efficiency of the pumping plant. Careful selection and upkeep of the plant's components are paramount to highly efficient operation.

3.3.1 Energy efficient irrigation

The energy required in crop irrigation is a function of the level of pumping required.

Pumping energy requirements, in turn, depend on several main factors:

1. The amount of vertical lift from the depth of the water source to the height of the application.
2. The water pressure required for the type of irrigation (e.g. high pressure for hydraulic gun sprinklers and low pressure for micro- and surface irrigation).
3. Depth of irrigation.
4. Frequency of irrigation.

5. The water system efficiency of the pumping system (i.e. the combination of water source, pumping plant and distribution system).

Therefore, energy may be saved in irrigation by addressing these five points. First of all, is it possible to reduce the depth of the water source? Are there any water sources at higher elevations? Second, is it feasible to choose an irrigation method with lower pressure, such as micro-irrigation? Low-pressure systems eliminate the need for additional booster pumps. Third, can the depth of irrigation be reduced? Is the crop watered extensively? Is it watered too heavily in particular areas? For example, many surface irrigation systems that rely on gravity for water distribution end up watering the high side of fields more than necessary. Fourth, is the field watered more frequently than required? Care must be taken to know the correct amount of water required, and at what interval, for a given crop. Last, and very important, is the pumping system operating with high efficiency? Pumping system efficiency is related to the proper match between the pumping plant, water source, and distribution system. It is also a function of components, including the motor or engine, the drive shaft, and the pump assembly.

Good design, as well as maintenance of the pumping system (e.g. pump repair and distribution system leak repair), can go a long way towards improving efficiency. Some of the primary efficiency measures associated with irrigation use are given below. It is important to remember that any conservation of water also conserves pumping energy.

Choose the appropriate crop for a given soil: It is best to plant the types of crops that use water most effectively for a given soil type.

Use only the required amount of water: Table 3.1 lists some approximate annual amounts of water required for various crops. It is important to use only the amount of water necessary for a particular crop, soil type, and region. Over-watering and nonuniform watering—in which some areas are over-watered while others are watered correctly—result in wasted water and pumping power.

Apply water at proper intervals for the soil type: Sandy soils generally require more frequent irrigation than heavy soils such as clay, because sandier soils drain more quickly. Therefore, it is usually best to apply the same quantity of water during a given application to a sandy soil as would be applied to a clay soil, but the applications should be more frequent for sandy soils. Heavy frequent irrigation wastes water and energy, and heavy infrequent irrigation is not as productive for sandy soils.

Eliminate unnecessary water loss: Water can be lost through leaks in pipes, valves, fittings, connections, and so on. Make sure that leaks are repaired in a timely manner to reduce water and pumping energy loss.

Table 3.1: Representative water requirements of selected crops.

Type of crop	Approximate annual water requirement (acre-feet)
Alfalfa	3 to 4.5
Barley	Dryland to 1
Beans	1.25 to 1.75
Beets	2 to 3
Chile peppers	1.5 to 2
Corn	1 to 2
Cotton	3 to 3.5
Grapes	2.5 to 3.5
Grain, sorghums	1.5 to 2
Lettuce	1
Onions	1.5
Orchard, fruit	2 to 3.5
Peanuts	1
Pecans	3 to 4
Permanent pasture	3 to 4.5
Potatoes	2.75 to 3.5
Tomatoes	2 to 3
Wheat Dryland to	2

Consider night time irrigation: Irrigate at night if feasible. Night time irrigation will save significant water and energy compared with day time irrigation, since less water will evaporate. Night time irrigation is particularly attractive for crops that are sprinkler irrigated.

Increase pump and engine or motor efficiency: Pump and engine or motor efficiency can be increased by good operation and maintenance practices. In addition, it may be feasible to replace inefficient equipment with high-efficiency models.

Minimise water pressure if possible: Pressurise the water only to the point required by the given irrigation system.

Choose the most appropriate irrigation method: Evaluate the crop, soil, and region carefully when determining the most efficient and cost-effective irrigation method to use. The three primary methods are surface irrigation, micro-irrigation, and sprinkler irrigation.

Improve surface irrigation efficiency: Surface irrigation requires the least amount of pumping energy because the water flows by gravity along the field. However, a surface irrigation system can be inefficient in terms of water loss if not carefully designed. There are several types of surface irrigation, including furrow, flood and surge. For furrow and flood systems, care must be taken to prevent soil from being over-saturated at the high point of the field due to percolation, and to reduce tail-water at the low point of the field. Surge systems apply water intermittently to a furrow with the use of surge valves. This improves irrigation efficiency over conventional furrow and flood systems. The use of gated pipes (i.e. pipes with openings called 'gates' on their sides) will improve efficiency over the use of header ditches in surface irrigation systems. Automated gated pipe systems are also available. In addition, the capture and reuse of tail-water will reduce water loss, though not without the small expense of pumping the water back to the top of the field. Since the pumping power required for surface pumping is considerably smaller than required for pumping groundwater, tail-water reuse in usually justified. It is also important to reduce seepage and general water loss by lining ditches with concrete or plastic, preventing rodent burrowing, and controlling weeds.

Consider using micro-irrigation: Micro-irrigation systems, such as bubblers and drip emitters, apply water near the bases of individual plants. This reduces seepage and evaporation losses compared with systems that water the entire soil surface or row. If used correctly, micro-irrigation systems are very water-efficient and require low pumping power. However, they are most cost-effective for applications in which the bubblers and emitters can remain in place, for example in orchards or vineyards. In some cases, they are feasible for vegetable crops, if the crops are of high value.

Choose an efficient sprinkler system: If a sprinkler system is the most viable approach for a given crop and conditions, consider selecting a low-energy sprinkler. The Low Energy Precision Application (LEPA) center pivot system is an improvement over conventional sprinkler systems. It uses lower pressure water, and applies the water between 20 and 38 centimeters (8 and 15 inches) above the soil. This reduces pumping energy, as well as evaporation losses. The system also incorporates circular farming and furrow dikes to prevent water runoff loss.

Use furrow dikes: Furrow dikes are installed with specialised diking equipment. The equipment creates mounds in furrows to contain water and eliminate runoff. Though very effective at reducing water consumption, some farmers object to the additional labour involved to create, and later remove, the dikes.

Measure soil water content: Devices such as evaporation pans, gypsum blocks, and tensiometers can be used to measure soil moisture and to aid in irrigation scheduling. In addition, soil moisture can be judged by the 'feel' method. Another valuable method for scheduling, using computers is based on the water budget approach.

Use a laser for levelling fields: A laser is an excellent tool for levelling fields. Laser levelling systems often consist of a rotating laser beam that is attached to a command post. As the laser rotates, the command post transmits a signal to a receiver located on the scraper of the tractor in the field. The signal from the laser controls the work of the scraper to produce a very level field. Level fields enable a more uniform distribution of water.

Design wells, pumps, and distribution systems carefully: An irrigation well should be designed and constructed with care by experts to ensure the well's success as a water resource. In addition, the pump and well should be designed in concert to match the irrigation requirements. It is also important to size the piping and distribution system to handle the flow efficiently.

Choose efficient pumps: Irrigation pumps are electric or fossil-fuel fired. To minimise energy use, select the most efficient pump available that also meets acceptable economic criteria. In an increasing number of applications, photovoltaic water pumps are viable. Photovoltaic systems are highly efficient, but may not be economically practical for large irrigation needs. However, as energy costs rise, and as photovoltaic technology matures, these systems are becoming more feasible.

Employ load management practices: Although load management will not necessarily save pumping energy for the irrigator, it will reduce the burden on utilities during peak demand hours. Many utilities offer the incentive of reduced rates for consumers who shift loads to off-peak hours. This could also translate to energy savings for irrigators in some cases; for example, if they convert to night time watering, irrigators will also benefit from reduced water loss to evaporation.

3.4 Energy conservation tips while using irrigation pumpsets

One of the largest single users of energy is agriculture, and within this field, the energy used for irrigation is quite significant. The requirement of energy for pumping groundwater, can be reduced if the power for this purpose is used most efficiently, which implies designing an efficient well assembly and having an efficient pumping unit and using the pumped water in the most efficient manner in the field.

A preliminary analysis of energy used in irrigation suggests that 30 to 50% of the total energy used could be saved through improved technology, i.e. more efficient valves pumps, proper design and construction of well and by better use of the water that is applied. Improper selection of pump, prime mover and well assembly, inadequate maintenance and faulty operation of the unit and other factors also contribute to the low efficiency of the system.

3.4.1   Causes of low efficiencies

Causes of low efficiencies are discussed below:

1. Undersized pipes: If pipes of smaller diameter are used, the initial cost will be less but the frictional head loss and the operational cost will be more. On the other hand, if pipes of larger diameter are used, the initial cost will be more but the frictional head loss and thereby the operational cost will be less. The optimum diameter of the pipe will have minimum total cost i.e. the initial and the operational cost. The farmers, while selecting the pipe size, give consideration to the initial cost only without bothering about the extra operational cost which they have to pay every year by way of increased energy bill. It is a general practice that with 100×100 mm pump, the suction and delivery pipes of 100 mm diameter are used. The velocities in the suction and delivery pipes should generally be lower than that at the entry and exit of the pump.

2. High delivery point: Large number of pumping units have extraordinary high delivery point. This is especially true for diesel units. Excessive height of delivery pipe causes extra energy consumption.

3. Poor fittings: The fittings provided by most of the farmers are very poor resulting in large losses and leakage. Head losses in a poor quality foot valves are high. Similarly the head loss in the sharp bends are also high. The farmers are mostly ignorant about the operational quality of the components.

4. Inefficient pumps: Field study has indicated that average efficiency of the pumps operated by electric motors is 47% and about two third of the pumps are operating at efficiencies less than 50%. Similarly the average efficiency of the pumps with diesel engines is 56% and more than half have efficiency less than 60%. Some of the pumps are operating at efficiencies less than the optimum efficiency. This is due to improper selection of pumps and mismatching prime movers and due to inferior quality of the pumps being marketed. The selection of the pumps should be governed by the characteristic curves. At the normal operating condition, the efficiency should be maximum.

5. Faulty prime movers: The prime movers of the pumps should be of proper size. Generally farmers go in for higher capacity motors and diesel engines. Generally the foundations and the belt transmissions are in poor shape. The farmers should be properly guided to choose the suitable sized prime mover, provide proper foundation and belt transmission, and to select the pump so as to take advantage of high efficiency of the pump.

Select right type of pump sets:

1. The various pumps sets available in the market require different quantities of diesel for pumping water. Therefore, it is important to choose an ISI mark pump. There is a potential of about 25% to 35% improvement in the efficiency of these pump sets.

2. The pump you select should be suitable for your well and your water requirements. It is not necessary that the pump that is good for your neighbouring farm is good for you too. If you give the following information to an expert, he will be able to help you choose the right pump:

   (a) The depth of your well.

   (b) The area of your field.

3. One should also select the right engine which makes the pump run at the right speed. For this, consult an expert. The engine you use should have sufficient Horsepower (HP) to operate the pump. The expert can calculate the amount of power needed for your engine. It is always better and beneficial to go in for a well-known and good quality engine. Look for the ISI mark of quality on the engine.

4. To ensure a high level of operational efficiency of your diesel powered engine, ensure the following:

   (a) Engine should not emit too much smoke.

   (b) Use the correct grade of lubricant recommended by the manufacturer.

   (c) Engine should be fitted with an oil filter.

   (d) Engine should have an air filter which should be cleaned regularly.

   (e) Engine jacket cooling water should be warm.

Foot valves:

1. Farmers can save about 10% to 30% diesel by simply using a foot-valve with larger area of openings than the one with narrow area of openings.

2. An efficient low friction ISI mark foot-valve, though slightly costlier, pays back fast the extra cost by saving a lot of diesel.

3. The larger valve helps to save electricity/diesel because less fuel and power is needed to draw water from the well.

Pipeline:

1. A rigid PVC pipeline, with bigger diameter saves energy.

   (a) More diesel is required to pump water through small diameter pipes because it offers higher friction. If the pipe is bigger than the pump flange size, a reducer must be used.

(b) How a 20% decrease in diameter increases the friction 3 times, if in place of 100 mm pipe, an 80 mm pipe is used, the loss due to friction for drawing the same quantity of water will be three times more, which will cause higher fuel consumption.

(c) Also pipes made of rigid PVC cause lower frictional loss as compared to pipes made of conventional galvanised iron. Such pipes thus help save fuel.

2. The fewer the no. of bends and fittings in a pipe, more the electricity saves.

3. The pipeline arrangement that has many bends and unnecessary fittings cause higher diesel consumption. Each bend in an 80 mm diameter pipeline leads to as much friction loss as an additional pipe length of 3 meters. Therefore, the fewer the number of bends and fittings in a pipe, the more the saving of diesel.

4. Sharp bends and L-joints in the pipe can lead upto 70% more frictional loss than standard bends.

5. Use good quality PVC suction pipe to save energy and save electricity up to 20 %.

Transmission:

1. Do not use belts that are old and worn out. Such belts can slip or snap anytime, causing loss in the transmission of power and hence increased fuel consumption.

2. Check points for efficient transmission.

3. Reduce the number of joints in the belt.

4. Check and adjust belt tension frequently.

5. Check alignment of the pump with the engine.

A farmer can save 15 litres of diesel every month simply by reducing the pipe height by 2 m. The pump works more efficiently when it is not more than 10 ft. above the water level of the well.

Apply oil and grease to pump set regularly as recommended by the manufacture.

To improve the power factor and voltage use ISI marked shunt capacitor of right capacity with motor. This will also save the electricity.

Switch off the light of well in the day time.

4
Energy conservation in food industry

4.1   Introduction

Food industry plays a significant role in country's economic development. Processed agricultural products account for about 30% of the processing industry's, out of which 90% is produced by the food industry.

The food industry is a complex and global collective of diverse businesses that supply most of the food consumed by the world population.

Agriculture is the process of producing food, feeding products, fibre and other desired products by the cultivation of certain plants and the raising of domesticated animals (livestock).

Food processing includes the methods and techniques used to transform raw ingredients into food for human consumption. Food processing takes clean, harvested or slaughtered and butchered components and use them to produce marketable food products. There are several different ways in which food can be produced. Modern food production is defined by sophisticated technologies. These include many areas. Agricultural machinery, originally led by the tractor, has practically eliminated human labour in many areas of production. Biotechnology is driving much change, in areas as diverse as agrochemicals, plant breeding and food processing.

4.2   Bakery industry

The bakery industry comprises mainly of bread, biscuits, cakes and pastries manufacturing units. Bread is the product of baking a mixture of flour, water, salt, yeast and other ingredients. The basic process involves mixing of ingredients until the flour is converted into a stiff paste or dough, followed by baking the dough into a loaf. Mixing has two functions: (i) to evenly distribute the various ingredients and (ii) allow the development of a protein (gluten) network to give the best bread possible. Each dough has an optimum mixing time, depending on the flour and mixing method used.

Rising (fermentation): Once the bread is mixed, it is then left to rise (ferment). As fermentation takes place, the dough slowly changes from a rough dense mass lacking extensibility and with poor gas holding properties into a smooth, extensible dough with good gas holding properties. The yeast cells grow, the gluten protein pieces stick together to form networks, and alcohol and carbon dioxide are formed from the breakdown of carbohydrates (starch, sugars) that are found naturally in the flour. The yeast uses sugars in much the same way as we do, i.e. it breaks sugar down into carbon dioxide and water. Enzymes present in yeast and flour also help to speed up this reaction. The energy which is released is used by the yeast for growth and activity. In a bread dough where the oxygen supply is limited, the yeast can only partially breakdown the sugar. Alcohol and carbon dioxide are produced in this process known as alcoholic fermentation. The carbon dioxide produced in these reactions causes the dough to rise and the alcohol produced mostly evaporates from the dough during the baking process. During fermentation, each yeast cell forms a centre around which carbon dioxide bubbles are formed. Thousands of tiny bubbles, each surrounded by a thin film of gluten, form cells inside the dough piece.

Kneading: Any large gas holes that may have formed during rising are released by kneading. A more even distribution of both gas bubbles and temperature also results. The dough is then allowed to rise again and is kneaded, if required by the particular production process being used.

Second rising: During the final rising, the dough again fills with more bubbles of gas, and once this has proceeded far enough, the doughs are transferred to the oven for baking.

Baking: The baking process transforms an unpalatable dough into a light, readily digestible, porous flavourful product. During baking the yeast dies at 46°C, and so does not use the extra sugars produced between 46–75°C for food. These sugars are then available to sweeten the bread crumb and produce the attractive brown crust colour. As baking continues, the internal loaf temperature increases to reach approximately 98°C. The loaf is not completely baked until this internal temperature is reached. Weight is lost by evaporation of moisture and alcohol from the crust and interior of the loaf. Steam is produced because the loaf surface reaches 100°C. As the moisture is driven off, the crust heats up and eventually reaches the same temperature as the oven. Sugars and other products, some formed by breakdown of some of the proteins present, blend to form the attractive colour of the crust. These are known as 'browning' reactions, and occur at a very fast rate above 160°C. They are the principal causes of the crust colour formation.

Cooling: In bakeries, bread is cooled quickly when it leaves the oven. The crust temperature is over 200°C and the internal temperature of the crumb is about 98°C. The loaf is full of saturated steam which also must be given time to evaporate. The whole loaf is cooled to about 35°C before slicing and wrapping can occur without damaging the loaf. A moist substance like bread loses heat through evaporation of water from its surface. The rate of evaporation is affected by air temperature and the movement of cool air around the loaf. In a bakery there are special cooling areas to ensure efficient cooling takes place before the bread is sliced and wrapped.

4.2.1 Energy savings in bakery industry

Bakeries all over the world have many ways in manufacturing process to reduce energy bills which includes thermal and electrical energy. Apart from the basic production line there are utilities such as air compressors, air conditioning, refrigeration, hot water generators, boilers, chillers, effluent treatment plant, conveyors, generators, air washers, cold rooms and lightening. Not only can a bakery reduce energy costs and greenhouse gas emissions, but it can impact its bottom line. Here are few tips and ideas for energy savings in bakery plant.

Motors

1. Eliminate improper cabling.
2. Check alignment.
3. Use synchronous motor to improve power factor.
4. Provide proper ventilation.
5. Maintain high power factor as lower power factor reduces the motor efficiency and also reduces the efficiency of electrical distribution. Lower power factor results when motors of plant runs below the full load ratings.

Lighting

1. Use low wattage lightings (CFL).
2. Use occupancy sensor.

Air compressor

1. We can save energy by having VFD installed for motor.
2. Screw compressors have far superior performance then other type of compressors.
3. Select compressors which have 'motor stop arrangement' in which motors stops on longer unload period.
4. Have cooling arrangement for incoming air.

Air conditioning/refrigeration

1. Proper insulation of pipes of chilled water and copper piping for refrigerant can results in energy savings.
2. Temperature can be set as per requirement and not going for extremes can help in reducing energy consumptions.
3. Avoid over sizing and select the refrigeration cap as per the load.

4. Consider gas powered refrigeration equipment to minimise the electrical load.
5. Connect refrigerated water load in series.

Oven

1. Waste heat generated can be utilised for heating water and pre-heat in take air recuperates and heat exchangers can be installed to use waste heat from flue gases.
2. Insulation are required for various industrial application such as ovens, furnaces, heaters, boilers, storage tanks, piping's, steam chilled water pipelines, etc. These insulations are made up of mineral wool, silica, alumina and mulite. Insulation can save heat losses considerably. While selecting insulation material, check thermal properties, physical and chemical specifications and repair damaged insulation.

Pumps

1. Schedule pumps to turn off whenever possible by putting up level sensors, float valves or pressure switches.
2. Avoid recirculation through bypass lines.
3. Minimise throttling.

Cooling towers

1. Use two-speed or variable-speed drives for cooling tower fan control if the fans are few. Stage the cooling tower fans with on-off control.
2. Turn off unnecessary cooling tower fans when the loads are reduced.
3. Control cooling tower fans based on leaving water temperatures.
4. Periodically clean plugged cooling tower water distribution nozzles.

Electrical

1. Provide separate lighting transformer for control of lighting systems.
2. Locate substation near the load centers to minimise energy losses in cables and also improve voltage levels.
3. Install capacitors with automatic power factor control panel to maintain a power factor of not less than 0.9. Install capacitors near the load points wherever possible or at the sub distribution board. The best practical method would be to install capacitors at larger capacity motors and at the substation for all other loads.
4. Identify under-loaded motors and examine the possibility of replacing them with the appropriate capacity motor or alternatively fix retrofitting devices that are available to save energy.

Fuels

1. Fuel consumption can be reduced by installation of fuel efficient burners. With advent of new technologies, one can get fuel efficient burners (for example, dual stage high pressure Weishaupt burners). Its a monoblock burner with inbuilt blowers in the burner.
2. Other important factors of combustion is air -fuel ratio. All manufacturers of burner provides exact ration of air fuel.
3. Regular maintenance of burners also helps in reducing fuel consumption.

4.2.2 Combined heat and power (CHP)

CHP provides an opportunity to reduce the overall energy consumption in facilities by generating electricity onsite and recovering waste heat from the electrical generation for the production process. When a facility obtains its electricity from the local utility and generates thermal energy through the combustion of natural gas, the energy conversion process is only 33% efficient. However, using CHP to produce electric energy onsite can result in 80% efficiency. CHP processes convert waste heat or steam into electrical power. The food industry produces biomass waste, which could be used as an alternative fuel source. CHP which requires a large capital investment, is an attractive opportunity for food processing facilities that have high energy intensity, a flat year-round load profile, and high thermal to electric ratios. Even facilities with an electrical demand less than 5 megawatts can benefit from CHP systems.

4.2.3 Waste heat recovery from biscuit oven stacks

Biscuit manufacturing involves high temperatures for baking. Various type of ovens are available for baking bakery products. Heat recovery can be done for biscuit oven which has 5–6 zones basically in indirect oven. As temperatures are very high (400–500°C) on feed end these stacks are then joined to a common duct which collects heat from these stacks with the help of blower and then these flue gases is passed onto the last zone which requires low temperature. This temperature is as low as 180–200°C can be attained by circulating flue gases collected in common duct from these stacks. This eliminates need of one burner all together at colouring zone or last zone. Fuel is saved by implementing this kind of arrangement.

4.3 Brewery industry

Energy consumption is equal to 3–8% of the production costs of beer, making energy efficiency improvement an important way to reduce costs, especially in times of high energy price volatility.

Energy-efficient technologies often include additional benefits, such as increasing productivity or achieving future of current environmental goals thus reducing the regulatory 'burden'.

4.3.1   Brewing process

Beer is an industrial product. A brewery is literally a beer factory in which the brewer takes advantage of and manipulates natural processes to create the perfect growth medium for yeast. On the surface the brewing process is simple. But if we look a little deeper we will find that there is a complex set of chemical reactions at work in the creation of beer. An overview of the brewing process is shown in Fig. 4.1.

Milling: The first step of the process is crushing the malt. This breaks apart the grains, exposing the starchy ball inside and making it accessible to the brewer. The grains are only lightly crushed, leaving the hulls intact to serve as a filter bed for the *lautering* process later on.

Mashing: Mashing is the process by which the brewer extracts fermentable sugars from the grain. Basically it consists of steeping the grains in water at temperatures between 140° and 160° Fahrenheit for a period of 60–90 minutes at a thickness similar to porridge. This activates naturally occurring enzymes in the grain that convert the grain starches into sugars, like maltose, that yeast can metabolise. This process occurs in a vessel called a *mash tun.*

Lautering: During *lautering*, the fermentable sugars are rinsed from the grains. As the sugary liquor from the mash, now called *wort*, is slowly drained from the bottom of the *mash tun*, heated water is pumped in from the top at the same slow rate. As the water flows through the grains, it raises the temperature to about 170° Fahrenheit, making the sugars more soluble and easier to remove. As mentioned above, the intact grain hulls form a natural filter, removing bits of grain and proteins from the *wort*. In some brewing systems, lautering occurs in the *mash tun*. In others, the entire mash is pumped to a special *lauter tun.*

Boiling: From the *lautering* stage, the wort is pumped to the *kettle* where it is boiled. A vigorous boil is maintained for sixty minutes or more. During the boil, the *wort* is sterilised and concentrated to the proper sugar density, haze causing proteins are removed (*hot break*), and light caramelisation occurs that deepens the flavour and colour of the beer. One of the most important things to occur during the boil is the addition of hops. Hops for bittering are added early in the boil, while those for flavour and aroma are added later.

Cooling: Following the boil, the *wort* is pumped through a heat exchanger to cool as quickly as possible to fermentation temperature. Rapid cooling minimises the danger of bacterial contamination and causes more haze causing proteins to precipitate out of the *wort* (*cold break*).

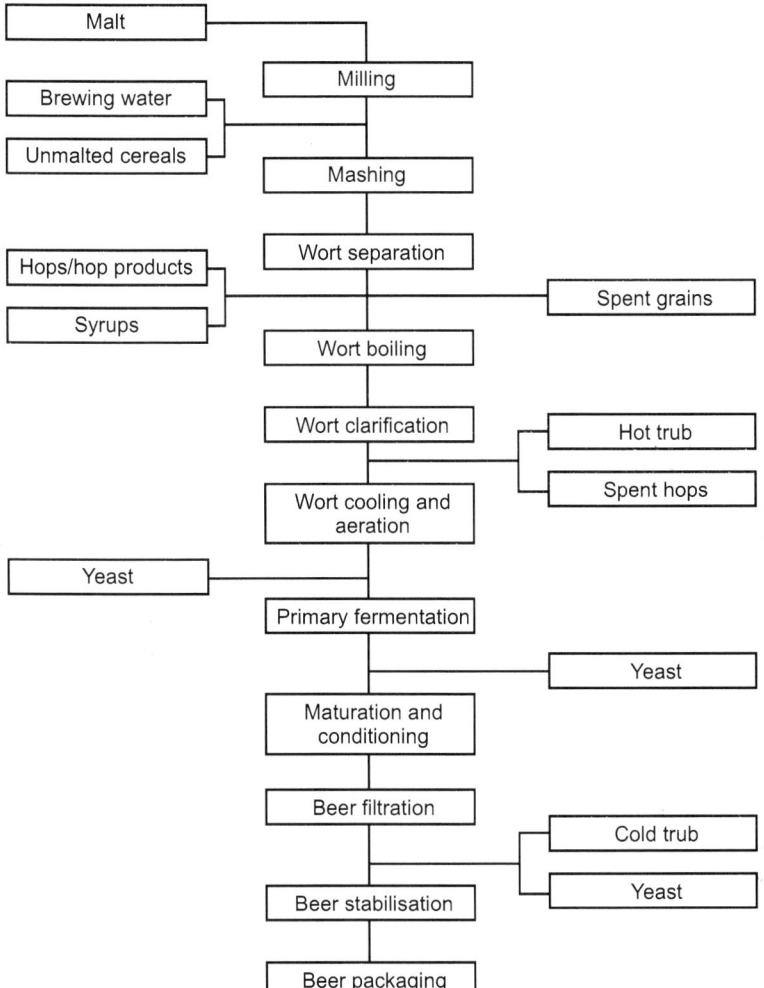

Figure 4.1: Schematic overview of the brewing process (input flows are indicated on the left side and output flows on the right side).

Fermentation: Yeast is added once the *wort* has reached the desired temperature for fermentation. Brewers call the addition of yeast *pitching*. Once the yeast has been *pitched*, the *wort* can properly be called *beer*. Fermentation can last a few days or a few weeks depending on the strain of yeast and the strength of the beer. During the process, the yeast is reproduce and then metabolise the sugars, making CO_2, alcohol and a host of other flavourful and aromatic compounds that add complexity to the beer. During the height of fermentation, the beer is capped by a thick creamy foam called *kreusen*. Once the available sugars have been consumed, the yeast cells clump together or *floc* and fall to the bottom of the fermenter.

Conditioning: When fermentation is complete, the beer is removed from the yeast and pumped to a conditioning or *bright tank* where it is stored at near freezing temperatures that cause most of the remaining yeast to drop out of suspension. Hops can be added at this point as well, a step known as *dry hopping* that lends the beer additional hop flavour and aroma. Once the beer is clear it is ready to filter and package.

4.3.2   Energy used in brewery

Within the breweries sector, natural gas and coal account for about 60% of the total primary energy and are mainly used for boiler fuel and for onsite electricity generation. Net electricity consumption, including losses, are 36% of primary energy requirements. Note that electricity expenditures account for 56% of the total, even though primary consumption is 36%. It is estimated that total energy expenditures account for roughly 3–8% of total production costs, based on data from the United Kingdom for breweries of similar design to those in the U.S.

Uses and sources of electricity in the brewer sector are as follows:

1. Boiler/hot water/steam generation.
2. Process cooling/refrigeration.
3. Machine drive (pumps, compressors, motors).
4. Facility HVAC.
5. Lighting.

The largest uses are in machine drives for pumps, compressed air and brewery equipment (46%), and for process cooling (32%). The onsite combined production of heat and power accounts for 22% of total electricity sources; a significant share especially when compared to other industries in the U.S. The vast majority of thermal energy is consumed in the form of steam at the brew house and for pasteurisation.

Energy intensity reflects the amount of energy required per unit of output or activity. Further, energy intensities will also vary depending on the size of the brewery. Large breweries (greater than 500,000 hL annual production) use less energy per unit output.

4.3.3   Energy efficiency improvement

This section discusses the energy efficiency opportunities identified for breweries. A brewery's choice of technology and the operational parameters of that technology are determined in large part by the standards for product quality and uniformity. In addition, this list is not presumed to be exhaustive and additional opportunities may exist. Although we have focused on energy efficiency measures, it is important to note that the reduction of raw materials needed or the reduction of product loss results in the indirect reduction in energy use. For example, the reduction of beer wastes can reduce the need for processing an equivalent amount of raw materials, thereby resulting in energy savings. While materials efficiency measures can also reduce energy consumption in breweries they are not specifically addressed in this section.

A variety of opportunities, may exist within breweries to reduce energy consumption while maintaining or enhancing quality and productivity of the plant.

Energy efficiency measures:
1. Boilers and steam distributions.
2. Meshing and lanter tun.
3. Wort boiling and cooling.
4. Motors and systems using motors.
5. Fermentation.
6. Refrigeration and cooling.
7. Processing.
8. Packaging.
9. Other utilities.

Energy efficiency measures in boilers and steam distribution include maintenance and controls, steam or heat recovery and process integration.

Improvements in motor and the systems that use them include downsizing of motors to match load requirements, introducing variable speed drives (VSD) and upgrading to better designed motors, pumps, and compressors. All of these measures have short payback periods of fewer than three years.

Refrigeration and cooling system improvements include, reducing the compressor size to match the load, improved operations and maintenance, insulation of the cooling lines and improved design. Energy savings for these measures range from 0.25 to 2 kWh/hL and many have additional benefits.

Systems modifications on some cooling systems often simplify it, by eliminating a step in the cooling process and reduce water use with direct ammonia cooling systems, there is always concern about the potential for ammonia leakage directly into the product.

'Other' utility efficiency measures also achieve non-energy benefits in addition to large energy savings. Lighting retrofits can increase productivity and the attractiveness of the workplace.

The most fuel intensive stage of brewing is the brew house given the high steam requirements for mashing and wort boiling. Electricity use is high because of its demand in operation of the cooling systems for fermentation and operation of machine drives in the plant. In the brew house, it is possible to recover low-grade heat from the mash or the hot water tank used in the mashing. Another possibility for saving energy during mashing is through the use of compression filters instead of plate filters.

Within the wort boiling and cooling step, many opportunities exist for recovering thermal energy for use in other brewery operations, either by minimising evaporation requirements (evaporation requirements are dictated by taste), recovering heat from vapour condensate or more efficient heat recovery (from wort cooling). Developments in kettle and brew house technology include thermal or mechanical vapour recompression, low pressure wort boiling, high gravity brewing, wort stripping and other system configurations.

High gravity brewing is a more common technology incorporated in the majority of breweries in the U.S., perhaps because of its many advantages. It increases brewing capacity with more efficient use of plant facilities, may improve product quality (better consistency and character have been reported, although the impact on flavour is an obvious concern), increases flexibility of beer type and productivity reduces water use lowers labour and cleaning costs and defers capital expenditures.

Wort stripping provides shorter cooking times for the wort and significant reductions in evaporation requirements with no changes in colour foam stability. Heat recovery from wort cooling systems when combined with water reuse in mashing bottle washing or cleaning, can reduce water use. Breweries using vapour condensers in the Netherlands and in the U.S. reported reduced water and operations and maintenance costs in addition to energy savings. Users have found that thermal vapour recompressors reduce the need for a circulation pump and reduce boiling times. Manufacturers claim mechanical vapour recompressors reduce aroma emissions almost entirely provide a gentler boiling process and save in steam in many cases.

Energy efficiency measures in the fermentation step include the use of immobilised yeast heat recovery systems and carbon dioxide (CO_2) recovery systems. While immobilised yeast technologies are currently being perfected they have been tested at a pilot plant and were found to reduce yeast reactor time significantly, i.e. from weeks to hours.

In addition, immobilised yeast technology has been found to improve process quality control and reduce materials through the reuse of yeast and the reduction in Kieselguhr required for filtration in the process. Implementing immobilised yeast technologies however may affect the flavour of the beer. Heat recovery systems may be expensive and limited data exists for these systems. Vendors estimate the payback period for carbon dioxide recovery systems to be 2 to 3 years from energy savings alone. Anheuser-Busch, however, estimates paybacks to be longer for CO_2 recovery systems for U.S. breweries, based on lower cost of domestic CO_2. CO_2 recovery systems are fairly common for large breweries but advances in the technology are making them more attractive for medium and small breweries.

In addition to the energy savings, manufacturers claim that compared to the traditional technology used; CO_2 recovery technology saves CO_2, requires less capital has much lower operation and maintenance costs, eliminates recirculation pumps and saves 50% of the water in scrubbing systems. Part of beer conditioning is used in removing all remaining unwanted bacteria before bottling usually through pasteurisation. Improvements in pasteurisation include tunnel or flash pasteurisation and heat recovery. Heat recovery in pasteurisation has been reported to save 1 kBtu/barrel of primary energy. Flash pasteurisation has been reported to reduce energy by two thirds compared to tunnel pasteurisation, a primary energy savings of 6 to 14 kBtu/barrel. In addition to energy savings flash pasteurisation has been found to require less space and coolant, lower initial investment and lower operation and maintenance costs. Since flash pasteurisation is integrally linked to the purchase and use of sterile filling technology, however the use of flash pasteurisation includes significant additional costs associated with sterile filtration requirements.

An alternative to pasteurisation is the use of sterile filtration in cross flow membrane filtration technologies. Though limited data exists for energy savings from oscillating microfiltration systems, investigations found potential 15–40% savings compared to standard steady-flow microfiltration. This technology is new to the brewery industry, but is being investigated for its potential energy savings as well as savings on disposal costs and reduction of waste. However some believe current cross flow membrane filtration systems may require as much extra energy as they save.

The processing stage is also the stage where alcohol is removed for non-alcoholic beer. The use of membranes is seen to be the most promising technology in the long term for production of non-alcoholic beer and has significant energy savings as well.

In packaging, the final stage of brewing, energy savings can be attained through heat recovery from bottle washing as well as from cleaning efficiencies. Researchers found energy savings at 6 kBtu/barrel for heat recovery washing. With paybacks of three years or less. In addition to the energy savings in heat recovery, water use was reduced by 40%.

Energy conservation in dairy industry

5.1   Introduction

Dairy products include milk and any of the foods made from milk, including butter, cheese, ice cream, yogurt, and condensed and dried milk. Milk has been used by humans since the beginning of recorded time to provide both fresh and storable nutritious foods. In few countries almost half the milk produced is consumed as fresh pasteurised whole, low-fat, or skim milk.

Cow milk (bovine species) is by far the principal type used throughout the world. Other animals utilised for their milk production include buffalo (in India, China, Egypt, and the Philippines), goats (in the Mediterranean countries), reindeer (in northern Europe), and sheep (in southern Europe). This section focuses on the processing of cow milk and milk products unless otherwise noted. In general, the processing technology described for cow milk can be successfully applied to milk obtained from other species.

5.2   Processing

5.2.1   Pasteurisation

Pasteurisation is most important in all dairy processing. It is the biological safeguard which ensures that all potential pathogens are destroyed. Extensive studies have determined that heating milk to 63°C (145°F) for 30 minutes or 72°C (161°F) for 15 seconds kills the most resistant harmful bacteria. In actual practice these temperatures and times are exceeded, thereby not only ensuring safety but also extending shelf life.

Most milk today is pasteurised by the continuous high-temperature short-time (HTST) method (72°C or 161°F for 15 seconds or above). The HTST method is conducted in a series of stainless steel plates and tubes, with the hot pasteurised milk on one side of the plate being cooled by the incoming raw milk on the other side. This 'regeneration' can be more than 90% efficient and greatly reduces the cost of heating and cooling.

There are many fail-safe controls on an approved pasteuriser system to ensure that all milk is completely heated for the full time and temperature requirement. If the monitoring instruments detect that something is wrong, an automatic flow diversion valve will prevent the milk from moving on to the next processing stage.

Higher temperatures and sometimes longer holding times are required for the pasteurisation of milk or cream with a high fat or sugar content.

Pasteurised milk is not sterile and is expected to contain small numbers of harmless bacteria. Therefore, the milk must be immediately cooled to below 4.4°C (40°F) and protected from any outside contamination.

The shelf life for high-quality pasteurised milk is about 14 days when properly refrigerated.

Extended shelf life can be achieved through ultra pasteurisation. In this case, milk is heated to 138°C (280°F) for two seconds and aseptically placed in sterile conventional milk containers. Ultra pasteurised milk and cream must be refrigerated and will last at least 45 days. This process does minimal damage to the flavour and extends the shelf life of slow-selling products such as cream, eggnog, and lactose-reduced milks.

Ultrahigh-temperature (UHT) pasteurisation is the same heating process as ultra pasteurisation (138°C or 280°F for two seconds), but the milk then goes into a more substantial container—either a sterile five-layer laminated 'box' or a metal can. This milk can be stored without refrigeration and has a shelf life of six months to a year. Products handled in this manner do not taste as fresh, but they are useful as an emergency supply or when refrigeration is not available.

5.2.2 Separation

Most modern plants use a separator to control the fat content of various products. A separator is a high-speed centrifuge that acts on the principle that cream or butterfat is lighter than other components in milk. (The specific gravity of skim milk is 1.0358, specific gravity of heavy cream 1.0083.) The heart of the separator is an airtight bowl with funnel like stainless steel disks. The bowl is spun at a high speed (about 6000 revolutions per minute), producing centrifugal forces of 4000 to 5000 times the force of gravity. Centrifugation causes the skim, which is denser than cream, to collect at the outer wall of the bowl. The lighter part (cream) is forced to the centre and piped off for appropriate use.

An additional benefit of the separator is that it also acts as a clarifier. Particles even heavier than the skim, such as sediment, somatic cells, and some bacteria, are thrown to the outside and collected in pockets on the side of the separator. This material, known as 'separator sludge,' is discharged periodically and sometimes automatically when buildup is sensed.

Most separators are controlled by computers and can produce milk of almost any fat content. Current standards generally set whole milk at 3.25% fat, low-fat at 1 or 2%, and skim at less than 0.5%. (Most skim milk is actually less than 0.01% fat.)

5.2.3   Homogenisation

Milk is homogenised to prevent fat globules from floating to the top and forming a cream layer or cream plug. Homogenisers are simply heavy-duty, high-pressure pumps equipped with a special valve at the discharge end. They are designed to break up fat globules from their normal size of up to 18 micrometers to less than 2 micrometers in diameter (a micrometer is one-millionth of a meter). Hot milk (with the fat in liquid state) is pumped through the valve under high pressure, resulting in a uniform and stable distribution of fat throughout the milk. Two-stage homogenisation is sometimes practiced, during which the milk is forced through a second homogeniser valve or a breaker ring. The purpose is to break up fat clusters or clumps and thus produce a more uniform product with a slightly reduced viscosity. Homogenisation is considered successful when there is no visible separation of cream and the fat content in the top 100 millilitres of milk in a one-litre bottle does not differ by more than 10% from the bottom portion after standing 48 hours. In addition to avoiding a cream layer, other benefits of homogenised milk include a whiter appearance, richer flavour, more uniform viscosity, better 'whitening' in coffee, and softer curd tension (making the milk more digestible for humans). Homogenisation is also essential for providing improved body and texture in ice cream, as well as numerous other products such as half-and-half, cream cheese and evaporated milk. Milk processing operations are shown in Fig. 5.1.

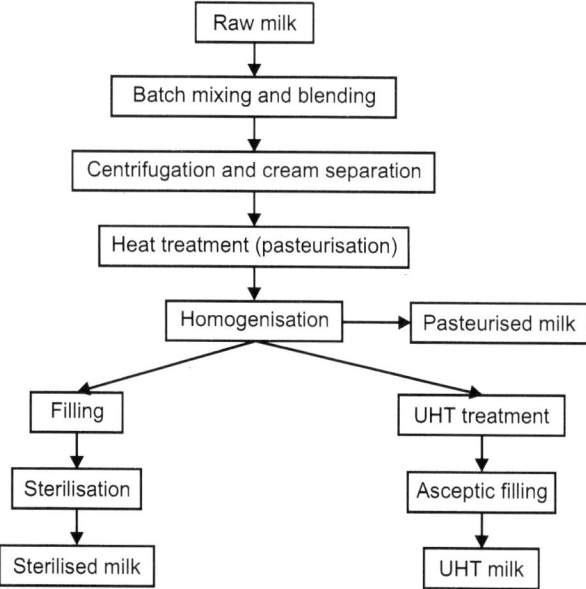

Figure 5.1: Milk processing operations.

5.3 Energy consumption

Dairy processing facilities consume considerable amounts of energy. Typically, approximately 80% of the energy requirements are for thermal uses to generate hot water and produce steam for process applications (e.g. pasteurisation, evaporation, and milk drying) and cleaning purposes. The remaining 20% is used as electricity to drive processing machinery, refrigeration, ventilation, and lighting.

In addition to recommendations to increase energy efficiency the following industry-specific measures are recommended:

1. Reduce heat loss by:
   (a) Using continuous, instead of batch, pasteurisers.
   (b) Partially homogenising milk to reduce the size of heat exchangers.
   (c) Using multistaged evaporators.
   (d) Insulating steam, water, and air pipes/tubes.
   (e) Eliminating steam leakage and using thermostatically controlled steam and water blending valves.
2. Improve cooling efficiency by:
   (a) Insulating refrigerated room/areas.
   (d) Installing automatic door closing (e.g. with microswitches) and applying airlocks and alarms.
3. Employ heat recovery for both heating and cooling operations in milk pasteurisers and heat exchangers (e.g. regenerative countercurrent flow).
4. Investigate the means to recover waste heat, including:
   (a) Recovering waste heat from refrigeration plant, exhaust and compressors (e.g. to preheat hot water).
   (d) Recovering evaporative energy.
   (e) Employing heat recovery from air compressors and boilers (e.g. waste gas exchanger).

5.4 Energy optimisation

As already discussed, pasteurised milk process in a diary plant consumes large amount of energy consisting of electricity and fuel.

5.4.1 Methodology

There is no tailor made proposed methodology, as the energy audit role changes with the organisation type. The aim of representing the methodology for milk dairy plant is to control the wastage and losses of the complete manufacturing cycle.

The methodology of energy optimisation for milk dairy plant processes is shown in Fig. 5.2. The first step of methodology is the identification of main work center or functions of the dairy plant.

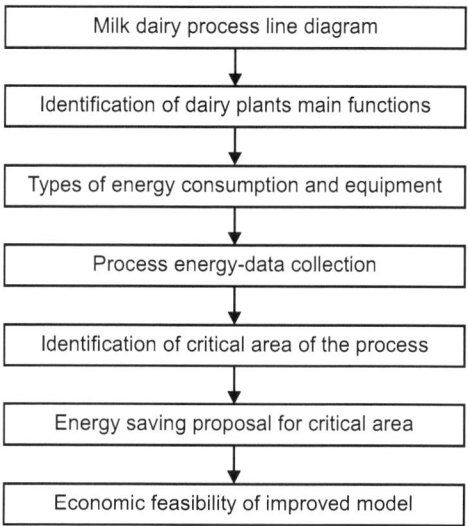

Figure 5.2: Methodology for milk dairy plant processes energy optimisation.

The energy audit is conducted on selected function's selected equipment only. The data of energy consumption and energy efficiency are collected. An audit team consisting of qualified and experienced electrical and mechanical engineers inspects the plant. However, it may be mentioned that each team should accompanied by an expert. The energy audit identifies the dairy plant areas where wastage of any type of energy found. To developed improved model of manufacturing system, the audit approach determines the critical areas and energy wastage and the advanced manufacturing techniques improved them. The proposed alternative designs for heat treatment, e.g. heating and cooling process during pasteurisation process by help of plate type heat exchanger. The existing system is then compared with the improved system on basis of various alternatives and on criterion of energy consumed, cost, noise generated and location feasibility.

5.4.2   Energy optimisation in milk dairy plant

The role of energy optimisation changes from plant to plant. Saving of energy is as important as environment saving. In different industries different energy conversion systems may be employed as every type of energy conversion system poses its own detrimental effect on the environment. Sometimes the harmful effluent developed by these systems causes halt to these systems. For example, the effluent of power station, industries, urban development's mining of coal or transportation of oil by sea, etc. Each type of energy conversion system has its own limitation regarding effluent and pollution. Both air and water pollution causes environment losses in the form of animal and human life losses and agricultural crop and forest losses. The industrial and nuclear wastages pose a danger to ecological system. They destroy the earth atmosphere, which in turn result in change in wind pattern, rainfall, life in coastal area and temperature increase.

The energy optimisation when applied to modern milk dairy plant means to develop a methodology for reducing energy different forms wastages and losses in a manufacturing system of dairy plant. It helps to develop capable manufacturing system with optimum energy efficiency for different products and grades. The selection and performance control of any energy conversion system requires proper planning and energy audit provide information for energy wise use only. The each equipment appropriateness and correct way of operations is determined and implemented for efficient and judicious use of energy. No matter how well-designed the plants is, the day to day output shows variation from unit to unit. The process on-line control is very important and essential and can be developed with continuous monitoring and improvement.

5.4.3 Importance of energy audit approach

The Energy Audit approach is a key approach for systematic decision making in process management. It quantifies the energy uses according to its various functions. It attempts to balance the total energy inputs with the output or the uses. The role of energy audit changes from preliminary audit to detailed audit. The detailed audit goes beyond quantitative estimates to the energy cost and saving and includes engineering applications and recommendations. The preliminary audit can be used as control tool to take feedback of the implemented projects and form basis of next project of improvements. The energy conservation and maximisation strategies for a process industry like dairy plant are cost effective, which conserve the environment automatically.

5.4.4 Other methods of energy conservation in milk dairy plan

By cogeneration system

Dairy industry requires both electricity and process heat. There is a chance of cogeneration. The cogeneration system shows that relative fuel saving and heat rate of cogeneration is less than the heat rate of power plant. Hence cogeneration is seen to be feasible, but it has following limitation: The power generation is so small hence, it is difficult to get the generator of such a small capacity. Even if we get the generator then scale of pay is very high hence, it is not that affordable.

The other methods are as follows:

1. By economiser to heat feed water for boilers, so use waste heat.
2. By proper integration of chilling system to the main process by load balancing.
3. To use waste heat of air compression outlet to heat boiler inlet water.
4. By proper loading of electrical motors and lightening system.
5. By using variable frequency drives in air compressors.
6. By proper maintenance of equipments.
7. By using flat plate heat exchanger to preheat milk of chilling process raw milk.

Using a plate type heat exchanger

In this alternative, it is suggested to utilise the energy wasted in the form of heat carried by the heated milk going into the chiller unit. The suggested alternative plans to use this heat and the same is used to preheat the milk coming from the supply and hence reduce the heating load required for the purpose of milk processing. This can be done by using a plate type heat exchanger. The suggested alternative design to use this heat and the same is used to preheat the milk coming from the fresh supply and hence reduce the heating load required for the purpose of milk processing in chilling unit. The energy audit is useful method and for any process industry like milk dairy plant. By the help of this technique the energy loss and wastages are easily identified and improved system or model can be developed.

In the pasteurisation process, milk is heated to kill pathogens and then cooled down again for preservation. Heat can be recovered by using heat exchangers, so that the incoming cold milk is heated by the outgoing pasteurised milk, which is in turn cooled. Transferring heat this way leads to substantial energy efficiency improvements.

6

Energy conservation in chemical process and allied industries

6.1   Introduction

The chemicals industry in industrialised nations produces a wide variety of chemicals ranging from commodity industrial chemicals used to make other products to specialty chemicals tailored for unique applications. As a major user of raw materials, both for energy consumption and as feedstocks, the chemicals industry can significantly impact the demand of non-renewable resources. Chemical industry is the fountain head for the manufacture of a variety of chemical products beginning with products like soda ash, paper, plastics, petrochemicals, petroleum products, organic and fine chemicals, fertilisers, pharmaceuticals and some speciality chemicals like isotopes (heavy water for example). All these products are manufactured using very large number of processes and operations. The level of processing and the energy consumption depends on the complexity of the process involved. Thus, the chemicals industry is a significant target when it comes to focusing on energy efficiency improvement in the context of decreasing the environmental burden of the industrialised nations.

The processes in the chemicals industry may be run in continuous or batch operation mode. The continuous operation mode is typical for the production of commodity chemicals (basic, bulk chemicals). This sector can be characterised by large plants, low profit margins, and high energy consumption. On the other hand, the batch processes used mainly in the sector of specialty and fine chemicals (SFC), are typically manufactured in lower volumes than basic chemicals, provide higher profit margins and have less cyclicality in their business cycle. Specialty chemicals are usually chemical substances (e.g. adhesives and dyes, catalysts, coatings, plastic additives) that are derived from basic chemicals and are more technologically advanced products than basic chemicals. The higher value-added to these chemicals is due to the fact that they cannot easily be duplicated by other producers or are shielded from competition by patents. Energy efficiency in the specialty chemicals sector has not been extensively promoted in the past due to the relatively high profit margins. Notwithstanding the higher margins (as compared with the basic chemicals which are produced in continuous mode), the energy saving motivation nowadays in the specialty chemicals sector is growing with increasing energy prices and along with a developing environmental consciousness of the specialty chemicals industry. Few of the specialty chemicals producers decided to actively support the aims of the Kyoto Protocol, which has the objective of reducing the emissions of greenhouse gases. Few successful companies announced reduction of energy requirement per tonne of product by up to 22% in recent years as a result of implementation of energy saving measures. Moreover, the life cycle of specialty chemicals creates an additional driving force for the optimisation of the process including energy efficiency, particularly as the patent protection approaches its expiration as shown in Fig. 6.1.

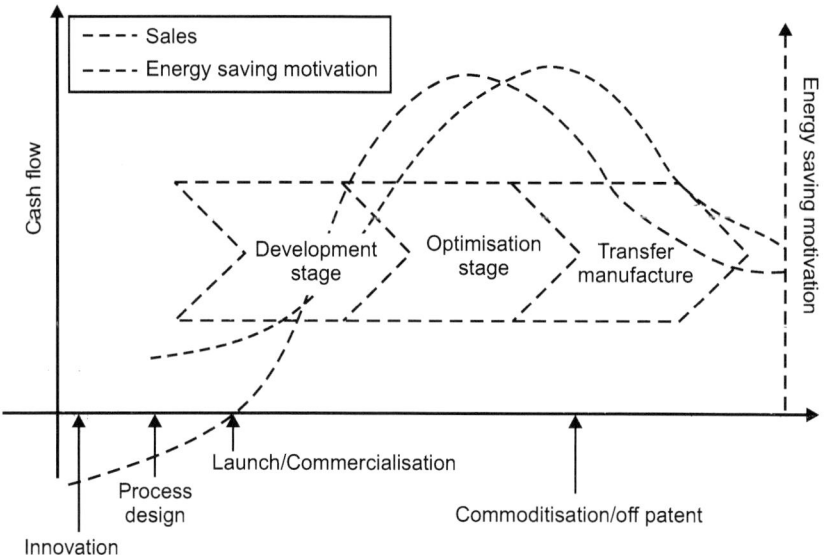

Figure 6.1: Energy saving in the context of the life cycle of the specialty chemicals.

The life cycle of the specialty chemicals starts with the innovation and process design stage. When designing a process for specialty chemicals, priority tends to be given to the product rather than to the process, therefore, the energy saving motivation is relatively low at this stage. This is because the unique function of the product must be protected. The process is likely to be small scale and operating costs tend to be less important than with commodity chemical processes. The capital cost of the process will be low relative to commodity chemical processes because of the scale. The time to market of the product is also likely to be important with specialty chemicals, especially if there is patent protection. If this is the case, then anything that shortens the time from basic research, through to product testing, pilot plant studies, process design, construction of the plant to product manufacturing will have an important influence on the overall project profitability. After the launch of the production in the plant scale, the optimisation of the process becomes an important issue. In particular, some of the energy systems might by oversized, resulting in high energy consumption and an unreliable control of the process parameters. Moreover, this can lead to quality variation, which is highly undesirable in the sector of specialty chemicals. Therefore, the motivation for process optimisation, including energy saving exponentially rises in this stage of the life cycle. Energy saving motivation peaks after the specialty chemical product has passed the maturity stage, and its sales and profitability start to decline. The main reason is to keep the production profitable and competitive as long as possible by cutting the production costs. When the critical point in the profitability of the product is reached, the manufacture is transferred to the low-cost destination. In the specialty chemicals sector, energy represents approximately 10% of the overall cost, making it an important item in the inter- and intra-enterprise competition. This percentage becomes more significant when it is considered that 60% of the overall cost is fixed and owes to raw materials cost, thus leaving 40% for potential cost savings. Moreover, the significant boost of energy prices in recent years has intensified the efforts of the manufacturers for more energy efficient strategies in plant operation. Although some energy losses along chemical production processes are unavoidable according to fundamental laws of thermodynamics, a major part can be viewed as potential for embracing efficient technologies and practices. For example, the losses of the energy utility used for heating in chemical batch processes can reach more than 50% of the overall heating utility consumption for particular process units. Therefore, the energy optimisation potential is high in these units. The practical implementation of energy optimisation is shown in Fig. 6.2. First, the energy saving initiative starts and the plant efficiency is estimated by statistical models or benchmarking, so the energy saving goals can be defined. Then, the optimisation cycle starts with data acquisition, modelling, analysis and finally optimisation, which influences the process data. Operational and design problems in heat-transfer equipment are shown in Table 6.1.

6.2 Energy conservation through technology

In considering the chemicals industry as a whole, there are three main areas for improving energy use: physical separation, energy recovery and product integration.

6.2.1 Physical separation technologies

Dramatic improvements in energy use can result from changes in the physical principles embodied in certain unit operations, especially in physical separation.

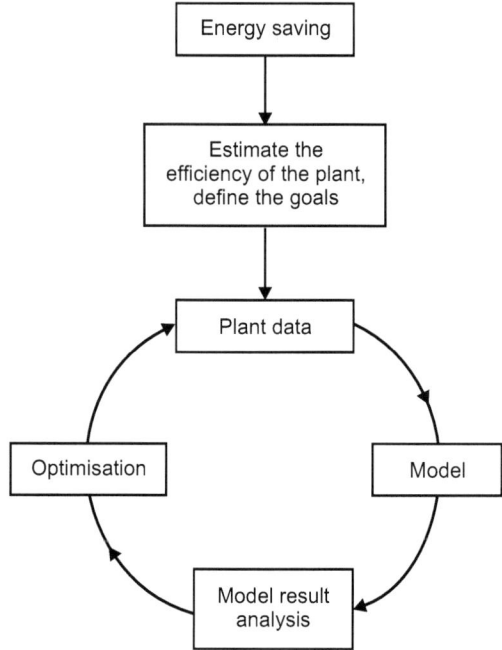

Figure 6.2: Energy optimisation cycle.

Table 6.1: Operational and design problems in heat-transfer equipment.

Common problems	Measures to overcome problems
Steam traps	
Faulty operation	Monitor
Leaking traps	Maintenance repair
Mismatch between steam line pressure and trap operating range	Use proper application and sizing
Steam tracing	
Leaks	Maintenance repair
Unnecessarily high steam temperature	Substitute another fluid for steam
Heat exchangers	
Fouling	Maintenance repair
Higher than necessary temperature separation between fluid streams	Design for low-temperature differences by increasing heat-transfer surface area

By far, the most widespread technique of chemical separation used today for mixtures of liquids is distillation. This is an energy intensive process, especially as practiced in the former days of cheap fuel. Already, incremental improvements in the process, retrofitted to existing installations, have achieved significant (e.g. 25%) savings in many plants. Further improvements of comparable magnitude can be expected during the next few years through redesign and add-on units, though generally at higher costs. Steam distillation columns provide opportunities for heat recovery in larger, integrated systems.

Alternative approaches to liquid separation include vacuum distillation, freeze crystallisation, and liquid-liquid (solvent) extraction. Dramatic increases in the cost effectiveness of turbo compressors and advances in vacuum pumps and cryogenic technology. However, the most promising technique seems to be liquid-liquid extraction, a process using a solvent with high affinity for one component of the mixture but immiscible with the remaining components. With this technique, separation involves two steps: decanting and closed-loop evaporation/condensation of the solvent.

Dehydration ('drying') using steam heat is another energy-intensive separation operation that can be dramatically improved in many cases. A technique of squeeze-drying wet solids or fabrics (prior to steam drying) can be adapted from technologies already developed in the paper industry, Separation (prior to disposal by incineration) of oily wastes or oil-soluble contaminants from water mixtures can be accomplished by using specially treated cellulose that has an affinity for oil. The oil-soaked cellulose can subsequently be burned, or squeeze-dried and then recycled.

6.2.2   Technologies for energy recovery and conservation

This category includes both heat recovery *per se* and improved utilisation of energy embodied in high-pressure gases or steam.

A variety of engineering schemes are available to recover waste heat from boilers and exothermic reactors. A 'bare burner' boiler, operating with excess air to ensure complete fuel combustion, will typically produce stack gases at 600°F with 6.2% oxygen, a stack gas heat loss of 19%, and an overall thermal efficiency of 78%. Modest improvements in efficiency could be achieved by more precise monitoring of stack temperatures, fuel and air intake, and closed-loop process control.

More significant improvements could result from using the heat of the stack gases either to preheat intake air or intake water via an 'economiser.' Overall efficiency of 85%, with stack gas temperatures reduced to 350°F, is readily achievable by either technique.

Many older plants used steam-driven vacuum jets instead of electric-or turbine-driven vacuum pumps because of lower capital costs. However, in a typical application, the vacuum pump is up to four times more efficient. For example, 80,000 Btu per hour are typically used for the electric pump versus 300,000 Btu per hour for the steam jet. Most existing steam jet-driven vacuum systems will probably be replaced older ones except in those situations of low pressure and low flow where they will continue to have an economic advantage.

6.2.3   Production integration technologies

Integration is a strategy for justifying energy and waste recovery that would not otherwise be economically justified. The simplest example is cogeneration of electricity and steam. Most firms in the chemicals industry have several applications of cogeneration under active consideration—in some cases, based on the use of process wastes as fuel.

Cogeneration opportunities exist to produce electricity or mechanical shaft power as a by-product of existing steam systems. For instance, in one plant an existing steam boiler produced 300- and 40-psi steam (as needed in the plant). By modifying the boiler to produce steam at 800 psi and 800°F, and interposing a turbo generator (with exhausts at 300 and 40 psi), enough electricity to supply the plant was generated.

Since, utility electricity normally requires 10,000 Btu to produce 1 kWh of electricity, and this operation used 4200 Btu to produce 1 kwh of electricity, there was a net energy savings of 5800 Btu/kWh.

Potential savings from production integration extend far beyond the case of cogeneration, however. Production of intermediates, such as ethylene and butadiene, is increasingly being integrated into petroleum refining complexes. This trend will be accelerated by the shift toward heavier cracking feedstocks such as heavy gas oil or fuel oil because of the greater importance of coproduction.

Integration of the production of ethylene, propylene, and a wide range of petrochemicals from a naphtha-based (aromatics-based) scheme is a strong possibility. Another option would be to integrate ethylene and acetylene production with ammonia and/or methanol. Ethylene/acetylene coproduction will become increasingly attractive as distillate prices rise and heavier feedstocks are used, and will undoubtedly result in some downstream process switching as acetylene again becomes competitive with ethylene as a feedstock for acrylates, vinyl acetate, and vinyl chloride.

In addition, the following five specific trends would occur in the chemicals industry,

1. Ethylene feedstocks will switch from gaseous.
2. Feeds (ethane and propane) to liquid feeds (naphtha and gas oil).
3. More chlorine will be produced from the diaphragm cell and less from the mercury cell.
4. Ammonia and methanol production from coal via synthesis gas will become more prevalent.

5. Acetylene production will move toward the crude oil, submerged-flame process.

6. Less phosphoric acid will be produced in electric arc furnaces.

While some of these trends were drawn purely on economic grounds, some are the result of the increasingly cautious attitude toward the use of limited feedstock resources by the industry. For example, by moving toward liquid feeds, an olefin plant will increase its feedstock flexibility because liquid feedstocks require additional vapourising equipment and this equipment can be used for a variety of liquid feeds. This chapter discusses the 'energy conservation aspects' of various 'chemical process industries' such as– petroleum refinery, fertiliser, sugar, pulp and paper, chlorine and caustic soda.

6.3   Petroleum refinery

Petroleum is a complex mixture of organic liquids called crude oil and natural gas, which occurs naturally in the ground and was formed millions of years ago. Crude oil varies from oilfield to oilfield in colour and composition, from a pale yellow low viscosity liquid to heavy black 'treacle' consistencies.

Crude oil and natural gas are extracted from the ground, on land or under the oceans, by sinking an oil well and are then transported by pipeline and/or ship to refineries where their components are processed into refined products. Crude oil and natural gas are of little use in their raw state; their value lies in what is created from them: fuels, lubricating oils, waxes, asphalt, petrochemicals and pipeline quality natural gas.

An oil refinery is an organised and coordinated arrangement of manufacturing processes designed to produce physical and chemical changes in crude oil to convert it into everyday products like petrol, diesel, lubricating oil, fuel oil and bitumen. As crude oil comes from the well it contains a mixture of hydrocarbon compounds and relatively small quantities of other materials such as oxygen, nitrogen, sulphur, salt and water. In the refinery, most of these non-hydrocarbon substances are removed and the oil is broken down into its various components, and blended into useful products. Natural gas from the well, while principally methane, contains quantities of other hydrocarbons- ethane, propane, butane, pentane and also carbon dioxide and water. These components are separated from the methane at a gas fractionation plant.

6.3.1   Petroleum hydrocarbon structures

Petroleum consists of three main hydrocarbon groups:

Paraffins: These consist of straight or branched carbon rings saturated with hydrogen atoms, the simplest of which is methane (CH_4) the main ingredient of natural gas. Others in this group include ethane (C_2H_6), and propane (C_3H_8).

Hydrocarbons: With very few carbon atoms (C_1 to C_4) are light in density and are gases under normal atmospheric pressure. Chemically paraffins are very stable compounds.

Naphthenes: Naphthenes consist of carbon rings, sometimes with side chains, saturated with hydrogen atoms. Naphthenes are chemically stable, they occur naturally in crude oil and have properties similar to paraffins.

Aromatics: Aromatic hydrocarbons are compounds that contain a ring of six carbon atoms with alternating double and single bonds and six attached hydrogen atoms. This type of structure is known as a benzene ring. They occur naturally in crude oil, and can also be created by the refining process.

The more carbon atoms a hydrocarbon molecule has, the 'heavier' it is (the higher is its molecular weight) and the higher is its the boiling point. Small quantities of a crude oil may be composed of compounds containing oxygen, nitrogen, sulphur and metals. Sulphur content ranges from traces to more than 5%. If a crude oil contains appreciable quantities of sulphur it is called a sour crude; if it contains little or no sulphur it is called a sweet crude.

6.3.2 Refining process

Every refinery begins with the separation of crude oil into different fractions by distillation. Basic refinery operations are shown in Fig. 6.3.

The fractions are further treated to convert them into mixtures of more useful saleable products by various methods such as cracking, reforming, alkylation, polymerisation and isomerisation. These mixtures of new compounds are then separated using methods such as fractionation and solvent extraction. Impurities are removed by various methods, e.g. dehydration, desalting, sulphur removal and hydrotreating. Refinery processes have developed in response to changing market demands for certain products. With the advent of the internal combustion engine the main task of refineries became the production of petrol. The quantities of petrol available from distillation alone was insufficient to satisfy consumer demand. Refineries began to look for ways to produce more and better quality petrol.

Two types of processes have been developed:
1. Breaking down large, heavy hydrocarbon molecules.
2. Reshaping or rebuilding hydrocarbon molecules.

Distillation (fractionation)

Because crude oil is a mixture of hydrocarbons with different boiling temperatures, it can be separated by distillation into groups of hydrocarbons that boil between two specified boiling points. Two types of distillation are performed: atmospheric and vacuum.

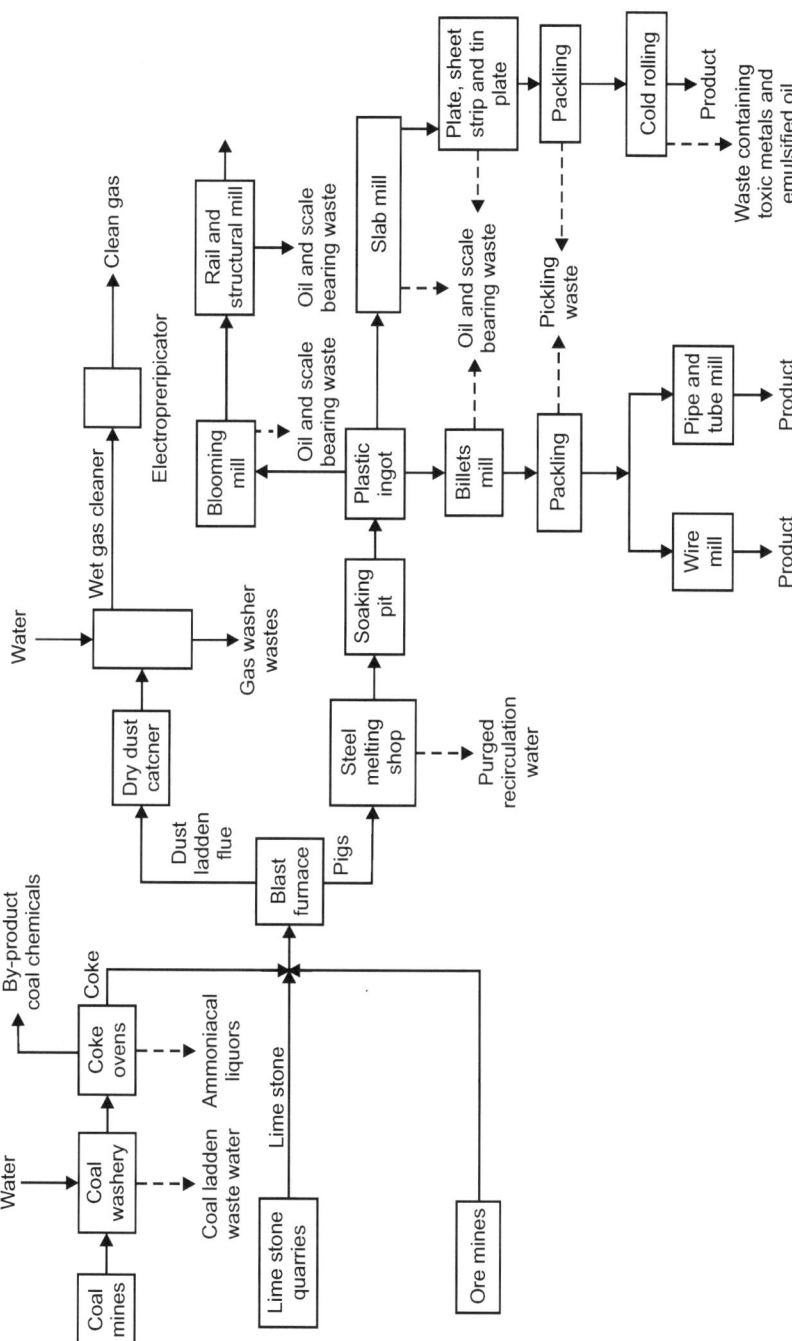

Figure 6.3: Basic refinery operations.

Atmospheric distillation takes place in a distilling column at or near atmospheric pressure. The crude oil is heated to 350–400°C and the vapour and liquid are piped into the distilling column. The liquid falls to the bottom and the vapour rises, passing through a series of perforated trays (sieve trays). Heavier hydrocarbons condense more quickly and settle on lower trays and lighter hydrocarbons remain as a vapour longer and condense on higher trays. Liquid fractions are drawn from the trays and removed. In this way the light gases, methane, ethane, propane and butane pass out the top of the column, petrol is formed in the top trays, kerosene and gas oils in the middle, and fuel oils at the bottom. Residue drawn of the bottom may be burned as fuel, processed into lubricating oils, waxes and bitumen or used as feedstock for cracking units.

To recover additional heavy distillates from this residue, it may be piped to a second distillation column where the process is repeated under vacuum, called vacuum distillation. This allows heavy hydrocarbons with boiling points of 450°C and higher to be separated without them partly cracking into unwanted products such as coke and gas.

The heavy distillates recovered by vacuum distillation can be converted into lubricating oils by a variety of processes. The most common of these is called solvent extraction. In one version of this process, the heavy distillate is washed with a liquid which does not dissolve in it but which dissolves (and so extracts) the non-lubricating oil components out of it. Another version uses a liquid which does not dissolve in it but which causes the non-lubricating oil components to precipitate (as an extract) from it.

Other processes exist which remove impurities by adsorption onto a highly porous solid or which remove any waxes that may be present by causing them to crystallise and precipitate out. Figure 6.4 shows distilling crude and product disposition.

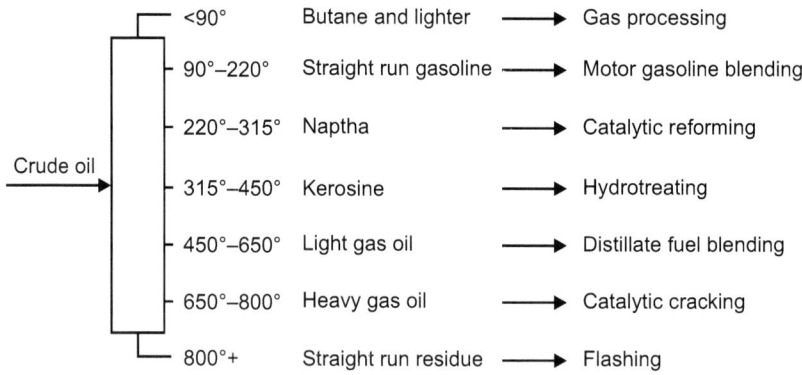

Figure 6.4: Distilling crude and product disposition.

6.3.3   Reforming

Reforming is a process which uses heat, pressure and a catalyst (usually containing platinum) to bring about chemical reactions which upgrade naphthas into high octane petrol and petrochemical feedstock. The naphthas are hydrocarbon mixtures containing many paraffins and naphthenes. This naphtha feedstock comes from the crudes oil distillation or catalytic cracking processes, but overseas it also comes from thermal cracking and hydrocracking processes. Reforming converts a portion of these compounds to isoparaffins and aromatics, which are used to blend higher octane petrol.

1. Paraffins are converted to isoparaffins.
2. Paraffins are converted to naphthenes.
3. Naphthenes are converted to aromatics.

For example,

Heptane	$\xrightarrow{\text{Catalyst}}$	Toluene	+	Hydrogen
C_7H_{16}		C_7H_8	+	$4H_2$
Cyclohexane	$\xrightarrow{\text{Catalyst}}$	Benzene	+	Hydrogen
C_6H_{12}		C_6H_6	+	$3H_2$

6.3.4   Cracking

Cracking processes break down heavier hydrocarbon molecules (high boiling point oils) into lighter products such as petrol and diesel. These processes include catalytic cracking, thermal cracking and hydrocracking.

For example,

A typical reaction:

$$C_{16}H_{34} \xrightarrow{\text{Catalyst}} C_8H_{18} + C_8H_{16}$$

Catalytic cracking is used to convert heavy hydrocarbon fractions obtained by vacuum distillation into a mixture of more useful products such as petrol and light fuel oil. In this process, the feedstock undergoes a chemical breakdown, under controlled heat (450–500°C) and pressure, in the presence of a catalyst-a substance which promotes the reaction without itself being chemically changed. Small pellets of silica-alumina or silica-magnesia have proved to be the most effective catalysts.

The cracking reaction yields petrol, LPG, unsaturated olefin compounds, cracked gas oils, a liquid residue called cycle oil, light gases and a solid coke residue. Cycle oil is recycled to cause further breakdown and the coke, which forms a layer on the catalyst, is removed by burning. The other products are passed through a fractionator to be separated and separately processed.

Fluid catalytic cracking uses a catalyst in the form of a very fine powder which flows like a liquid when agitated by steam, air or vapour. Feedstock entering the process immediately meets a stream of very hot catalyst and vapourises. The resulting vapours keep the catalyst fluidised as it passes into the reactor, where the cracking takes place and where it is fluidised by the hydrocarbon vapour. The catalyst next passes to a steam stripping section where most of the volatile hydrocarbons are removed. It then passes to a regenerator vessel where it is fluidised by a mixture of air and the products of combustion which are produced as the coke on the catalyst is burnt off. The catalyst then flows back to the reactor. The catalyst thus undergoes a continuous circulation between the reactor, stripper and regenerator sections.

The catalyst is usually a mixture of aluminium oxide and silica. Most recently, the introduction of synthetic zeolite catalysts has allowed much shorter reaction times and improved yields and octane numbers of the cracked gasolines. Thermal cracking uses heat to break down the residue from vacuum distillation. The lighter elements produced from this process can be made into distillate fuels and petrol. Cracked gases are converted to petrol blending components by alkylation or polymerisation. Naphtha is upgraded to high quality petrol by reforming. Gas oil can be used as diesel fuel or can be converted to petrol by hydrocracking. The heavy residue is converted into residual oil or coke which is used in the manufacture of electrodes, graphite and carbides.

Hydrocracking can increase the yield of petrol components, as well as being used to produce light distillates. It produces no residues, only light oils. Hydrocracking is catalytic cracking in the presence of hydrogen. The extra hydrogen saturates, or hydrogenates, the chemical bonds of the cracked hydrocarbons and creates isomers with the desired characteristics. Hydro-cracking is also a treating process, because the hydrogen combines with contaminants such as sulphur and nitrogen, allowing them to be removed.

Gas oil feed is mixed with hydrogen, heated, and sent to a reactor vessel with a fixed bed catalyst, where cracking and hydrogenation take place. Products are sent to a fractionator to be separated. The hydrogen is recycled. Residue from this reaction is mixed again with hydrogen, reheated, and sent to a second reactor for further cracking under higher temperatures and pressures. In addition to cracked naphtha for making petrol, hydrocracking yields light gases useful for refinery fuel, or alkylation as well as components for high quality fuel oils, lube oils and petrochemical feedstocks.

Following the cracking processes, it is necessary to build or rearrange some of the lighter hydrocarbon molecules into high quality petrol or jet fuel blending components or into petrochemicals. The former can be achieved by several chemical process such as alkylation and isomerisation.

6.3.5   Alkylation

Olefins such as propylene and butylene are produced by catalytic and thermal cracking. Alkylation refers to the chemical bonding of these light molecules with isobutane to form larger branched-chain molecules (isoparaffins) that make high octane petrol.

Olefins and isobutane are mixed with an acid catalyst and cooled. They react to form alkylate, plus some normal butane, isobutane and propane. The resulting liquid is neutralised and separated in a series of distillation columns. Isobutane is recycled as feed and butane and propane sold as liquid petroleum gas (LPG).

For example,

Isobutane	Catalyst	Butylene		Isooctane
C_4H_{10}	$+$	C_4H_8	$\rightarrow$	C_8H_{18}

6.3.6   Isomerisation

Isomerisation refers to chemical rearrangement of straight-chain hydrocarbons (paraffins), so that they contain branches attached to the main chain (isoparaffins).

This is done for two reasons:

1. They create extra isobutane feed for alkylation.
2. They improve the octane of straight run pentanes and hexanes and hence make them into better petrol blending components.

Isomerisation is achieved by mixing normal butane with a little hydrogen and chloride and allowed to react in the presence of a catalyst to form isobutane, plus a small amount of normal butane and some lighter gases. Products are separated in a fractionators. The lighter gases are used as refinery fuel and the butane recycled as feed.

Pentanes and hexanes are the lighter components of petrol. Isomerisation can be used to improve petrol quality by converting these hydrocarbons to higher octane isomers. The process is the same as for butane isomerisation.

6.3.7   Polymerisation

Under pressure and temperature, over an acidic catalyst, light unsaturated hydrocarbon molecules react and combine with each other to form larger hydrocarbon molecules.

Such process can be used to react butenes (olefin molecules with four carbon atoms) with iso-butane (branched paraffin molecules, or isoparaffins, with four carbon atoms) to obtain a high octane olefinic petrol blending component called polymer gasoline.

6.3.8 Hydrotreating and sulphur plants

A number of contaminants are found in crude oil. As the fractions travel through the refinery processing units, these impurities can damage the equipment, the catalysts and the quality of the products. There are also legal limits on the contents of some impurities, like sulphur, in products.

Hydrotreating is one way of removing many of the contaminants from many of the intermediate or final products. In the hydrotreating process, the entering feedstock is mixed with hydrogen and heated to 300–380°C. The oil combined with the hydrogen then enters a reactor loaded with a catalyst which promotes several reactions:

1. Hydrogen combines with sulphur to form hydrogen sulphide (H_2S)
2. Nitrogen compounds are converted to ammonia
3. Any metals contained in the oil are deposited on the catalyst
4. Some of the olefins, aromatics or naphthenes become saturated with hydrogen to become paraffins and some cracking takes place, causing the creation of some methane, ethane, propane and butanes.

6.3.9 Sulphur recovery plants

The hydrogen sulphide created from hydrotreating is a toxic gas that needs further treatment. The usual process involves two steps:

1. The removal of the hydrogen sulphide gas from the hydrocarbon stream.
2. The conversion of hydrogen sulphide to elemental sulphur, a non-toxic and useful chemical.

Solvent extraction, using a solution of diethanolamine (DEA) dissolved in water, is applied to separate the hydrogen sulphide gas from the process stream. The hydrocarbon gas stream containing the hydrogen sulphide is bubbled through a solution of diethanolamine solution (DEA) under high pressure, such that the hydrogen sulphide gas dissolves in the DEA. The DEA and hydrogen mixture is heated at a low pressure and the dissolved hydrogen sulphide is released as a concentrated gas stream which is sent to another plant for conversion into sulphur.

Conversion of the concentrated hydrogen sulphide gas into sulphur occurs in two stages.

Combustion of part of the H_2S stream in a furnace, producing sulphur dioxide (SO_2) water (H_2O) and sulphur (S).

$$2H_2S + 2O_2 \rightarrow SO_2 + S + 2H_2O$$

Reaction of the remainder of the H_2S with the combustion products in the presence of a catalyst. The H_2S reacts with the SO_2 to form sulphur.

$$2H_2S + 2O_2 \rightarrow 3S + 2H_2O$$

As the reaction products are cooled the sulphur drops out of the reaction vessel in a molten state. Sulphur can be stored and shipped in either a molten or solid state.

6.3.10   Refineries and the environment

Air, water and land can all be affected by refinery operations. Refineries are well aware of their responsibility to the community and employ a variety of processes to safeguard the environment. The processes described below are used by the refinery in managing the environmental aspects of refining.

6.3.11   Air

Preserving air quality around a refinery involves controlling the following emissions:

1. Sulphur oxides
2. Hydrocarbon vapours
3. Smoke
4. Smells

Sulphur enters the refinery in crude oil feed. To deal with this refineries incorporate a sulphur recovery unit which operates on the principles described above. Many of the products used in a refinery produce hydrocarbon vapours. The escape of vapours to atmosphere are prevented by various means. Floating roofs are installed in tanks to prevent evaporation and so that there is no space for vapour to gather in the tanks. Where floating roofs cannot be used, the vapours from the tanks are collected in a vapour recovery system and absorbed back into the product stream. In addition, pumps and valves are routinely checked for vapour emissions and repaired if a leakage is found. Smoke is formed when the burning mixture contains insufficient oxygen or is not sufficiently mixed. Modern furnace control systems prevent this from happening during normal operation.

Smells are the most difficult emission to control and the easiest to detect. Refinery smells are generally associated with compounds containing sulphur, where even tiny losses are sufficient to cause a noticeable odour.

6.3.12   Water

Aqueous effluent's consist of cooling water, surface water and process water. The majority of the water discharged from the refinery has been used for cooling the various process streams. The cooling water does not actually come into contact with the process material and so has very little contamination. The cooling water passes through large 'interceptors' which separate any oil from minute leaks, etc., prior to discharge.

Rainwater falling on the refinery site must be treated before discharge to ensure no oily material washed off process equipment leaves the refinery. This is done first by passing the water through smaller 'plant oil catchers', which treat rainwater from separate areas on the site, and then all the streams pass to large 'interceptors' similar to those used for cooling water. The rainwater from the production areas is further treated in a Dissolved Air Flotation (DAF) unit. This unit cleans the water by using a flocculation agent to collect any remaining particles or oil droplets and floating the resulting flock to the surface with millions of tiny air bubbles. At the surface the flock is skimmed off and the clean water discharged.

Process water has actually come into contact with the process streams and so can contain significant contamination. This water is treated in the 'sour water treater' where the contaminants (mostly ammonia and hydrogen sulphide) are removed and then recovered or destroyed in a downstream plant. The process water, when treated in this way, can be reused in parts of the refinery and discharged through the process area rainwater treatment system and the DAF unit. Any treated process water that is not reused is discharged as waste to the sewerage system. This trade waste also includes the effluent from the refinery sewage treatment plant and a portion of treated water from the DAF unit. As most refineries import and export many feed materials and products by ship, the refinery and harbour authorities are prepared for spillage from the ship or pier.

6.3.13 Land

The refinery safeguards the land environment by ensuring the appropriate disposal of all wastes.

Within the refinery, all hydrocarbon wastes are recycled through the refinery slops system. This system consists of a network of collection pipes and a series of dewatering tanks. The recovered hydrocarbon is reprocessed through the distillation units.

Wastes that cannot be reprocessed are either recycled to manufacturers (e.g. some spent catalysts can be reprocessed), disposed of in EPA-approved facilities off-site, or chemically treated on-site to form inert materials which can be disposed to land-fill within the refinery.

6.4 Energy conservation in oil refineries

Large amounts of heat energy are required during the refining process, so the refineries are equipped with heaters, boilers and other facilities. These facilities burn petroleum gas, a by-product of fuel oil refining, releasing CO_2, SO_x, NO_x and other gases.

Energy is the life blood of refinery and its uninterrupted supply is crucial to the operation and profitability of the facility but also, and crucially, to it safety. Moreover, the factors that influence the refining margins are the refinery capacity utilisation, energy consumption, prices of crude oil and products, and the refinery complexity. Due to the fact that the oil prices are an undependable factor for the refineries, the financial results could be improved mainly by reducing costs.

6.4.1   Refining energy costs

As mentioned above, the crude refining consists of many processes, which require certain amount of energy in order to achieve the production of dedicated products. In general, the technological processes in the refineries consume energy between 6% and 15% of the crude oil intake. The wide range of this use depends on the level of conversion and complexity of the refinery. The simplest the refinery (mainly distillation and products treatment), the lower the specific energy consumption. In the opposite, the higher the crude conversion (the presence in the technological scheme of conversion processes like catalytic cracking, thermal cracking, hydrocracking, etc.), the higher the energy consumption. Besides, some sources assume that the refinery age also affects the energy consumption due to the different approach and focus towards energy efficiency in the past decades.

On the other hand, the energy consumed by the processes in a refinery comprises between 60% and 80% from the operating expenses or approximately 6% to 7% related to the crude oil costs.

Reviewing the refinery processes and their energy consumption it could be summarised that the energy giants in a typical refinery are: the atmospheric and vacuum distillation with share of around 40% of the total energy consumption. This could be explained by the fact that the whole amount of crude oil is processed by these units, and the crude oil is separated into different fractions by means of rectification, which is known to be highly energy intensive. Other processes, such as catalytic cracking and reforming are also energy-intensive, but on the other hand, the released excessive energy could be utilised by producing steam and hydrogen for further use.

Energy consumption of the main refinery technological processes are given below:

1. Coking
2. Catalytic hydrotreating
3. Alkylation
4. Catalytic reforming
5. Fluid catalytic cracking

6. Vacuum distillation

7. Atmospheric distillation

Therefore, refiners desperately need to optimise the energy use by applying not only reduction measures but more important efficiency measures.

6.4.2   Need for energy efficiency

Due to all mentioned factors about the energy consumption in the world, industry sector and refining particularly, the increased regulatory requirements about the products quality and emissions, as well as the need of being competitive, it is easy to understand that improvement of energy efficiency in refining is a question of surviving. Some sources have also concluded the great importance of the energy efficiency due to the high environmental costs and low refining margins. On the other hand, the need for emissions reduction due to the climate changes also raises the importance of the energy efficiency. Furthermore, the energy efficiency is considered in several different dimensions, such as legal, environmental, social, economic and financial and technological. The latter is related to integration of improved technologies at all stages of energy consumptions: improved products and processes, applied equipment, procedures, knowledge, as well as energy efficiency orientation and development of new more efficient technologies.

6.4.3   Energy efficiency barriers

Stressing the importance of the energy efficiency it is worthwhile to summarise the main obstacles for its development and improvement. The understanding of the barriers is useful for their overcoming and avoiding, when an energy efficiency programme is to be implemented.

The main barriers to energy efficiency are:

1. Lack of time.
2. Lack of priorities for investment.
3. Different priorities for investment.
4. Lack of capital needed.
5. Cost of possible production failures.
6. Technical risk of production disruptions.
7. Insufficient information for energy consumption of the purchased equipment.

Few researchers relate the energy efficiency barriers with the information, the availability of which is limited and with the invisibility of the energy efficiency. Furthermore, the challenges for energy efficiency incorporate technical, economic, organisational (behavioural), informational, knowledge, financial and policy factors. Typically, the main obstacles in a refinery to improve its energy efficiency could be related with the human behaviour in the process operation's level. The energy efficiency measures often are perceived with resistance because of a high risk for the reliability and stability of the technological processes.

On the other hand, few administrative layers in the refinery also reveal resistance to suggested energy efficiency improvements even when clear evidences are available. The suggested reason for this is again the human nature to oppose to changes. The determination of all barriers of the energy efficiency improvement and implementation, as well as the roots of those obstacles is a field for further investigation.

6.4.4   Approaches for energy efficiency improvement

Probably one of the most broadly discussed area is the possibilities, techniques, technologies, etc., for improvement of the energy efficiency. Broad energy saving opportunities are given below:

1. Process and process control.
2. Furnaces, boilers and steam distribution.
3. Motors (pumps, fans, etc.).
4. Air compression.
5. Process operations and equipment.
6. Industrial buildings.
7. Energy management.

Other sources observe technologies with huge potential for energy consumption reduction such as variable speed drive for motors, high efficient motors, improved compressed-air systems, waste heat recovery, etc. In several literature sources it was suggested that the refiners should improve their energy efficiency by implementing more efficient heat exchange and heat integration, improved process control, special approach to huge energy consuming equipment, reduction of fouling in the equipment, better pipe insulation, waste heat recovery cogeneration for steam and electricity production, improved steam use, implementation of advanced process control systems, etc.

The fouling in heat exchanger in the refineries is regarded as a huge problem because of the reduced heat transfer capacity, which leads to large growth of the energy consumption. On the other hand, it is a reason for unpredicted shut downs and respectively more financial losses for the refineries. Prediction tools for fouling could lead to energy savings of 2% of total fuel in a refinery. Therefore, dedicated tools for the heat exchangers efficiency incorporated in energy management systems is suggested to be useful and leading to cost saving in energy and maintenance.

6.4.5   Energy audits

In order to implement and successfully use energy efficiency system, it is absolutely necessary to provide an energy audit. Energy efficiency check reveals the possible areas for energy savings. Purpose of such audit is to understand the current energy consumption and production and use this database for improvement evaluations with further calculations of the achieved effect on savings. In addition, the benefits of an energy audit are as decrease in energy use, resulting in reduction of operating costs, lowering the emissions of greenhouse gases, increase of profitability and overall performance, etc.

Three types of energy audits are: (i) the *preliminary energy audit* (walk-through audit) is timely limited and aims to identify the energy costs, and priorities the energy consumption optimisation possibilities, (ii) the *targeted energy audit* provides more depth analysis of one or more areas of the project and (iii) the *detailed energy audit* is a comprehensive analysis aiming to generate a detailed action plan and develop feasible options for energy use reduction.

6.4.6   Energy management systems

The main elements of the energy management system are identified to be:

1. Management responsibility.
2. Energy policy.
3. Energy planning.
4. Implementation.

It highlights on the management involvement during all the stages of the system and on the need of development of procedures, regular audits and indicators of efficiency.

Nevertheless, a worthwhile energy management system should provide its users with feedback of achieved results, optimal targets, as well as comparison with previous periods. This is needed for the users in order to understand the real meaning of the observed energy consumption and to undertake corrective measures. This monitoring tool provides continuous feedback to the operators and visually presents the energy saving potentials. Further useful element of this system is the calculation of the energy savings into financial values.

Thus, dedicated energy management systems are vital to develop and implement in order to achieve comprehensive monitoring of the energy resources use, to provide relevant targets, to present real-time feedback of the consumption, and to visualise the potential of energy savings. Investments in this area, as well as change in people's perception and attitude towards energy efficiency are required in order to improve energy efficiency.

6.5 Fertiliser industry

Fertiliser is a substance added to the soil to improve plants growth and yield. Fertilisers can be classified as nitrogenous fertilisers, phosphatic fertilisers and complex fertilisers (Fig. 6.5). Plants may be producing only nitrogenous fertilisers, like urea, ammonium sulphate, ammonium nitrate and ammonium chloride, or only phosphatic fertilisers like super phosphates; there are plants where complex fertilisers containing both nitrogen and phosphates, like ammonium phosphate and ammonium sulphate phosphate, are produced. Also, some fertiliser units are only involved in mining and undertake no other activity.

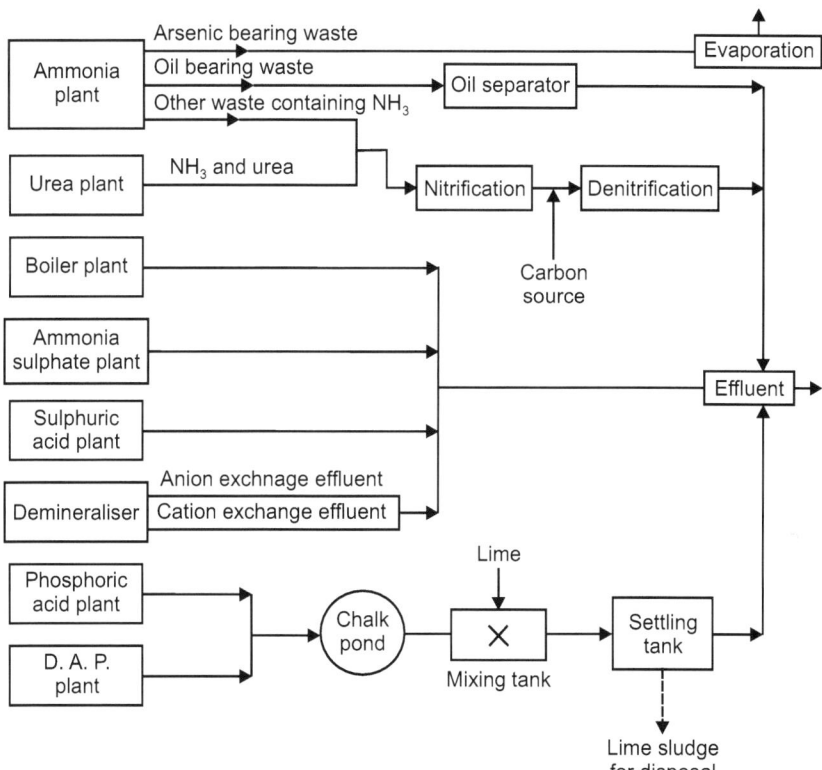

Figure 6.5: Flow diagram for effluent treatment of a complex fertiliser plant.

6.5.1 Manufacturing process

Ammonia is the principal intermediate in the manufacture of all nitrogenous fertilisers. So, except when the by-product ammonia is available from a coke-oven, raw materials for nitrogenous fertiliser production are carbonaceous material, which are required for making ammonia. Thus, all nitrogenous fertiliser plants essentially have an ammonia production unit and a reactor where the synthetic ammonia is reacted with other chemicals to produce the final product. The plant may have auxiliary units to produce reacting chemicals also.

Steps in manufacture

Basic process steps in the manufacture of urea from carbonaceous raw materials like naphtha are:

1. Reaction of the carbonaceous material with steam and air to form a mixture of hydrogen and carbon monoxide, known as synthesis gas.
2. Reaction of carbon monoxide with steam over a catalyst to form more hydrogen and carbon dioxide.
3. Separation and purification of carbon dioxide.
4. Removal of residual carbon monoxide from gas mixture.
5. Synthesis of ammonia by reacting hydrogen and nitrogen over a catalyst (Nitrogen is supplied as air in an earlier step).
6. Synthesis of urea by treating ammonia with carbon dioxide in a reactor at higher temperature and pressure.

Plants using by-product ammonia from other manufacturing plants (like coke oven) have to produce carbon dioxide separately for the production of urea. Ammonium sulphate may be produced by reacting anhydrous ammonia with sulphuric acid, usually obtained as a by-product from other manufacturing plants. Ammonium sulphate may also be manufactured from gypsum or from calcium sulphate sludge, obtained from the phosphatic fertiliser plant, using ammonia and carbon dioxide obtained from ammonia plant. In this process, calcium sulphate is reacted with ammonium carbonate solution to produce ammonium sulphate.

Ammonium nitrate is produced when ammonia is reacted with nitric acid. Normally, the required quantity of nitric acid is produced in the same plant, by oxidising ammonia.

Super phosphate is produced by merely mixing the phosphate ore (commonly known as phosphatic rock) with sulphuric acid to convert the phosphate to 'monocalcium phosphate'. The by-product calcium sulphate of this process may be used in the manufacture of ammonium sulphate.

Ammonium phosphate is made by treating phosphoric acid with ammonia; the phosphoric acid production process involves the following steps:

1. Dissolving phosphate rock in enough sulphuric acid.
2. Holding the mixture until the calcium sulphate crystals grow to adequate size.

3. Separation of the phosphoric acid and calcium sulphate by filtration.

4. Concentration of acid to the desired level.

Fertiliser production is one of the most energy intensive processes. Energy is consumed in the form of natural gas, associated gas, naphtha, fuel oil, low sulphur heavy stock and coal. The choice of the feedstock is dependent on the availability of feedstock and the plant location.

Production of ammonia has greatest impact on energy use in fertiliser production. It accounts for 80% of the energy consumption for nitrogenous fertiliser. The feedstock mix used for ammonia production has changed over the past. The shift towards the increased use of natural/associated gas and naphtha is beneficial in that these feedstocks are more efficient and less polluting than heavy fuels like fuel oil and coal. Furthermore, capacity utilisation in gas based plants is generally higher than in other plants. Therefore, gas and naphtha present the preferred feedstocks for nitrogenous fertiliser production.

Energy intensity in fertiliser plants has decreased over time. This decrease is due to advances in process technology and catalysts, better stream sizes of urea plants and increased capacity utilisation. Capacity utilisation is important as losses and waste heat are of about the same magnitude no matter how much is actually produced in a plant at a specific point of time.

The production of phosphatic fertiliser requires much less energy than nitrogenous fertiliser. Depending on the fertiliser product, energy consumption varied from negative input for sulphuric acid to around 1.64 GJ/T of fertiliser for phosphoric acid. For sulphuric acid, the energy input is negative since more steam (in energy equivalents) is generated in waste heat boilers than is needed as an input. The age of the technology, the scale of the plant and management practices have a large impact on energy efficiency of the overall process. Energy savings potential are highest for naphtha and fuel oil/LSHS based plants especially for plants built before 1980.

The age of the technology, the scale of the plant and management practices have a large impact on energy efficiency of the overall process. Energy savings potential are highest for naphtha and fuel oil/LSHS based plants especially for plants built before 1980. A closer look at all pre-1980 plants reveals that with current installed capacity of 5.09 MT, energy savings in the pre-1980 ammonia plants alone would account for up to 11 Pcal per year. This means, with a share of less than 50% in total installed capacity energy savings in these plants alone would be more than 75% of total possible energy savings.

6.5.2 Categories for energy efficiency improvement

The following factors have been identified as affecting energy consumption in ammonia-urea plants: capacity utilisation, type of feedstock, technology employed and vintage of the plant. Since ammonia-urea plants built more recently already reach world best efficiency levels. For an older ammonia plant a typical revamp would include the following: (a) capacity increase, (b) energy-saving, (c) reduction in raw material and utility consumption, (d) reduction in environmental impact, (e) improved safety and reliability and (f) improved control systems. All of this would directly or indirectly benefit energy consumption.

Specifically, improvements to the energy efficiency of various process components of ammonia (and to a lower extent of urea) include improvements in the reforming, CO_2 removal, synthesis, and purge gas recovery (all of these can further broken down into process optimisation, maximisation of heat recovery, and the fine-tuning of process parameters), and energy savings from development of new and better catalysts and improved metallurgy leading to superior and more efficient process equipment.

6.5.3 Energy intensity of fertilisation

Inorganic fertilisers are major consumers of energy in the agricultural sector. In the United States, inorganic fertilisation accounts for about a third of total energy input to crop production. In contrast to tractors, irrigation pumps, and other types of equipment, fertilisers are indirect energy consumers. That is, the bulk of energy use associated with fertilisers is not consumed directly at the agricultural site, but indirectly during its production, packaging, and transportation to the site. Additional energy is then used on-site during fertiliser application.

Most fertiliser energy use is attributable to the production of nitrogen fertilisers with natural gas. Natural gas is the principal energy resource for creating anhydrous ammonia, a key nitrogen fertiliser. The natural gas provides a source of hydrogen in the synthesis of ammonia by the Haber process. Over 90% of nitrogenous fertilisers contain ammonia and/or other fertiliser elements derived from ammonia (e.g. ammonium nitrate, sodium nitrate, calcium nitrate, ammonium sulphate, ammonium phosphates, and urea).

Producing ammonia is a very energy intensive process; it requires about 1090 to 1250 m^3 of natural gas to produce 1 metric ton of anhydrous ammonia. Natural gas is also used in other ways in the fertiliser industry. For example, natural gas as a fuel provides the process heat for producing other types of fertilisers. It is estimated the natural gas supplies between 70 and 80% of all energy for fertiliser production.

Each of the three primary nutrients in inorganic fertilisers has a different set of energy requirements during its lifecycle. However, these requirements can be separated into four main stages: production, packaging, transportation, and application.

Organic fertilisers also have energy requirements. In fact, transportation and application energy demands are often higher for organic fertilisers since they are less nutritious per unit weight. Some of the requirements for processed organic fertilisers may include:

1. Collection of organic waste.
2. Loading and transportation of waste to a processing plant.
3. Unloading and putting waste into windrows.
4. Turning and irrigation of windrows to expedite composting.
5. Collection, loading, and transportation of composted waste from processing plant to field.
6. Unloading waste for storage.
7. Loading and applying waste to field by farm equipment.

Organic material can be applied directly to a field, but it will then take longer to decompose, delaying the availability of nutrients to plants. In addition, 'fresh' application will often result in a temporary nitrogen deficiency in the soil, as micro-organisms use the nitrogen for cell production during the decomposition process.

Organic fertilisers that are processed obviously require more energy that those that are directly applied. Therefore, there is a point at which the energy expenditure per unit weight of nutrients in transporting and composting the organic material exceeds the energy requirement of inorganic fertilisers.

6.5.4   Energy-efficient fertilisation practices

Since natural gas is such a critical resource in fertiliser production, natural gas price fluctuations have a dramatic effect on fertiliser costs. As energy costs continue to rise, and the demand for fertilisers increases, this effect is becoming more pronounced. The implementation of energy-efficiency measures in the production and use of fertilisers will help curb the effects of rising gas costs, as well as the effects of energy costs in general. One of the most obvious areas of energy consumption to address in the fertiliser industry is the production of anhydrous ammonia for nitrogenous fertilisers. Efficiency strides have been made since the inception of the original Haber process in the early 1900s, but more improvements are still possible. This area of energy consumption is the responsibility of fertiliser manufacturers.

6.5.5   Measures to increase the efficiency of ammonia production

1. Replace process equipment with high-efficiency models.
2. Improve process controls to optimise chemical reactions.

3. Recover process heat.

4. Maximise the recovery of waste materials.

The next area of energy consumption to address is the application of fertilisers to crops. Measures to improve the use of fertilisers are the responsibility of the farmer. Therefore, the efficient use of fertilisers is more controllable by farmers (and is thus more directly applicable to farmers) than is the efficient production of fertilisers.

Barriers to energy efficiency improvement

Although integrating energy savings measures and technologies in all plants would lead to net savings both in terms of energy and overall costs, only few measures have been or are currently being implemented in various fertiliser sector. Barriers to energy efficiency improvement are of both general and firm/process specific nature. A 5–6 year payback period is one of the barriers to energy-saving investment. Energy policies in general and price-based policies in particular can help overcome these barriers in giving proper incentives and correcting distorted prices. Appropriate provisions should be made in the retention pricing scheme to further encourage investment in energy conservation projects. On a long term basis, substantial further investments in energy efficiency technologies for existing and new plants have to be made. Therefore, sectoral policies should be devoted to the promotion of such investments. A stable foreseeable policy environment would substantially help firms to reduce the risk of taking large investments.

6.5.6   Trends in ammonia technologies

Ammonia is produced by the synthesis of elemental hydrogen and nitrogen. Hydrogen is separated from water and nitrogen from air using the energy of fossil fuels such as naphtha, fuel oil or natural gas. The process is energy intensive at every step and involves very high operating temperature, pressure and is highly corrosive. Special material are needed to fabricate process equipment and pipelines. In order to take advantage of economy of scale of operations and reduce overall energy consumption, new generation ammonia plants are designed for large capacities.

Two prominent technologies to process ammonia are: the partial oxidation of heavy hydrocarbons, vacuum residue and coal, and steam reforming of lighter hydrocarbons and natural gas.

In the conventional partial oxidation process, the feedstock along with steam and oxygen react in a gasifier to produce carbon monoxide (CO) and hydrogen (H_2) at $1200°C$. After cooling and cleaning the gas, H_2S is stipped off; CO is converted to CO_2 and stripped off in a CO_2 removal unit to yield a H_2 stream. It is then washed with liquid nitrogen in the LNG wash unit to condense and remove impurities to yield an ammonia-synthesis gas containing hydrogen and nitrogen in the 3:1 ratio.

6.5.7 Waste-heat recovery

The waste-heat recovery system is associated with flue gas from the reformer furnace and process gas from the secondary reformer. It generates high-pressure steam in specially designed boilers. These boilers and steam super heaters operations under very high heat flux on the one side and corrosive water on the other. Consequently, meticulous care is needed to avert failures. Proper material selections and stringent water quality are two proactive loss prevention methods. Various material grades of SA 213, ASTM A312, ASTM A335 and ASTM A351 provide useful service life in this section. The order part of the combustion air preheater–the trail end of the flue-gas-heat-recovery train–is more likely to corrode due to sulphur dioxide (SO_2) condensing from the flue gas. In this area, cast iron or glass will resist the acid attack. Carbon steel preheater tubes, joined with 1.5 to 2.0 m of SS 304 tubes at the cold end of the tube sheet, can ensure reasonable service life. Typically, the flue-gas temperature to the stack is maintained above the dew point of SO_2 to prevent condensation. During startup and shutdown, condensation of SO_2 will occur. The high-pressure feed water heaters are prone to leaks at the tube-to-tube sheet joints. Lining or over-laying the tube sheet with material such as, alloy 60 and using tube materials with 1 CR-0.5 Mo can mitigate this risk considerably.

6.6 Sugar industry

Energy consumption and its impact on the efficiency and environmental issues has become of great importance recently. Sugar is the main energy consumer among food product industry and has an old technology.

The industry sector plays a significant role in global energy consumption. Sugar is one of the most energy consuming industries. Since energy production is extensively based on using fossil fuels, the environmental issues will become of great importance. The economical and environmental issues and obligations cause the industry to move toward better design conditions. Therefore, there is a need to pursue a new policy to force producers to undertake energy-efficient practices to establish sustainable production systems without disrupting the natural resources.

6.6.1 Major processes involved in a sugar mill

The sugar manufacturing process normally comprises.

1. Juice extraction
2. Juice clarification and evaporation

3. Crystallisation
4. Centrifuging
5. Drying and packing

The energy demand in sugar processing is determined by four interlinked energy intensive process stages: extraction, juice purification, evaporation and crystallisation. The energy system is composed of a boiler house, multiple-stage evaporator, and a process heating subsystem. Flow diagram of sugar production process is shown in Fig. 6.6.

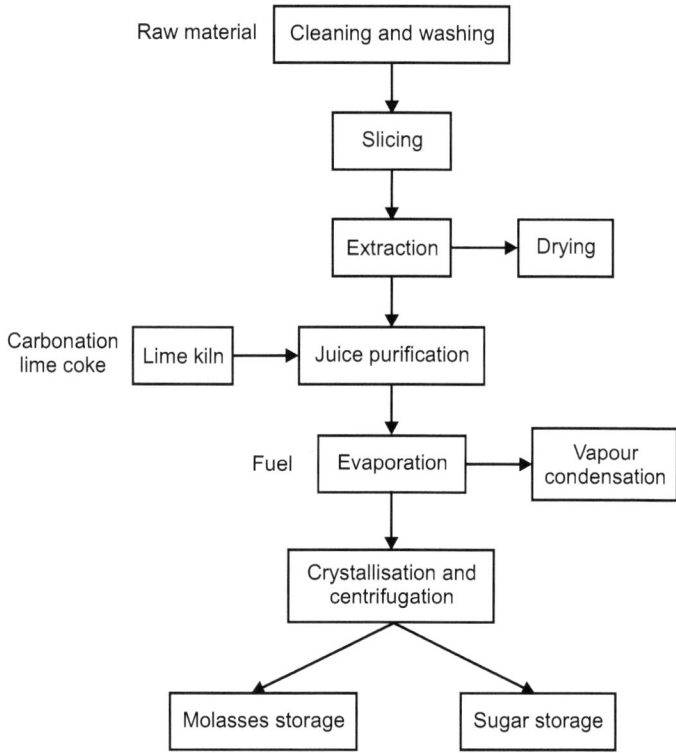

Figure 6.6: Flow diagram of sugar production process.

A multiple-stage evaporator system is used in sugar processing, in order to improve energy efficiency. The crystallisation process is biggest energy consumer of vapour extracted during evaporation.

6.6.2 Power saving

During the milling of cane the power is consumed by the following units:

1. Bagasse

2. Friction between shafts and bearings of the rollers.
3. Friction between bagasse and trash plate.
4. Friction of scrapers and toe of the trash plate against the rollers, to which should be added the work of dislodging the bagasse at these points.
5. In driving of the intermediate carriers.
6. Gears

The sugar factory owner has to necessitate care in selection of the hydraulic loading at the mills, selection of mill bearings, designing and setting of the trash plate, designing of the intermediate carrier gears, etc., such that all these result in minimum consumption of power during milling operation. However, automation for controlling the milling operations may result in improved sugar extraction and reduced energy cost. The areas of automation can be feeding, control of mill speeds, application of maceration water for juice extraction, etc.

6.6.3   Further reduction in steam consumption

It is possible to reduce further the consumption of steam through the use of thermo-compressors, mechanical vapour recompressors, liquid heat-exchangers for heating of juice, etc.

6.6.4   Automation

The automatic control system of the evaporator should be designed in such way that the syrup, leaving the evaporators has a predetermined consistency. This design should also take care of the time-lag between the admittance of steam and exist of syrup. The control of steam should also be linked with the flow of vapours for heating and boiling of juice.

The maintenance of consistent higher brix of syrup at the evaporators is essential in view of the multiple effect advantage at this station compared to the other stations like juice heating and pan boiling.

6.6.5   Sugar crystallisation

Energy requirement by vacuum pans

Through proper controls, energy can be saved at this stage which is estimated to be 15–18% of the total energy required in a sugar plant. This could be reduced to about 10–12% by taking remedial measures. The automation of vacuum pans can be of great advantage in reducing both the mechanical/electrical energy as also the heat energy requirement in this area of operation. Molasses conditioners should be installed in line to feed from the storage tank to the vacuum pans. The inline conditioning of the molasses avoids the chances of molasses cooling during the storage.

Energy requirement by crystallisers

The massecuite when discharged from the vacuum pans is received, stored and cooled in a set of cooling crystallisers, the massecuite is subjected to either air cooling in case of high grade massecuite or water cooling in case of low grade massecuite. The degree of supersaturation further increases during cooling, resulting in deposition of more sugar on the available/existing crystal surface. The energy requirement of crystallisers can be reduced by installing continuous vertical crystalliser which will then dispense with the use of a large battery of batch crystallisers. A large number of factories are already making use of the continuous crystallisers and their use should be spread to as many more factories as possible.

Sugar centrifugals

The energy consumption at centrifugals is also related to the selection of the process. The methods being used for cooling, conveying and grading of sugar can be modified to save energy. This primarily may require the use of fluidised bed drier for cooling and conveying. The design of the graders can also be modified so that a fewer number of graders can handle the required through put.

6.6.6 Steam generation

The factors responsible for efficient use and generation of steam are:
1. Steam pressure.
2. Steam temperature.
3. Design of boilers.
4. Design of accessories.
5. Design of furnaces.
6. Use of heat recovery units.
7. Bagasse drying.

The commonly used furnaces in the sugar factory boilers are: (i) step grate furnace, (ii) horse shoe furnace and (iii) spreader stroker furnace.

Boiler operation

The automation in boilers can be of immense use in achieving a higher boiler efficiency and hence reduced energy costs. The automation in boilers can include control of CO_2 and/or O_2 %, steam flow, feed control, excess air control and excess O_2 monitoring system, etc.

Boiler efficiency

Boiler efficiency can be measured either by the direct or indirect method. The indirect method measures the losses around a boiler and is an instantaneous measure of boiler efficiency. It is an accepted method used by boiler manufacturers to assess boiler performance. The limitation of this technique is that it is a snapshot of boiler performance. The method is usually reserved for measurement when the boiler is burning coal. Catch sampling of coal and ash may introduce errors into the measurement. The measurement of excess combustion air is also a problem. The direct method assesses boiler efficiency by relating energy in steam output as a percentage of energy in fuel and boiler feed water. The limitation of this technique is that it requires accurate massing of fuel, boiler feed water, steam and blow down losses. This method is more suited to efficiency reporting over longer periods for which bagasse and coal massing can be more accurately determined.

Another factor which affects energy efficiency is the quality of cane being processed. Sugar factories have limited control over this aspect. Higher fibre per cent cane will make more bagasse available for combustion whilst higher ash and higher moisture reduce the calorific value of bagasse.

6.6.7   Power generation

Many sugar factories in the world are producing or co-generating additional power for transfer to the grid or for use in the ancillary industry. Almost all factories are equipped with turbo generators which are generally of the back pressure type. Live steam from the boilers is used for operating turbo generators set for producing power. Exhaust steam is used for this process.

6.6.8   Use of non-conventional energy in sugar industry

Solar energy and bioenergy can be used in the sugar industry for energy conservation. Water to be fed to boilers can be preheated through solar water heaters. Bioenergy in terms of methane gas can be generated from sugar/distillery effluents which could supplement the fuel consumption by 50% in oil-fired or coal-fired boilers. Energy can also be produced from bagasse and other by-products of sugarcane industry.

6.6.9   Energy audits and measurements in a sugar factory

An energy conservation project should be initiated with an energy audit of a sugar factory to achieve the most economical reduction of energy costs in order to save the maximum energy at the lowest cost. The manner in which the total energy utilisation at the plant is managed between the different stages is determined by energy measurements, to be able to draw up an energy balance for the plant.

The energy audit comprises:

1. Acquaintance with the energy systems of the factory.

2. Information collection regarding energy utilisation, production, etc., to base these energy balance.
3. Planning and execution of energy measurements.
4. Information collection regarding the energy systems development to serve as a base for drawing up energy saving measures.

6.6.10 Energy saving in a sugar plant

The various means through which the Indian sugar industry is attempting to save energy and/or fuel are:

1. Efficient production of steam.
2. Efficient use of mechanical/electrical energy.
3. Efficient use of steam.

This can be further elaborated as under:

1. Maximum generation of steam per unit bagasse or fuel.
2. Maximum generation of power set unit of steam.
3. Minimum consumption of power.
4. Minimum consumption of process steam.
5. Minimum line losses both on account of steam and power transmission.
6. Maximum heat recovery through reclamation of a hot condensate, flashing, etc.
7. Minimum usage of chilled water.

6.6.11 Electrical saving potential

The major consumers of the electricity are motors and drivers. Using controllers on the fan's speed, appropriate use of electrical motors, and also utilising high efficient motors will result in significant electrical saving.

6.6.12 Thermal saving potential

The major form of energy used in sugar plants is the thermal energy. The following potential are found in the sugar factories:

1. Replacing old technologies with new, efficient ones: replacing motors in the mills, CO_2 and cane pumps, steam furnace, water pump, and dryers to reach maximum efficiency, using ejectors instead of vacuum pumps, convert shell and tube evaporators into plate and frame ones.
2. Installing monitoring and control equipments.
3. Improving insulation including valves, pipes and tubes and other connections.

4. Heat recovery from furnaces, evaporators, and boiler stacks.

5. Improving combustion process: Exact adjust of furnace performance, install gas analyser on the output gas stream from boiler, install flow meters to calculate the input water to the boiler, correct the ratio of fuel to air at the inlet of the furnace.

6. Increasing the efficiency of evaporators: Adding the additional facilities to increase the evaporator's efficiency, increase the stages, increase the concentration and temperature of input solution to first stage, vacuum the last section of evaporators, correct the piping system for steam.

7. Using high efficient electrical motors.

8. Modify capacity.

9. Installing pre heaters for the input juice and air.

10. Modifying dryers.

11. Improvement of performance of other equipments: Recycle the condense water to the boiler, avoid the steam leakage, correct the operation of pumps and electric motors.

Due to the huge usage of steam applying cogeneration of heat and power which typically incorporates a steam boiler and a steam turbine will lead to reduce the waste.

Identification of utilisation possibilities for waste and by-products

Several possibilities exist for utilising waste from the sugar production process by the selection of solutions that offer energy saving potentials. Energy minimisation in the sugar plant is one of the most important issues within the context of sustainability.

6.6.13 Technologies of the 21st century in the sugar industry

Various new and futuristic technologies and processes that could be used in the sugar industry are:

1. Cane handling: Computerised cane weighing machines should be used. The present mechanical weighing scale can be replaced by electrical scales.

2. Cane preparation: Many sugar factories use heavy duty shredders to achieve the preparatory index above 90% and this should be evaluated under Indian conditions.

3. Milling juice extraction technologies: Mis Taxmaco Ltd., Calcutta has collaborated with Walkers of Australia for a new design of a constant ratio 5-roller mill. This design of the mill has capability to give higher

milling efficiency and low energy consumption. It has already been commissioned in a sugar factory in Tamil Nadu. The indigenous development in the field has not been lagging behind. Mis WIL, Pune, have developed self setting 3-roller mills and a milling tandem of this design have already been commissioned in a sugar factory in Tamil Nadu. The possibility of adopting diffusers in the cane sugar industry should be re-evaluated from the view point of capital cost, maintenance cost, energy consumption and extraction efficiencies.

4. Low pressure extraction system: The power requirement of the system is claimed to be about 0.6 kW per TCD.

5. Cane sugar separation technology: The industry is rapidly replacing the steam turbine drives with variable speed DC drive, which results in fuel saving and gives a better flexibility in operational control. Hydraulic motors for operating the mills can be a breakthrough for the sugar industry in India in near future. Modifications in design of other accessories like imbibition equipment, etc., would also bring about saving in energy and costs.

6. Juice treatment: Use of hydrogen peroxide for treatment and removal of colour in syrups has been found highly successful in Karnataka. This can be adopted for a better removal of colour with reduced sugar losses. The possibility to use short retention clarifier should be evaluated to avoid losses due to inversion. Use of membrane filters and bag filters also offer good potential for achieving better quality of filtration in future. Extensive studies have been conducted for clarification of juice by membrane filtration.

7. Evaporation: The combination of extensive vapour bleeding, condensate flashing and recirculation, use of thermo-compressors/MVR's, etc., can reduce the process steam consumption to about 30% on cane against the present 45% in the Indian sugar industry.

8. Sugar crystallisation: Use of continuous vacuum cooling crystallisation system developed by a French sugar group (Beghisay) can improve the performance of crystallisation section.

9. Centrifugals: To save energy, Mis Krupp, Germany, have developed a new design of continuous centrifugals where two centrifugal baskets are mounted on a common shaft and driven by a common motor.

10. Sugar handling: A firm in Bangalore has developed a sugar drier cum conveyor on the principle of fluidised-bed drier. It has been installed in Karnataka. The size of this equipment is 5 m × 1.2 m × 5 m which can handle up to 150 bags of sugar/hour.

11. Steam generation: The sugar industry can opt for modem design of boilers and new furnaces based on following principles:
    (a) Boiler with membrane wall construction.
    (b) Furnace design for mixed fuel.
    (c) Fluidised-bed furnaces.

    Ponni sugars at Erode has installed a Igni fluid make boiler of 25 T/h capacity and 45 kg/cm^2 pressure which is very compact in design and is highly efficient.

    It can use coal, sugar pith and lignite and other alternate fuels too.

    The concept of tap power directly from coal is under development of BHEL and BARC, Bombay. It is estimated that this technology can offer an overall efficiently of 60% against 30% at present through steam at turbo generators.

12 Power generation: The conventional method of producing power through back pressure turbines can be replaced with production of power by extraction-condensing turbines or by topping turbines. In this manner additional power can be produced for use in other industries or for bleeding to state grid.

    Automation at various sectors of sugar plant shall result in improved productivity, reducing sugar losses and energy cost. The researchers affirms that if the above mentioned improvements are incorporated, there could be 30% energy conservation in the sugar industry.

There are a number of other actions and precautions that can help to minimise energy and heat losses in a sugar plant.

These include:

1. All pipe lines, tanks which conveyor receive hot juices, steam, bled vapours, etc., should be properly lagged for avoiding radiation and condensation losses. Similarly, all heat-exchangers should be lagged.

2. Pumps and motors should be carefully selected. It has been seen that many factories install over size pumps to cater to any future requirements during expansion. This is an incorrect approach and results in a recurring loss of power.

3. The type of pumps should be selected after matching the duty requirements with the performance curves of the pumps to achieve maximum efficiency. The use of screw pumps, wherever required, would also result in saving of power'.

4. The slat and rake type conveyers should be replaced with belt conveyers, if possible.

5. The power factor of the electric distribution system should be maintained above 0.95 through use of capacitors, wherever possible.

6. Hot condensate should be collected and suitably recycled to the boilers or to the process. A proper recycling of the condensate can totally eliminate the use of cold water in the process, and at the boilers. The flashing of condensates should also be recovered.

7. A proposal to generate electric power at 3 kV or 11 kV should also be considered and high tension motors at all major consuming ends, to reduce transmission losses.

6.7   Pulp and paper industry

Probably half of the fibre used for paper today comes from wood that has been purposely harvested. The remaining material comes from wood fibre from sawmills, recycled newspaper, some vegetable matter and recycled cloth. Coniferous trees, such as spruce and fir, used to be preferred for papermaking because the cellulose fibres in the pulp of these species are longer, therefore making for stronger paper. These trees are called 'softwood' by the paper industry. Deciduous trees (leafy trees such as poplar and elm) are called 'hardwood.' Because of increasing demand for paper, and improvements in pulp processing technology, almost any species of tree can now be harvested for paper.

Some plants other than trees are suitable for paper-making. In areas without significant forests, bamboo has been used for paper pulp, as has straw and sugarcane. Other materials used in paper manufacture include bleaches and dyes, fillers such as chalk, clay, or titanium oxide, and sizings such as rosin, gum, and starch.

6.7.1   Manufacturing process

Making pulp: Several processes are commonly used to convert logs to wood pulp. In the mechanical process, logs are first tumbled in drums to remove the bark. The logs are then sent to grinders, which break the wood down into pulp by pressing it between huge revolving slabs. The pulp is filtered to remove foreign objects. In the chemical process, wood chips from de-barked logs are cooked in a chemical solution. This is done in huge vats called digesters. The chips are fed into the digester and then boiled at high pressure in a solution of Paper sodium hydroxide and sodium sulphide. The chips dissolve into pulp in the solution. Next the pulp is sent through filters. Bleach may be added at this stage, or colourings. The pulp is sent to the paper plant. Flow diagram of paper mill is shown in Fig. 6.7.

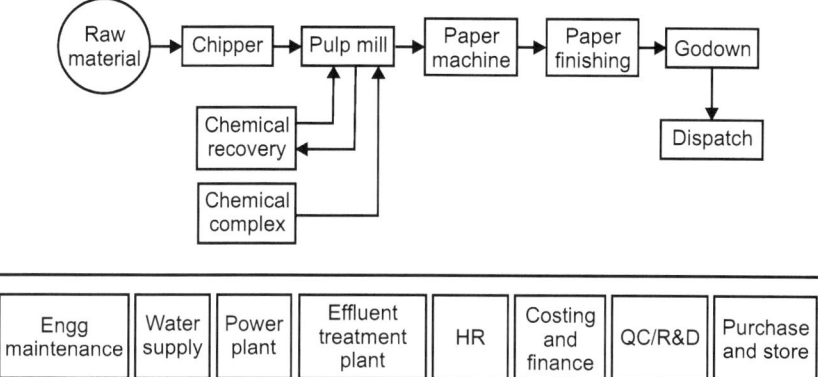

Figure 6.7: Flow diagram of paper mill.

Beating: The pulp is next put through a pounding and squeezing process called, appropriately enough, beating. Inside a large tub, the pulp is subjected to the effect of machine beaters. At this point, various filler materials can be added such as chalks, clays, or chemicals such as titanium oxide. These additives will influence the opacity and other qualities of the final product. Sizings are also added at this point. Sizing affects the way the paper will react with various inks. Without any sizing at all, a paper will be too absorbent for most uses except as a desk blotter. A sizing such as starch makes the paper resistant to water-based ink (inks actually sit on top of a sheet of paper, rather than sinking in). A variety of sizings, generally rosins and gums, is available depending on the eventual use of the paper. Paper that will receive a printed design, such as gift wrapping, requires a particular formula of sizing that will make the paper accept the printing properly.

Pulp to paper: In order to finally turn the pulp into paper, the pulp is fed or pumped into giant, automated machines. One common type is called the Fourdrinier machine, which was invented in England in 1807. Pulp is fed into the Fourdrinier machine on a moving belt of fine mesh screening. The pulp is squeezed through a series of rollers, while suction devices below the belt drain off water. If the paper is to receive a water-mark, a device called a dandy moves across the sheet of pulp and presses a design into it.

The paper then moves onto the press section of the machine, where it is pressed between rollers of wool felt. The paper then passes over a series of steam-heated cylinders to remove the remaining water. A large machine may have from 40 to 70 drying cylinders.

Finishing: Finally, the dried paper is wound onto large reels, where it will be further processed depending on its ultimate use. Paper is smoothed and compacted further by passing through metal rollers called calendars. A particular finish, whether soft and dull or hard and shiny, can be imparted by the calendars. The paper may be further finished by passing through a vat of sizing material. It may also receive a coating, which is either brushed on or rolled on. Coating adds chemicals or pigments to the paper's surface, supplementing the sizings and fillers from earlier in the process. Fine clay is often used as a coating. The paper may next be supercalendered, that is, run through extremely smooth calendar rollers, for a final time. Then the paper is cut to the desired size.

Environmental concerns

The number of trees and other vegetation cut down in order to make paper is enormous. Paper companies insist that they plant as many new trees as they cut down. Environmentalists contend that the new growth trees, so much younger and smaller than what was removed, cannot replace the value of older trees. Efforts to recycle used paper (especially newspapers) have been effective in at least partially mitigating the need for destruction of woodlands, and recycled paper is now an important ingredient in many types of paper production. The chemicals used in paper manufacture, including dyes, inks, bleach, and sizing, can also be harmful to the environment when they are released into water supplies and nearby land after use.

6.7.2   Energy efficiency opportunities in pulp and paper industry

Energy efficiency improvement is an important way to reduce these costs and to increase predictable earnings, especially in times of high energy price volatility. There are a variety of opportunities available at individual plants in pulp and paper industry to reduce energy consumption in a cost-effective manner. The energy management practices applicable to the typical pulp and paper mill are in the following categories:

1. Energy management programmes and systems.
2. Steam systems.
3. Motor systems.
4. Pumps.
5. Fan systems.
6. Compressed air systems.
7. Raw material preparation.
8. Chemical pulping, bleaching and chemical recovery.

9. Mechanical pulping.

9. Papermaking.

10. Emerging energy efficient technologies.

Energy represents a significant cost to the pulp and paper industry. Energy costs are a sizeable fraction of operating costs, equal to roughly 20% of the industry's total cost of materials.

Electricity is used throughout the typical pulp and paper mill to power motors and machine drives, conveyors, and pumps, as well as building operations such as lighting and ventilation systems. The largest use of fuels is in boilers to generate steam for use in pulping, evaporation, papermaking, and other operations. Black liquor is the dominant fuel for boilers in the pulp and paper industry, followed by hog fuel and natural gas, and to a lesser extent, coal. Natural gas and oil are typically used in lime kilns. The major processes employed in the manufacture of pulp and paper products include raw materials preparation, pulping (chemical, semi-chemical, mechanical, and waste paper), bleaching, chemical recovery, pulp drying, and paper making.

The two by-products of the pulp and paper production process—black liquor and hog fuel (i.e. wood and bark)—meet over 50% of the industry's annual energy requirements. This self generation significantly reduces the industry's dependence on fossil fuel inputs and purchased electricity, with the added benefits of reduced raw material costs (i.e. avoided pulping chemical purchases) and reduced waste generation. Natural gas and coal comprise the majority of the remaining fuel used by the industry.

Black liquor, hog fuel, coal, and residual oils are used exclusively as boiler fuels to generate power and to produce steam for use in the various pulping and papermaking processes. Black liquor is combusted in a recovery boiler, which is designed for the dual purpose of generating steam and recovering inorganic smelt for regeneration into white liquor. Because of the low heat contents of black liquor and hog fuel, the efficiencies of boilers that combust these fuels are around 65%. Natural gas is also used as a boiler fuel, but it is also used in significant quantities for direct process heating in lime kilns and limited drying applications (e.g. coating and tissue drying).

In the manufacture of pulp, evaporation, cooking (includes digestion through washing for chemical pulps), and chemical preparation are the most energy intensive processes. Steam is used in significant quantities for nearly every process, but most notably in the evaporation, cooking, and bleaching processes for process heat. The sole use of direct fuel is the chemicals preparation process (in the lime kiln). Drying is by far the most energy intensive step associated with paper machine operations, accounting for roughly two-thirds of total papermaking energy use.

Energy efficiency improvement opportunities

Many opportunities exist within the U.S. pulp and paper industry to reduce energy consumption while maintaining or enhancing productivity. Ideally, energy efficiency opportunities should be pursued in a coordinated fashion at multiple levels within a facility. At the component and equipment level, energy efficiency can be improved through regular preventative maintenance, proper loading and operation, and replacement of older components and equipment with higher efficiency models (e.g. high efficiency motors) whenever feasible. At the process level, process control and optimisation can be pursued to ensure that production operations are running at maximum efficiency. At the facility level, the efficiency of space lighting and ventilation can be improved while total facility energy inputs can be minimised through process integration, where feasible. Lastly, at the level of the organisation, energy management systems can be implemented to ensure a strong corporate framework exists for energy monitoring, target setting, employee involvement, and continuous improvement.

At individual pulp and paper mills, the actual payback period and savings associated with a given measure will vary depending on facility activities, configuration, size, location, and operating characteristics.

The energy efficiency measures can be divided into three primary categories:

1. Cross-cutting measures, which are measures applicable to cross-cutting systems (e.g. motors and pumps) that are common across industries.

2. Process-specific measures, which are measures applicable to processes specific to the pulp and paper industry (e.g. pulping).

3. Emerging technologies, which are energy efficient technologies that hold significant promise for commercialisation and/or substantial market penetration in the next several years.

Table 6.2 provides a summary of some of the key cross-cutting measures.

Table 6.2: Summary of cross-cutting energy efficiency measures.

Energy management programmes and systems	
Energy management programmes	Energy teams
Energy monitoring and control systems	
Steam systems	
Boilers	
Boiler process control	Minimising blow down
Reduction of flue gas quantities	Blow down steam recovery
Reduction of excess air	Flue gas heat recovery
Improved boiler insulation	Burner replacement
Condensate return	

(Cont'd...)

Steam distribution systems

Steam distribution controls	Steam trap maintenance
Improved insulation	Steam trap monitoring
Insulation maintenance	Leak repair
Steam trap improvement	Flash steam recovery

Process integration and self-generation

Pinch analysis	Steam injected gas turbines
Cogeneration of heat and power (CHP)	Steam expansion turbines

Motor systems

Motor management plan	Adjustable-speed drives (ASDs)
Strategic motor selection	Power factor correction
Maintenance	Minimising voltage unbalance
Properly sized motors	

Pump systems

Pump system maintenance	Avoiding throttling valves
Pump system monitoring	Replacement of belt drives
Pump demand reduction	Proper pipe sizing
Controls	Precision casting, surface coating or polishing
High-efficiency pumps	Sealings
Properly sized pumps	Curtailing leakages through clearance reduction
Multiple pumps for variable loads	Adjustable-speed drives (ASDs)
Impeller trimming	Dry vacuum pumps

Fans

Maintenance	High efficiency belts (cog belts)
Properly sized fans	Duct leakage repair
Adjustable-speed drives (ASDs) and improved control	

Compressed air systems

System improvements	Improved load management
Maintenance	Pressure drop minimisation
Monitoring	Inlet air temperature reduction
Leak reduction	Controls
Turning off unnecessary compressed air	Properly sized pipe diameter
Modification of system *in lieu* of increased pressure	Heat recovery
Replacement of compressed air by alternative sources	Natural gas engine-driven compressors

Examples of cross-cutting efficiency measures

Steam systems: Steam systems are by far the most significant end use of energy in the U.S. pulp and paper industry. Over 80% of the energy consumed by the industry is in the form of boiler fuel. Energy efficiency improvements to steam systems therefore represent the most significant opportunities for energy savings in pulp and paper mills.

Motor systems: Motor-driven systems are by far the most significant consumer of electrical energy in a typical pulp and paper mill. Motor-driven systems accounted for around 90% of all the electricity used by the pulp and paper industry. Furthermore, pumps, fans, and materials processing equipment account for the majority (over 70%) of motor-driven systems electricity use in the typical paper mill. Other important uses of electricity in pulp and paper manufacturing include materials handling systems (e.g. conveyors) and compressed air systems.

Efficiency improvements to motor-driven systems can therefore lead to significant energy savings in most pulp and paper mills.

Examples of process-specific efficiency measures

Table 6.3 provides a summary of some of the process-specific energy measures.

Raw material preparation: The processes associated with raw material preparation are estimated to consume roughly 10% of the electricity use and 3% of the steam use in pulp manufacturing operations. One option for reducing the energy use associated with debarking is described below.

Cradle debarker: The cradle debarker is designed to remove bark from delimbed logs in a manner that reduces debarking energy use by up to 33%. Reportedly, a cradle debarker works in the following manner. Logs are loaded into a long trough that contains a series of horizontal and vertical conveyor chains, which are oriented at a slight angle to the path of the logs. The chains lift and drop the logs as they move along the trough; this action loosens and removes bark via compressive and shear forces that are generated between the logs in the trough. Additional reported benefits include less damage to logs leading to a greater wood recovery rate, decreased transportation costs through elimination of off-site debarking, and greater process control.

Chemical (Kraft) pulping: The vast majority (85%) of wood pulp is produced by chemical pulping processes. Furthermore, chemical (i.e. Kraft) pulping and its associated chemical recovery account for the vast majority of steam, electricity, and direct fuel used by the industry in the manufacture of pulp. Efficiency improvements to the chemical pulping process can therefore lead to significant energy savings across the industry.

Table 6.3: Summary of process-specific energy efficiency measures.

Raw material preparation	
Cradle debarkers	Automatic chip handling and screening
Replace pneumatic chip conveyors with belt conveyors	Bar-type chip screening
Use secondary heat instead of steam in debarking	Chip conditioning

Chemical Pulping

Pulping

Use of pulping aids to increase yield	Digester blow/flash heat recovery
Optimise the dilution factor control	Heat recovery from bleach plant effluents
Continuous digester control system	Improved browstock washing
Digester improvement	Chlorine dioxide (ClO_2) heat exchange

Bleaching

Heat recovery from bleach plant effluents	Chlorine dioxide (ClO_2) heat exchange
Improved brownstock washing	

Chemical recovery

Lime kiln oxygen enrichment	Improved composite tubes for recovery boiler
Lime kiln modification	Recovery boiler deposition monitoring
Lime kiln electrostatic precipitation	Quaternary air injection
Black liquor solids concentration	

Mechanical pulping

Refiner improvements	Increased use of recycle pulp
Refiner optimisation for overall energy use	Heat recovery from de-inking plant
Pressurised groundwood	Fractionation of recycled fibres
Continuous repulping	Thermopulping
Efficient repulping rotors	RTS pulping
Drum pulpers	Heat recovery in TMP

Papermaking

Advanced dryer controls	Waste heat recovery
Control of dew point	Vacuum nip press
Energy efficient dewatering–rewetting	Shoe (extended nip) press
Dryers bars and stationary siphons	Gap forming
Reduction of blowthrough losses	CondeBelt drying
Reduction air requirements	Air impingement drying
Optimising pocket ventilation temperature	

Two such measures are described below:

Digester heat recovery: In the Kraft chemical pulping process, steam is produced when hot pulp and cooking liquor is reduced to atmospheric pressure at the end of the cooking cycle. In batch digesters, steam is typically stored as hot water in an accumulator tank. In continuous digesters, extracted black liquor flows to a tank where it is flashed. Recovered heat from these processes can be used in other facility applications, such as chip pre-steaming, facility water heating, or black liquor evaporation. For black liquor evaporation, flash steam from batch digester blow (created by flashing from the hot water accumulator) or black liquor flash from a continuous digester can used for thermal energy in a multi-stage evaporator. This thermal energy will offset the need for steam generated by a boiler for black liquor evaporation. In chip steaming, the black liquor that is flashed in stages from continuous digesters can be used in two ways. Flash vapour from the first stage is normally used to heat the chips in the steaming vessel, while the flash vapour of the second stage can be used instead of live steam in the chip bin.

Black liquor solids concentration: Black liquor concentrators are designed to increase the solids content of black liquor prior to combustion in a recovery boiler. Increased solids content means less water must be evaporated in the recovery boiler, which can increase the efficiency of steam generation substantially.

There are two primary types in use today: Submerged tube concentrators and falling film concentrators.

In a submerged tube concentrator, black liquor is circulated in submerged tubes where it is heated but not evaporated; the liquor is then flashed to the concentrator vapour space, causing evaporation. Capital costs of the high solids concentrator will include concentrator bodies, piping for liquor and steam supplies, and pumps.

A tube type falling film evaporator effect operates almost exactly the same way as a more traditional rising film effect, except that the black liquor flow is reversed. The falling film effect is more resistant to fouling because the liquor is flowing faster and the bubbles flow in the opposite direction of the liquor. This resistance to fouling allows the evaporator to produce black liquor with considerably higher solids content (up to 70% solids rather than the traditional 50%), thus eliminating the need for a final concentrator. Martin and others estimate a steam savings of 0.8 GJ per tonne of pulp.

Papermaking: The papermaking process accounts for about half of the total steam, electricity, and direct fuel used by the pulp and paper industry. In particular, the drying stage of the paper machine accounts for the vast majority of thermal energy use in papermaking. Most energy saving opportunities for papermaking are therefore related to improving the efficiency of the drying process and recovering its waste heat for beneficial use. Four example measures are provided below.

Advanced dryer controls: Control systems are a well-known way to optimise process variables and thereby reduce energy consumption, increase productivity, and improve the quality of industrial processes. One example of a control system for dryers is Dryer Management System control software, which reportedly offers advanced control of dryer system set points and process parameters to reduce steam use and improve productivity.

Waste heat recovery: In the paper drying process, several opportunities exist to recover thermal energy from steam and waste heat. One mill replaced the dryers with stationary siphons in their paper machine and was able to achieve energy savings of 0.89 GJ/T due to improved drying efficiency. Heat can also be recovered from the ventilation air of the drying section and used for heating of the facilities.

Shoe (extended nip) press: After paper is formed, it is pressed to remove as much water as possible. Normally, pressing occurs between two felt liners pressed between two rotating cylinders. Extended nip presses use a large concave shoe instead of one of the rotating cylinders. The additional pressing area allows for greater water extraction, (about 5–7% more water removal) to a level of 35–50% dryness. Greater water extraction leads to decreased energy requirements in the dryer, which leads to reductions in steam demand. Furthermore, reduced dryer loads allow plants to increase capacity up to 25% in cases where production in dryer is limited.

Reduced air requirements: Air to air heat recovery systems on existing machines recover only about 15% of the energy contained in the hood exhaust air. This percentage could be increased to 60–70% for most installations with proper maintenance and extensions of the systems. Paper machines with enclosed hoods require about one-half the amount of air per tonne of water evaporated compared to paper machines with a canopy hoods. Enclosing the paper machine reduces thermal energy demands since a smaller volume of air is heated. Electricity requirements in the exhaust fan are also reduced.

Published estimates suggest steam savings of 0.76 GJ per tonne of paper and electricity savings of 6.3 kWh per tonne of paper by installing a closed hood and an optimised ventilation system.

6.8   Chlorine and caustic soda

Most chlorine is manufactured electrolytically by the diaphragm, membrane, or mercury cell process. In each process, a salt solution (sodium or potassium chloride) is electrolysed by the action of direct electric current which converts chloride ions to elemental chlorine. Chlorine is also produced in a number of other ways, for example, by electrolysis of molten sodium or magnesium chloride to make elemental sodium or magnesium metal; by electrolysis of hydrochloric acid; and by non-electrolytic processes.

Diaphragm cell technology: Currently in North America, most chlorine production is from diaphragm cell technology. The products of this type of cell are chlorine gas, hydrogen gas, and cell liquor composed of sodium hydroxide and sodium chloride solution. A nearly saturated sodium chloride solution (brine) enters the diaphragm cell anolyte compartment and flows through the diaphragm to the cathode section. Chloride ions are oxidised at the anode to produce chlorine gas. Schematic of a diaphragm cell shown in Fig. 6.8. Hydrogen gas and hydroxide ions are produced at the cathode. Sodium ions migrate across the diaphragm from the anode compartment to the cathode side to produce cell liquor containing 10% to 12% sodium hydroxide. Some chloride ions also migrate across the diaphragm resulting in the cell liquor containing about 16% sodium chloride. The cell liquor is typically concentrated to 50% sodium hydroxide by an evaporation process. The salt recovered in the evaporation process is returned to the brine system for reuse.

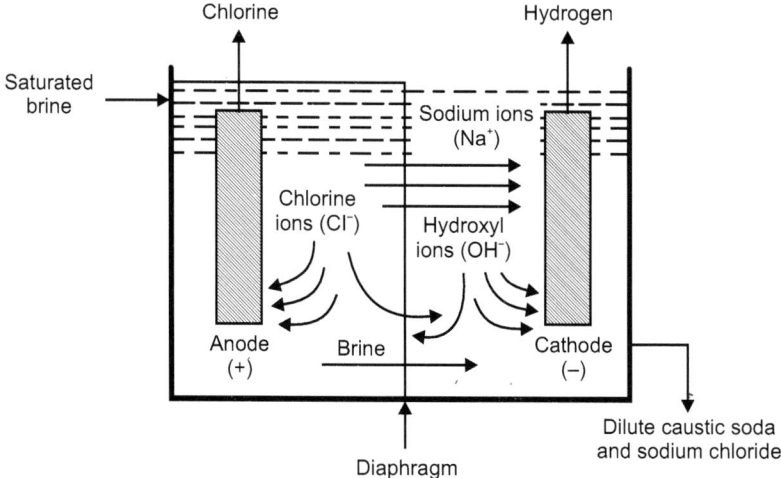

Figure 6.8: Schematic of a diaphragm cell.

Membrane cell technology: Membrane cell technology uses sheets of perfluorinated polymer ion exchange membranes to separate the anodes and cathodes within the electrolyser (Fig. 6.9). Ultra-pure brine is fed to the anode compartments, where chloride ions are oxidised to form chlorine gas. The membranes are cation selective resulting in predominantly sodium ions and water migrating across the membranes to the cathode compartments. Water is

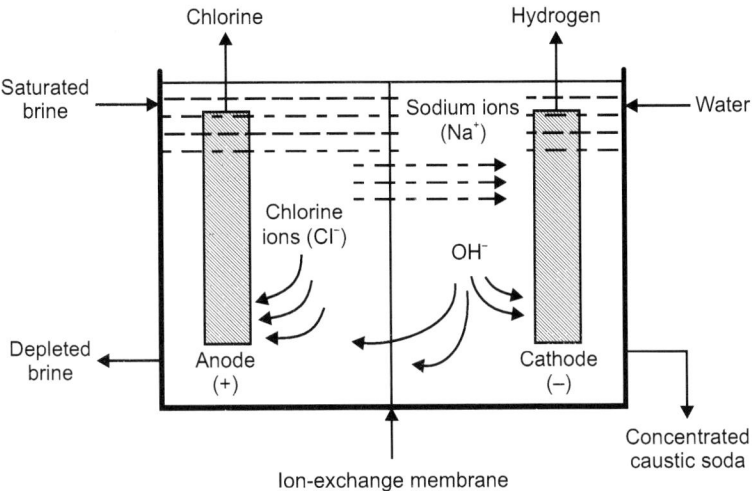

Figure 6.9: Schematic of a membrane cell.

reduced to form hydrogen gas and hydroxide ions at the cathodes. In the cathode compartment, hydroxide ions and sodium ions combine to form sodium hydroxide. Membrane electrolysers typically produce 30%–35% sodium hydroxide, containing less than 100 ppm of sodium chloride. The sodium hydroxide can be concentrated further, typically to 50%, using evaporators.

Mercury cell technology: Mercury Cell technology uses a stream of mercury flowing along the bottom of the electrolyser as the cathode. The anodes are suspended parallel to the base of the cell, a few millimeters above the flowing mercury (Fig. 6.10). Brine is fed into one end of the cell box and flows by gravity between the anodes and the cathode. Chlorine gas is evolved and released at the anode. The sodium ions are deposited along the surface of the flowing mercury cathode. The alkali metal dissolves in the mercury, forming a liquid amalgam. The amalgam flows by gravity from the electrolyser to the carbon-filled decomposer, where deionised water is added. The water chemically strips the alkali metal from the mercury, producing hydrogen and 50% sodium hydroxide. The mercury is then pumped back to the cell inlet, where the electrolysis process is repeated.

6.8.1   Energy conservation in chlor-alkali technology

Energy consumption can be reduced in membrane cells by significantly lowering the voltage required to overcome electrochemical polarisation. The cell voltage can be reduced by replacing hydrogen-evolving cathodes with 'zero-gap' oxygen-depolarised cathodes. Energy consumed is directly proportional to the total cell voltage therefore, the reduction in cell polarisation

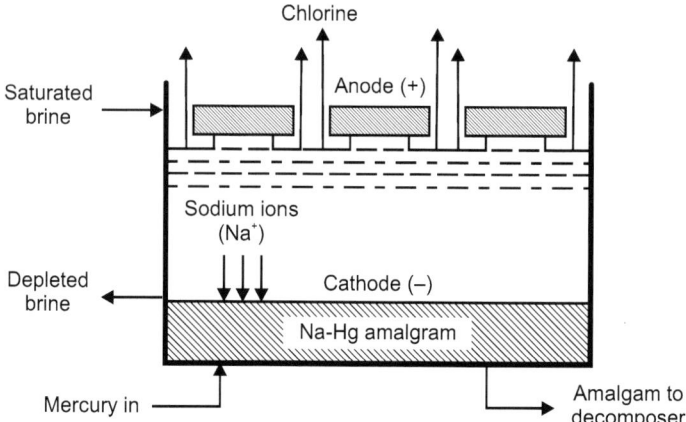

Figure 6.10: Schematic of a mercury cell.

voltage amounted to an electrical energy savings of up to 32%. Overall the oxygen-depolarised cathode cells including the energy required to produce O_2 provides an energy saving of nearly 28% over conventional membrane cells and 50% over diaphragm cells.

6.8.2   Barriers in energy conservation

1. Existing chlor-alkali membrane technology has been optimised to the extent that no further reduction of the cell voltage is expected from additional cell or membrane modifications.

2. Oxygen-supplied cathodes must satisfy two conflicting criteria, high gas permeability and low liquid permeability.

3. A stable interface between liquid and gas within the active layer of the cathode must be maintained in order to ensure long-term operation of the cell.

4. Existing chlor-alkali membrane cells cannot be simply retrofitted due to the different principles and operating conditions of the oxygen cathodes.

5. The presence of oxygen can result in accelerated corrosion of the cathode hardware and other cell components.

Pathways

The following pathways are taken to overcome the barriers related to an oxygen- depolarised cathode:

1. Identify factors leading to performance losses of an oxygen-depolarised cathode cell.

2. Study effects of cell design, operating conditions, and materials for cathode and other cell components on an oxygen- depolarised cathode.
3. Optimised cell components and operating conditions.

Results

The following results can be achieved.

1. Zero-gap chlor-alkali cells with oxygen-depolarised cathodes offered significant energy savings of as much as 38% at 0.31 amperes per square centimeter (A/cm^2).
2. Zero-gap cells with a modified anode structure generated caustic soda with current efficiency above 96% at standard industrial current densities ($\leq 0.4\,A/cm^2$). This matches or exceeds the current norms for conventional membrane cells.
3. Significantly reduced generation of peroxide, an unwanted by-product associated with the oxygen-depolarised cathode reaction.
4. Eliminated oxygen-depolarised cathode hardware corrosion problems by utilising silver-plated cathode hardware which exhibited excellent corrosion resistance both under open-circuit conditions and during electrolysis.

6.8.3 Energy requirements

Electricity is the largest energy source used for production of chlorine and sodium hydroxide

Electricity fuels the electrolysis process and represents the primary energy source. The amount of electricity required depends on the design of the cell, the design operating current, concentration of electrolytes, temperature, and pressure. Energy in the form of fuels or steam is used primarily for evaporation of the sodium hydroxide solution to a useable state.

Some fuels are also consumed in the production and purification of brine feedstock before it is sent to the electrolysis cell. For every category, energy use for process heat is distributed according to the various fuel types used throughout the industry.

Among the three types of chlorine cells, the mercury cell is the most energy-intensive, with electricity requirements of nearly 3600 kWhr per metric ton of chlorine. The membrane cell is the least energy-intensive in terms of both steam and electricity requirements. Steam requirements are less than half those of the diaphragm or mercury cell. Electricity requirements for the membrane cell are in the range of 2800 kWhr per metric ton of chlorine. The diaphragm cell is intermediate between these energy consumption ranges.

Thus, efficient operation of the cell is critical to optimised energy use and cost-effective production. Sources of energy losses in chlorine cells include anode or cathode overvoltage, too large a drop across the diaphragm, oxygen evolution on the anode, and failure to recover heat and energy from hydrogen, chlorine, and cell liquor streams.

A key consideration in membrane processes is the purity of the brine. Using very pure brine at an optimum flow rate minimises blockage through the membrane and allows sodium to penetrate freely. Brine purity is also important in mercury cells. Impurities tend to increase hydrogen by-product and reduce the current efficiency. Another issue is brine flow rate. Flow rates that are too high increase cell temperature and electrical conductivity of the medium. Brine rates that are too low create temperatures and high cell voltages that are higher than the most efficient voltage (3.1 to 3.7 volts).

Energy requirements for the manufacture of sodium carbonate from trona ore is consumed in the form of steam used for vacuum crystallisation of the trona solution to produce an initial 30% solids solution of sodium carbonate, and for calcining. Electricity is used for dissolution, clarifying, thickening, precipitation, dewatering, and calcining. Overall electricity use for this process is very low, about 127 Btu per pound of solid product.

6.8.4   Low energy consumption in chlor-alkali cells using oxygen reduction electrodes

Chlorine as one of the most important bulk chemicals in the world is produced by the electrolysis of brine. Chlorine is not only used in the day-to-day life but also is an essential part in the chemical building block, resulting in a myriad of reactions and products in the major plastic, pharmaceutical, inorganic and fine chemical and specialty industries. It is an energy intensive process, where electrical power consumption between 2100 and 3300 kWh t^{-1} Cl_2 is used depending on the operating parameters and the type of the process. There are three major processes based on mercury, diaphragm and membrane cells in use for the electrolysis.

The membrane cell technology as a recent advance since its introduction in 1970 has fewer exhausts to the environment and is relatively more efficient in the use of electric power. Despite the fact that the overall energy intensity, i.e. energy per unit output for the production of the chlorine/caustic soda has been reduced due to the successive introduction of ion exchange membrane cells instead of the 'unclean' mercury and diaphragm process, the issue of energy consumption is still a major issue. For example, a yearly production of over 4.2 million tons of chlorine in Japan by the membrane process requires a power consumption of 100 TW. Thus, there is a great incentive to look into alternative process parameters that have either to be modified or replaced so that less energy and lower capital-intensive alternatives be applied in order to increase productivity and competitiveness for the chlorine industry. The amount of electric energy needed for driving the electrode reaction depends mainly on the type of the electrolytic cell with their respective thermodynamic decomposition potential difference (DE). In the electrolytic process, as shown in for the overall reaction for every ton of chlorine produced, about 1.1 tons of caustic is generated and 28 kg hydrogen is evolved as a by-product.

$$2\,NaCl + 2\,H_2O \;\rightarrow\; Cl_2 + H_2 + 2NaOH$$

During the Hydrogen Evolution Reaction (HER) a high voltage input, i.e. 1.23 V more than the Oxygen Reduction Reaction (ORR) is required and thus, results in an inefficient means of energy consumption for producing the desired and undesired products. Furthermore, investments in equipment and energy input associated with the recovery, further purification, handling and storage of the hydrogen are also needed. Depending on the type of the process, the hydrogen has to pass through a series of systems, such as demisters, coolers, catalytic combustors (for removal of traces of oxygen), heat exchangers, blowers, compressors, etc. Although the value chain of hydrogen in the process does not correspond to the total energy input and investment costs necessary for its recovery, the hydrogen recovered has found applications in chemical processes, such as hydrogenation, catalytic reductions, hydrotreating and hydrocracking reactions, ammonia and methanol syntheses, etc. However, in most cases the hydrogen is used for heating or drying purposes, process steam generations or is simply flared-off.

In an earlier effort, a total voltage cut of 1.0 V was achieved by oxygen reduction electrodes, i.e. a reduction by ca 34% vis-à-vis a chlor-alkali cell based on the conventional membrane with HER. Thus, by using non-noble metal electrocatalysts as low-cost alternatives to Pt and Pt alloys and by constructing double-layer gas diffusion electrodes, high performance and stability could be obtained. Moreover, due to the thermodynamic relations and changes in entropy for each of the half-cell reactions, the ORR shows more exothermic reaction than the HER, implying that less heat can be supplied to the system.

6.8.5 Major energy saving areas having maximum potential in chlor-alkali sector

There is a huge possibility of saving energy in the chlor-alkali sector in different units or operations. The best possible zones where energy saving could be done in chlor-alkali sector are given below:

1. Optimisation of electrolysers for current consumption by monitoring cell voltages and replacing membranes in time.

2. Maximisation of utilisation of hydrogen generated in the cells (achieving more than 90% of utilisation.

3. Optimum liquefaction of chlorine with respect to usage.

4. Usage of vapours generated in manufacture of caustic flakes in caustic evaporation (48% lye manufacture).

5. Optimisation of steam consumption in concentrating 32% to 48% caustic by using multiple effect evaporators.

6. Heat recovery by provision of brine and chlorine recuperator for pre heating the feed brine towards the cell.

7. Burner design of caustic concentration unit (flakes production) to be done at available cell hydrogen pressure.

8. Use of waste steam in vapour absorption refrigeration unit to meet chilled water requirement in chlor-alkali plant.

9. Use of heat recovery system in each section of caustic evaporation unit. The heat from 50% Caustic product is recovered by preheating of 37% and 41% caustic by heat exchangers.

10. Introduction of VSDs (Variable Speed Drives) in all the critical pumps in the plant.

11. Use of Hypo jets by Hypo circulation for maintaining vacuum in Hypo unit. Conventionally pumps are used to create and maintain vacuum in the unit.

6.9   Lubrication

Many factors come into play when selecting a lubricant, including the projected life of a gearbox, its seals, and the desired performance of the gearbox within an application. As manufacturers continue to push the limits of machine performance to increase productivity and reduce downtime for greater customer satisfaction, suppliers are called upon to offer increasingly creative solutions. Interestingly, one of the most effective ways to achieve these performance levels is also one most frequently overlooked: proper lubrication.

Choosing the right lubricant can be an especially tricky task. Not only do industrial lubricants come in many varieties and formulations, but many industries also have their own industry regulations and standards. As a result, choosing the proper lubricant for an application is critical.

Typically, end users rely on their OEMs to determine the best lubricant, so it is important for OEMs to value lubricant as a machine element much in the same way that they value hardness of the gears, bearing selection, materials, and geometry. Like all of these other physical components of a gearbox, the right lubricant will allow the gearbox to achieve optimum performance. As a result, end users will enjoy the benefits of lower wear rates, lower operating temperatures, and best of all, greater energy efficiency.

For decades, lubricant suppliers have been developing and manufacturing specialty lubricants tailored to the requirements of industrial applications. There are general technical requirements that all lubricants must meet, such as reducing friction and wear, protecting against corrosion, dissipating heat and providing a sealing effect. But depending on the operating conditions and manufacturing processes in a plant, lubricants may also be expected to provide a host of additional properties.

Oils, greases, pastes, and waxes represent the most common categories of industrial lubricants. Typically, an oil lubricant contains 95% base oil (most often mineral oils) and 5% additives. Greases consist of lubricating base oils that are mixed with a soap to form a solid structure. Pastes contain base oils, additives and solid lubricant particles. Finally, lubricating waxes are comprised of synthetic hydrocarbons, water, and an emulsifying agent, which becomes fluid when a certain temperature level is exceeded.

6.9.1 Choosing the best lubricant

The key requirement for selecting the proper lubricant is the base oil viscosity. To select the appropriate viscosity, consider the following information about your application:

1. Operating speed (variable or fixed).
2. Specific type of friction (e.g. sliding or rolling).
3. Load and the environmental conditions.
4. Industry standards.

For example, some lubricants, like PAG (polyalkylene glycol) oils, are good for sliding friction but are not well suited for rolling friction. Likewise, PAO oils are used for rolling friction and can handle some sliding friction, whereas silicon and PFPE lubricants are typically used for extremely high temperatures.

Synthetics and mechanical applications

When synthetic oil is selected, it is generally to provide mechanical and chemical properties superior to those found in traditional mineral oils. Synthetic base oils have many benefits, including:

1. Low/high-temperature viscosity performance.
2. Decreased evaporative loss.
3. Reduced friction.
4. Reduced wear.
5. Improved efficiency.

6. Chemical stability.
7. Resistance to oil sludge problems.
8. Extended drain intervals.
9. Reduced operating costs resulting from less downtime.
10. Improved labour utilisation (less time required for lubrication and maintenance).
11. Measurable energy savings and increased output.

Despite their many benefits, synthetic lubricants are also known for one distinct disadvantage that is the cost. But the cost may be mitigated by extended change intervals, as synthetic and specialty lubricants can last five times longer or more than nonsynthetic lubricants when a high-quality base oil is used. The majority of oil lubricants, including many motor oils, are mineral oil distillates of crude oil (petroleum), while synthetic oil lubricants are also used. Synthetic oils, such as Polyalphaolefins (PAOs) or synthetic esters, are produced artificially from other compounds. Because of this, the composition is quite different from petroleum oil.

Their higher purity and uniformity provide for several enhanced properties, such as viscosity index, oxidation stability, and colour.

There are also semi-synthetic oils (also called synthetic blends), which are a blend of mineral and synthetic oil. This class of lubricants provides many of the benefits of synthetic oil at a fraction of the cost.

Most OEMs find that for ease of distribution, it is beneficial to use an H1 product because H1 synthetic gear oils are high-performance lubricants with the added benefit of being food-grade. Thus, they can be employed in both food and industrial environments. It is important to note that standards for food-grade gear oils are just as high as for other gear oils, and the synthetics perform better than standard mineral oils.

6.10 Conserve energy by adopting to total lubrication management

It is a simple fact that good machine lubrication can lead to energy savings and an improved corporate profitability. This ought to interest any plant management, who is looking for ways to reduce operating costs, and is especially significant at a time when operating in competitive global economy, besides energy conservation is a national cause.

6.10.1 Total lubrication management

This section describes how manufacturing plants can use 'Total Lubrication Management' (TLM) recommended best practices to reduce their energy consumption, emissions and operating costs– all at the same time. Electric utility bills of the plants are far larger than the maintenance and lubrication costs. So, while controlling or reducing maintenance and lubrication costs is important, reducing electric utility usage is critical. There are tremendous opportunities that exist to use an improved lubrication reliability programme to decrease plant energy costs, thereby increasing corporate profitability.

Energy for work

During conversions from one form of energy to another, some useable energy is lost. These energy losses can be extremely costly. The science of physics reveals that lubrication can play a role in reducing energy losses by reducing friction. Placed between two moving surfaces, a lubricant decreases the coefficient of friction. Naturally, this would also mean the more a lubricant decreases friction, the less energy a well lubricated machine consumes.

Lubricant formation

All lubricants consist of base oil of required viscosity, blended with special chemicals called 'Additives'. These additives are carefully selected by the oil suppliers, keeping in view the end use application–such as engine oils, gear oils, transmission oil, hydraulic oils, compressor oils, etc.

All lubricants are approved by OEMS after field tests under stringent test conditions. Lots of research work is done by oil companies and OEM before a lubricant is approved to be offered to users or reach the market. However, just buying an expensive lubricant also does not ensure maximum lubricant performance and energy savings. The lubricant must be the right one for the application and must be properly maintained for its quality in order for it to provide maximum machine performance.

6.10.2 Recommended parameters of TLM implementation

The basic recommended parameters of TLM implementation are as follows:

1. Select correct grade and viscosity of lubricants for the specific application and ensure this grade has OEM acceptance.

2. Store oil in good environment to keep oil uncontaminated in storage. Greases should be stored indoors to avoid day/night temperature fluctuations. This can lead to soap-oil separation, making grease unfit for usage. Good house-keeping at lubricants storage is the most important and is at fulcrum of entire activity.

3. Adopt colour coding to eliminate any possibility of mix-up in oils leading to contamination.

4. Use good and clean lubrication equipment to ensure feeding uncontaminated lubricants to machines.

5. Keep oil clean by providing 'Breathers' on machines oil sump and inspect oil filters on machines oil systems regularly.

6. Test oils regularly for oil condition and machine condition, i.e. contamination, additive depletion, wear debris and elemental analysis, etc.

7. Regular training to lubrication staff for correct lubrication techniques. All lubrication staff should be in skilled category.

8. Enforce excellent house-keeping at oil storage, handling and dispensing area.

9. Keep oil points at machine 'clean' to ensure that no dust or dirt particles go in the machine sump along with oil.

10. Adopt target based oil management system. Ensure that atleast 95% oil is drained out from the sump, before feeding new oil into the sump, failing which, may be adding new oil into 'muck' in the oil sump or machine system.

6.10.3   Lubrication and energy savings

It is possible to measure energy savings in a variety of ways, including production output, temperature changes or reduction in electrical energy consumption. Another measurement is maintenance costs and fuel consumption.

Production output

When using any mechanical equipment, it is possible to evaluate the equipment's energy efficiency by recoding its production output. For example, if a machine is capable of producing a certain number of parts in a given amount of time and the lubricant is kept clean as per recommended cleanliness standards and lubrication systems are improved. This shall be resulting in a higher volume of production in the same amount of time, than the machine has become more energy efficient and productive.

Temperature changes

Monitoring temperature changes is another way to optimise lubrication programme performance. Increased friction in a machine moving parts results in higher operating temperatures. Friction is a result of metal-to-metal contact that occurs between two surfaces moving relative to another. Even between highly machined surfaces, under microscopic view, asperity contact occurs. The greater the amount of metal to metal contact, the greater is the amount of friction. As a result, more energy is required to move the surfaces relative to one another. This friction results in higher electrical power costs. Lubricants and good lubrication system can reduce that friction. Therefore, when friction is reduced, less electricity is required to drive a gearbox, compressor, pump or other equipment, and this leads to energy conservation in the industry.

6.10.4   Electrical energy reduction

Tracking electrical consumption is a highly reliable way to evaluate improvements in plant energy use. In fact, various organisations have been able to document improvements in electrical energy efficiency after implementation of lubrication management programmes. Companies that upgrade their lubrication and reliability practices have been able to document a 5–10% reduction in power consumption, more than enough to pay for implementing good lubrication programme by professional service providers. Average documented savings were 10% in gear boxes, 12% in air compressors and 4% in electric motors. Electric motors power most plant machinery, including gearboxes, compressors, refrigeration systems, pumps, hydraulic systems and ball mills.

The following equation can determine the amount of electricity used by an electric motor:

1. $kW = V/1000 \times A \times 1.73$ (where V is volts and A is amperes).

2. $kW = \sqrt{3} \ VI. \ Cos \ \emptyset$ – corresponding Cos Ø at 0.9 which all the plants are meeting as per requirement of electricity board.

    Both are common metric measurements of electrical current measured using a voltmeter or ammeter. For a three-phase motor, 1.73 is a standard factor. Data logging equipment is available that allows one to measure and collect data for either amperes, volts or both. Yet, most electrical consumers pay for electricity by kilowatt-hour (kWh) per month. The following formula is commonly used to determine the electrical charge per month (ECM).

3. $ECM = kW \times h \times EC$ (where h is hours of service and EC is the electrical charge.)

Air compressors are an excellent source for energy savings. Compressed air is one of the most expensive uses of energy in a manufacturing plant, and approximately 70% of all manufacturers have a compressed air system. These systems power a variety of equipment, including machine tools, material handling and separation equipment and spray painting equipment. According to a study, compressed air systems account for 10% of all electricity and roughly 16% of industrial motor system energy use. This adds up to large amount of expenditure per year in energy costs. Energy audits conducted suggest that more than 50% of compressed air systems at industrial facilities have significant energy conservation opportunities.

Energy conservation is very important to industry, as important as conserving natural resources, reducing emissions and improving profitability. Governments and corporate management in the industry alike are looking for ways to reduce energy consumption. It is possible to make dramatic gains in energy efficiency by reducing friction, and one of the best ways to do that is to 'employ good lubrication practices', including the use of high-performance lubricants and the adoption of lubrication reliability best practices.

Lubricants influence energy efficiency mainly through reducing energy losses, which include churning losses and friction losses in hydrodynamic, elastohydrodynamic and boundary lubrication regimes. The total energy loss depends on lubricant viscosity and chemical composition. The best energy efficiency can be achieved when using all possibilities for improvement, including boundary friction modification and viscosity optimisation. Friction modifiers, being added to the oil additive package in low concentrations (about one to two per cent), reduce boundary friction. Such concentrations have very little effect on the oil viscometric behaviour. Therefore, these two ways of obtaining energy efficiency improvements can be considered and utilised almost independently.

Energy conservation in pharmaceutical industries

7.1 Introduction

The pharmaceutical industry develops, produces, and markets drugs or pharmaceuticals for use as medications. Pharmaceutical companies may deal in generic or brand medications and medical devices. They are subject to a variety of laws and regulations that govern the patenting, testing, safety, efficacy and marketing of drugs. The technology operated by the pharmaceutical industry–the chemical and industrial processes through which medicines are produced, packaged and shipped–seems to fit the constant returns to scale hypothesis almost perfectly.

7.2 Stages in production of bulk pharmaceuticals

There are three overall stages in the production of bulk pharmaceutical products: (i) R&D, (ii) conversion of natural substances to bulk pharmaceuticals and (iii) formulation of final products.

7.2.1 Conversion to bulk pharmaceutical substances

Bulk pharmaceutical substances are produced via chemical synthesis, extraction, fermentation, or a combination of these processes.

7.2.2 Chemical synthesis

Figure 7.1 shows a simplified diagram is the chemical synthesis process for pharmaceuticals. There are five primary stages in chemical synthesis: (i) reaction, (ii) separation, (iii) crystallisation, (iv) purification and (v) drying.

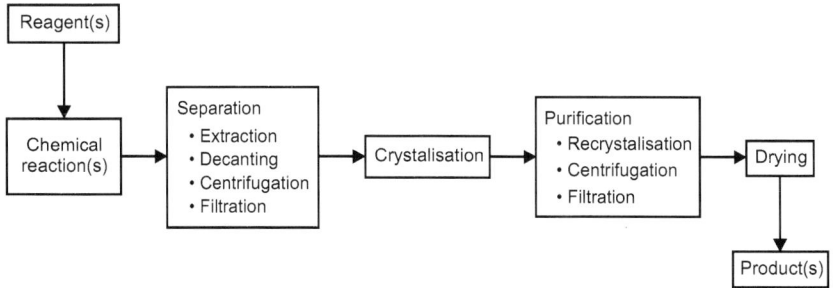

Figure 7.1: Simplified chemical synthesis diagram.

Reaction

In the reaction process, raw materials are fed into a reactor vessel, where reactions such as alkylations, hydrogenations, or brominations are performed. The most common type of reactor vessel is the kettle-type reactor. These reactors, which are generally made of stainless steel or glass-lined carbon steel, range from fifty to several thousand gallons in capacity. The reactors may be heated or cooled, and reactions may be performed at atmospheric pressure, at elevated pressure, or in a vacuum. Generally, both reaction temperature and pressure are monitored and controlled. Nitrogen may be required for purging the reactor, and some intermediates may be recycled back into the feed. Some reactions are aided via mixing action provided by an agitator. A condenser system may be required to control vent losses. Reactors are often attached to pollution control devices to remove volatile organics or other compounds from vented gases.

Separation

The main types of separation processes are extraction, decanting, centrifugation, filtration, and crystallisation.

Extraction: The extraction process is used to separate liquid mixtures. Extraction takes advantage of the differences in the solubility of mixture components. A solvent that preferentially combines with only one of the mixture components is added to the mixture. Two streams result from this process: the extract, which is the solvent-rich solution containing the desired mixture component, and the raffinate, which is the residual feed solution containing the non-desired mixture component(s).

Decanting: Decanting is a simple process that removes liquids from insoluble solids that have settled to the bottom of a reactor or settling vessel. The liquid is either pumped out of the vessel or poured from the vessel, leaving only the solid and a small amount of liquid in the vessel.

Centrifugation: Centrifugation is a process that removes solids from a liquid stream using the principle of centrifugal force. A liquid-solid mixture is added to a rotating vessel or centrifuge, and an outward force pushes the liquid through a filter that retains the solid phase.

The solids are manually scraped off the sides of the vessel or with an internal scraper. To avoid air infiltration, centrifuges are usually operated under a nitrogen atmosphere and kept sealed during operation. Filtration separates fluid/solid mixtures by flowing fluid through a porous media, which filters out the solid particulates. Batch filtration systems widely used by the pharmaceutical industry include plate and frame filters, cartridge filters, nutsche filters, and filter/dryer combinations.

Crystallisation

Crystallisation is a widely used separation technique that is often used alone or in combination with one or more of the separation processes described above. Crystallisation refers to the formation of solid crystals from a super-saturated solution. The most common methods of super saturation in practice are cooling, solvent evaporation, and chemical reaction.

The solute that has crystallised is subsequently removed from the solution by centrifugation or filtration.

Purification

Purification follows separation, and typically uses the separation methods described above. Several steps are often required to achieve the desired purity level. Recrystallisation is a common technique employed in purification. Another common approach is washing with additional solvents, followed by filtration.

Drying

The final step in chemical synthesis is drying the product (or intermediates). Drying is done by evaporating solvents from solids. Solvents are then condensed for reuse or disposal. The pharmaceutical industry uses several different types of dryers, including tray dryers, rotary dryers, drum or tumble dryers, or pressure filter dryers.

7.2.3   Product extraction

Active ingredients that are extracted from natural sources are often present in very low concentrations. The volume of finished product is often an order of magnitude smaller than the raw materials, making product extraction an inherently expensive process.

Precipitation, purification, and solvent extraction methods are used to recover active ingredients in the extraction process. Solubility can be changed by pH adjustment, by salt formation, or by the addition of an anti-solvent to isolate desired components in precipitation. Solvents can be used to remove active ingredients from solid components like plant or animal tissues, or to remove fats and oils from the desired product. Ammonia is often used in natural extraction as a means of controlling pH.

7.2.4   Fermentation

In fermentation, micro-organisms are typically introduced into a liquid to produce pharmaceuticals as by-products of normal micro-organism metabolism. The fermentation process is typically controlled at a particular temperature and pH level under a set of aerobic or anaerobic conditions that are conducive to rapid micro-organism growth. The process involves three main steps: (i) seed preparation, (ii) fermentation and (iii) product recovery.

Seed preparation: The fermentation process begins with seed preparation, where inoculum (a medium containing micro-organisms) is produced in small batches within seed tanks. Seed tanks are typically 1–10% of the size of production fermentation tanks.

Fermentation: After creating the inoculum at the seed preparation stage, the inoculum is introduced into production fermentors. In general, the fermentor is agitated, aerated, and controlled for pH, temperature, and dissolved oxygen levels to optimise the fermentation process. The fermentation process lasts from hours to weeks, depending on the product and process.

Product recovery: When fermentation is complete, the desired pharmaceutical by-products need to be recovered from the fermented liquid mixture. Solvent extraction, direct precipitation, and ion exchange may be used to recover the product. Additionally, if the product is contained within the micro-organism used in fermentation, heating or ultrasound may be required to break the micro-organisms cell wall. In solvent extraction, organic solvents are employed to separate the product from the aqueous solution. The product can then be removed from the solvent by crystallisation. In direct precipitation, products are precipitated out of solution using precipitating agents like metal salts. In ion exchange, the product adsorbs onto an ion exchange resin and is later recovered from the resin using solvents, acids, or bases.

7.2.5 Formulation of final products

The final stage of pharmaceutical manufacturing is the conversion of manufactured bulk substances into final, usable forms. Common forms of pharmaceutical products include tablets, capsules, liquids, creams and ointments, aerosols, patches, and injectable dosages.

To prepare a tablet, the active ingredient is combined with a filler (such as sugar or starch), a binder (such as corn syrup or starch), and sometimes a lubricant (such as magnesium sterate or polyethylene glycol). The filler ensures the proper concentration of the active ingredient; the purpose of the binder is to bond tablet particles together. The lubricant may facilitate equipment operation during tablet manufacture and can also help to slow the disintegration of active ingredients.

Tablets are produced via the compression of powders. Wet granulation or dry granulation processes may be used. In wet granulation, the active ingredient is powdered and mixed with the filler, wetted and blended with the binder in solution, mixed with lubricants, and finally compressed into tablets. Dry granulation is used when tablet ingredients are sensitive to moisture or drying temperatures. Coatings, if used, are applied to tablets in a rotary drum, into which the coating solution is poured. Once coated, the tablets are dried in the rotary drum; they may also be sent to another drum for polishing.

Capsules are first constructed using a mold to form the outer shell of the capsule, which is typically made of gelatin. Temperature controls during the molding process control the viscosity of the gelatin, which in turn determines the thickness of the capsule walls. The capsule's ingredients are then poured (hard capsules) or injected (soft capsules) into the mold.

For liquid pharmaceutical formulations, the active ingredients are weighed and dissolved into a liquid base. The resulting solutions are then mixed in glass-lined or stainless steel vessels and tanks. Preservatives may be added to the solution to prevent mold and bacterial growth. If the liquid is to be used orally or for injection, sterilisation is required. Ointments are made by blending active ingredients with a petroleum derivative or wax base. The mixture is cooled, rolled out, poured into tubes, and packaged.

Creams are semisolid emulsions of oil-in-water or water-in-oil; each phase is heated separately and then mixed together to form the final product.

7.3   Energy consumed in various processes

Table 7.1 shows that R&D and bulk manufacturing are typically the most important energy consuming activities in the pharmaceutical industry.

7.4   Energy efficiency opportunities for the pharmaceutical industry

A variety of opportunities exist within pharmaceutical laboratories, manufacturing facilities, and other buildings to reduce energy consumption while maintaining or enhancing productivity.

Although technological changes in equipment conserve energy, changes in staff behaviour and attitude can also have a great impact. Energy efficiency training programmes can help a company's staff incorporate energy efficiency practices into their day-to-day work routines. Personnel at all levels should be aware of energy use and company objectives for energy efficiency improvement. Often such information is acquired by lower-level managers but neither passed up to higher-level management nor passed down to staff. Energy efficiency programmes with regular feedback on staff behaviour, such as reward systems, have had the best results. Though changes in staff behaviour (such as switching off lights or closing windows and doors) often save only small amounts of energy at one time, taken continuously over longer periods they can have a much greater effect than more costly technological improvements. Other staff actions such as the closing of fume hood sashes could result in significant and

Table 7.1: Distribution of energy use in the pharmaceutical industry.

	Overall	Plug loads and processes	Lighting	Heating, ventilation and air conditioning (HVAC)
Total	100%	25%	10%	65%
R&D	30%	Microscopes Centrifuges Electric mixers Analysis equipment Sterilisation processes Incubators Walk in/reach in areas (refrigeration)	Task and overhead lighting	Ventilation for clean rooms and fume hoods areas requiring 100% make-up air chilled water hot water and steam
Offices	10%	Office equipment including computers, fax machines, photocopiers, printers Water heating (9%)	Task, overhead, and outdoor lighting	Space heating (25%) Cooling (9%) Ventilation (5%)
Bulk manufacturing	35%	Centrifuges Sterilisation processes Incubators Dryers Separation processes	Task and overhead lighting	Ventilation for clean rooms and fume hoods areas requiring 100% make-up air chilled water, hot water and steam
Formulation, Packaging and filling	15%	Mixers Motors	Mostly overhead, some task	Particle control ventilation
Warehouses	5%	Fork lifts Water heating (5%)	Mostly overhead lighting	Space heating (41%) Refrigeration (4%)
Miscellaneous	5%		Overhead	

immediate improvement. Establishing formal management structures and systems for managing energy that focus on continuous improvement are important strategies for helping companies manage energy use and implement energy efficiency measures.

7.5   Cogeneration

For industries like pharmaceutical manufacturing that have requirements for process heat, steam, and electricity, the use of combined heat and power (CHP) systems may be able to save energy and reduce pollution. Cogeneration plants are significantly more efficient than standard power plants because they take advantage of waste heat. In addition, transmission losses are minimised when CHP systems are located at or near the plant.

Often, utility companies will work with individual companies to develop CHP systems for their plants. In this scenario, the utility company owns and operates the plant's CHP system; therefore, the company avoids the capital expenditures associated with CHP projects, but gains the benefits of a more energy efficient source of heat and electricity.

In addition to energy savings, CHP systems also have comparable or better availability of service than utility generation. In the automobile industry, for example, typical CHP units are reported to function successfully for 95–98% of planned operating hours. For installations where initial investment is large, potential multiple small-scale CHP units distributed to points of need could be used cost effectively.

Currently, most large-scale CHP systems use steam turbines. Switching to natural gas-based systems will improve the power output and efficiency of the CHP system, due to increased power production capability. Although the overall system efficiency of a steam turbine-based CHP system (80%, HHV) is higher than that of a gas turbine-based CHP system (74%, HHV), the electrical efficiency of a gas turbine-based CHP system is much higher (27–37% for typical industrial scale gas turbines). The power-to-heat ratio of a steam turbine-based CHP system is very low (limited to about 0.2), limiting the output of electricity. The power-to-heat ratio of a gas turbine-based CHP system is much higher (between 0.5 and 1.0), producing more power for the same amount of fuel. This may improve the profitability of a gas-based CHP unit, depending on the price of power to the plant. Modern gas-based CHP systems have low maintenance costs and will reduce emissions of NO_x, sulphur dioxide, CO_2, and particulate matter from power generation considerably, especially when replacing a coal-fired boiler.

In general, the energy savings of replacing a traditional system (i.e. a system using boiler-based steam and grid-based electricity) with a standard gas turbine-based CHP unit is estimated at 20%–30%. The efficiency gain will be higher when replacing older or less maintained boilers.

Combined cycles (combining a gas turbine and a back-pressure steam turbine) offer flexibility for power and steam production at larger sites, and potentially at smaller sites as well. Steam-injected gas turbines (STIG) can absorb excess steam (e.g. due to seasonal reduced heating needs) to boost power production by injecting steam into the turbine.

The size of typical STIGs starts around 5 MW. This type of turbine uses the exhaust heat from a combustion turbine to turn water into high-pressure steam. This steam is then fed back into the combustion chamber to mix with the combustion gas.

The advantages of this system are as follows:

1. The added mass flow of steam through the turbine increases power by about 33%.
2. The machinery involved is simplified by eliminating the additional turbine and equipment used in combined cycle gas turbine.
3. The steam is cool compared to combustion gases helping to cool the turbine interior.
4. The system reaches full output more quickly than combined-cycle unit (30 minutes versus 120 minutes).

Additional advantages are that the amounts of power and thermal energy produced by the turbine can be adjusted to meet current power and thermal energy (steam) loads. If steam loads are reduced, the steam can then be used for power generation, increasing output and efficiency. Drawbacks include the additional complexity of the turbine's design. Additional attention to the details of the turbine's design and materials are needed during the design phase. This may result in a higher capital cost for the turbine compared to traditional models. The economics of a cogeneration system depend strongly on the local situation, including power demand, heat demand, power purchasing and selling prices, natural gas prices, as well as interconnection standards and charges, and utility charges for backup power (backup charges).

7.5.1 Trigeneration

Furthermore, new CHP systems offer the option of trigeneration, which provides cooling in addition to electricity and heat. Cooling can be provided using either absorption or adsorption technologies, which both operate using recovered heat from the cogeneration process (Fig. 7.2).

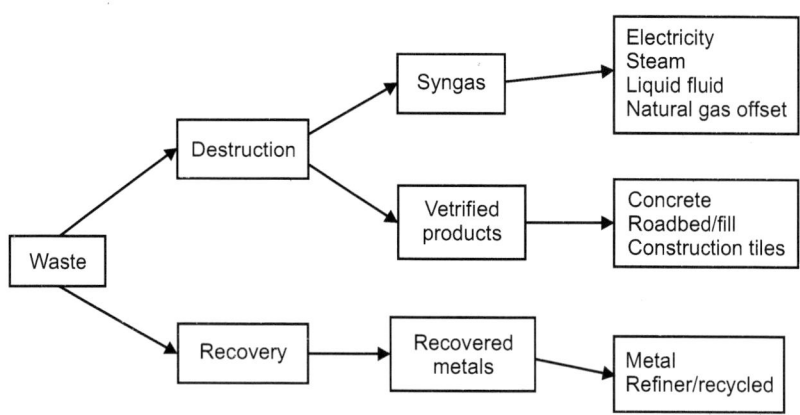

Figure 7.2: Recovering of heat from the cogeneration process.

Absorption cooling systems take advantage of the fact that ammonia is extremely soluble in cold water and much less so in hot water. Thus, if a water-ammonia solution is heated, it expels its ammonia. In the first stage of the absorption process, a water-ammonia solution is exposed to waste heat from the cogeneration process, whereby ammonia gas is expelled. After dissipating the heat, the ammonia gas (still under high pressure liquefies). The liquid ammonia flows into a section of the absorption unit where it comes into contact with hydrogen gas. The hydrogen gas absorbs the ammonia gas with a cooling effect. The hydrogen-ammonia mixture then meets a surface of cold water, which absorbs the ammonia again, closing the cycle. In contrast, adsorption cooling utilises the capacity of certain substances to adsorb water on their surface, from where it can be separated again with the application of heat. Adsorption units use hot water from the cogeneration unit. These systems do not use ammonia or corrosive salts, but use silica gel (which also helps to reduce maintenance costs). The thermal performance of absorption and adsorption systems is similar, with a coefficient of performance (COP) between 0.68 and 0.75. The capital costs of both systems are also comparable. However, the reliability of an adsorption unit is expected to be better and maintenance cost is expected to be lower.

7.5.2   Power recovery turbines

Steam is often generated at high pressures (typically at 120–150 psig), but often the pressure is reduced (to as low as 10–15 psig) to allow the steam to be used by different process. Typically, pressure reduction is accomplished through a pressure reduction valve, which does not recover the energy embodied in the pressure drop. This energy could be recovered by using a micro scale-back pressure steam turbine, which is produced by several manufacturers. Power recovery turbines are capable of producing 13.5 kWh/MBtu steam. The actual power generation on a particular site will vary depending on steam pressures and steam uses.

7.6   Improving energy efficiency in pharmaceutical manufacturing operations

7.6.1   HVAC

HVAC systems, which consist of dampers, supply and exhaust fans, filters, humidifiers, dehumidifiers, heating and cooling coils, ducts, and various sensors.

There are many energy efficiency measures that can be applied to HVAC systems, some significant opportunities are discussed below.

Non-production hours set-back temperatures: Setting back building temperatures (that is, turning temperatures down in winter or up in summer) during periods of non-use, such as weekends or non-production times, can lead to significant savings in HVAC energy consumption. Similarly, reducing ventilation in clean rooms and laboratories during periods of non-use can also lead to energy savings.

Adjustable speed drives (ASDs): Adjustable speed drives can be installed on variable-volume air handlers, as well as recirculation fans, to match the flow and pressure requirements of air-handling systems precisely. Energy consumed by fans can be lowered considerably since they are not constantly running at full speed. Adjustable speed drives can also be used on chiller pumps and water systems pumps to minimise power consumption based on system demand.

Heat recovery systems: Heat recovery systems reduce the energy required to heat or cool facility intake air by harnessing the thermal energy of the facility's exhaust air. Common heat recovery systems include heat recovery wheels, heat pipes, and run-around loops. For areas requiring 100% make-up air, studies have shown that heat recovery systems can reduce a facility's heating/cooling.

Improving HVAC chiller efficiency: The efficiency of chillers can be improved by lowering the temperature of the condenser water, thereby increasing the chilled water temperature differential. This can reduce pumping energy requirements. Another possible efficiency measure is installing separate high temperature chillers for process cooling. Sizing chillers to better balance chiller load with demand is also an important energy efficiency strategy.

7.6.2   Clean room HVAC

A recent study found that HVAC systems accounted for 36–67% of clean room energy consumption. Another recent study estimated the following energy distribution for clean room operation: 56% for cooling, 36% for heating, 5% for fans, and 3% for pumps.

The following measures can improve energy efficiency in clean rooms.

Reduce recirculation air charge rates

Improve air filtration quality and efficiency: High Efficiency Particulate Air (HEPA) filters and Ultra Low Penetration Air (ULPA) filters are commonly used in the pharmaceutical industry to filter make-up and recirculated air. The adoption of alternative filter technologies might allow for lower energy consumption. For example, new air filtration technologies that trap particles in the ultra-fine range (0.001–0.1 microns), a range for which current filter technologies are not effective, might reduce the energy necessary for reheating/recooling clean room air.

Use cooling towers: In many instances, water cooling requirements can be met by cooling towers in lieu of water chillers. Water towers can cool water much more efficiently than chillers and can therefore reduce the overall energy consumption of clean room HVAC systems.

Reduce clean room exhaust: The energy required to heat and cool clean room make-up air accounts for a significant fraction of clean room HVAC energy consumption. Measures to reduce clean room exhaust airflow volume can therefore lead to significant energy savings.

7.6.3 Boilers

Boilers and steam distribution systems are major contributors to energy losses at many industrial facilities; they are therefore an area where substantial efficiency improvements are typically feasible.

The following measures can improve energy efficiency in boilers.

Reduce flue gas quantities: Often excessive flue gas results from leaks in the boiler and/or in the flue. This reduces the heat transferred to the steam and increases pumping requirements. These leaks are often easily repaired. Savings amount to 2–5% of the energy formerly used by the boiler.

Reduce excess air: The more excess air is used to burn fuel, the more heat is wasted in heating this air rather than in producing steam. A rule of thumb often used is that boiler efficiency can be increased by 1% for each 15% reduction in excess air or 40°F (22°C) reduction in stack gas temperature.

Properly size boiler systems: Correctly designing the boiler system at the proper steam pressure can save energy by reducing stack temperature, reducing piping radiation losses, and reducing leaks in traps and other sources. In a study done in Canada on 30 boiler plants, savings from this measure ranged from 3–8% of the total gas consumption.

Properly insulate boiler

Perform regular maintenance: A simple maintenance programme to ensure that all components of a boiler are operating at peak performance can result in substantial savings. On average, the energy savings associated with improved boiler maintenance are estimated at 10%.

Reuse condensate: Reusing hot condensate in boilers saves energy, reduces the need for treated boiler feed water, and reclaims water at up to 100°C (212°F) of sensible heat. A Pfizer plant (in the year 2011) in Groton, Conn., upgraded their condensate recovery system and realised a 9% reduction in electricity consumption, and an 8% reduction in water consumption and wastewater discharge. As a result, Pfizer saved roughly $175,000 per year through avoided oil, gas and water purchases.

7.7 Water and energy conservation in the pharmaceutical industry

In particular, the pharmaceutical industry requires consistent, high-quality water for production and wastewater treatment to meet the demands of ever-stricter regulatory discharge limits. To meet these challenges, companies must question conventional thinking and typical approaches and explore new technologies and solutions in order to remain competitive.

Because of this increased focus on water and energy, companies are evaluating new technologies and integrated solutions to reduce water consumption and increase energy efficiency. These new technologies and solutions cannot, however, compromise the dependability and robustness demanded by the marketplace. There is perhaps no better example of the need for dependable water solutions and product safety than in the pharmaceutical industry.

For pharmaceutical applications, there are three main drivers that force the issues of utility conservation into product design: long-term lifecycle costs, regulatory requirements and conservation/corporate responsibility.

Operating cost reduction has become increasingly important for pharmaceutical and virtually every other industry to allow companies to operate as efficiently as possible. Reducing the cost of ownership on a water system is an important aspect of system design. Exploring new methods to reduce the cost of treating water, wastewater discharge, and utility expenses challenges decades-old system designs. Many cost savings techniques can actually offer enhanced reliability and performance of conventional water systems, while lowering cost of ownership.

Achieving regulatory requirements on a water system design usually includes two distinct areas of responsibility in a pharmaceutical application: maintaining minimum water quality standards for discharge or reuse within in the facility and meeting discharge volume from the facility. There are specific contaminant limits on the discharge of water into municipalities or other waste streams. Exceeding these limits can result in severe financial penalties or put a plant's operation at risk. In particular, pharmaceutical manufacturers must operate within strict national and local regulatory limits.

Finally, conservation of natural resources and public perception of the pharmaceutical manufacturers is critical for the image of these companies. Companies can receive negative public perception and ratings from regulatory bodies as well as shareholders, if not operating at an optimal level. More and more, companies want to be recognised for the efficient use of natural resources and a gentle global footprint. In addition, companies experience increased scrutiny and pressure from various entities including the public, regarding concerns about pharmaceuticals in drinking water. So, ensuring tight management of water supply is a serious and critical point for pharmaceutical manufacturers—both in the manufacturing process and in product development.

7.7.1 Microbial contamination is key to current technology system design

Water is used widely during the production of pharmaceutical products as a direct ingredient as well as indirect uses such as formulation, rinsing, sanitising and cleaning. Most high quality makeup water systems used within pharmaceutical production have some element of recirculation inherent in the system design. When water is not needed, water systems typically go into a recirculating standby mode to control microbial proliferation within the water and on the wetted surfaces of the water system. This conventional approach has been used for decades and has been considered the best '*in situ*' approach for controlling microbial contamination. Often, microbial contamination is the most difficult, and costly, aspect of the water system design. While chemical and organic impurities can usually be managed with little difficulty, proliferation of bacteria, viruses and other organisms can challenge even the best system designs. In an effort to build robustness into the water systems of a pharmaceutical facility, many of these systems are built with a 2 × 100% redundancy approach. This ensures the water system is always available when needed. All water systems require periodic maintenance monitoring and adjustments, and a redundant water system approach allows one train to be available while the other train is being serviced or maintained. This redundant water train will continually recirculate while in the standby mode, and will consume water and electricity, and produce wastewater, nearly 100% of the time since it is by nature a standby or backup system.

7.7.2 Technology developments reduce energy and water consumption

Recent technology developments for pharmaceutical water systems eliminate the need to continually recirculate water. A traditional water system constantly recirculates water, which consumes electricity for pumps, ultraviolet lights, instruments and other devices. Often the water must be heated or cooled to maintain adequate water temperature specifications. Additionally, certain unit processes such as reverse osmosis produce a waste stream during operation. A recirculating water system, even if it is not currently producing water for production, is producing waste that must be discharged or treated prior to discharge. These systems can quickly consume valuable raw water, consume electrical and steam utilities and produce a wastewater stream even if the water system is not being used in the production process. A sanitise/start/stop design eliminates most of this waste by shutting down the water system when not needed. While in the standby mode, the system will briefly turn on and receive periodic heat sanitisation during extended periods of non-use.

These sanitisation periods are quick and relatively low temperature (typically 60°C) rather than the more typical 80 to 85°C. The brief but frequent sanitisation periods are very effective at challenging microbial proliferation within a water system.

When the pharmaceutical production requires water, the system performs a brief pulse sanitisation just before water is sent to the manufacturing process. This ensures the water system is freshly sanitised for optimal water quality when it is needed. Standby or redundant trains can now sit relatively idle, receiving periodic pulse sanitisations to maintain water quality. Just like conventional water systems, the sanitise/start/stop design can be combined with chemical cleaning, chemical sanitisation and conventional heat sanitisation to address microbial contamination and biofilm formation within the water system. The economic and water volume savings from the sanitise/ start/stop approach can be dramatic, sometimes saving many millions of gallons of water per year. Large water systems, redundant trains, high raw water costs, high discharge water costs, and water discharge limitations can greatly impact the total savings. While the savings are greater on larger systems, even single-train, relatively small water systems with moderate water costs still result in rapid payback and significant savings.

8

Energy conservation in textile industry

8.1 Introduction

The textile industry touches the lives of all people in one or the other ways. Apparel, home textiles, technical textiles, industrial textiles, medical textiles, safety textiles, smart or intelligent textiles, there are variations for all - consumers, traders, manufacturers, technologists, engineers and others.

Textile industry has come a long way to be an organised industry from being a mere domestic industry. Starting with the industrial revolution, it has gained a state of supremacy with time. High production of wool, cotton and silk all over the world has given a boost to the textile industry in past years. Though the industry originated in U.K., the art of textile production passed to Europe and North America after mechanisation of textile manufacturing process in those areas. Asian countries also industrialised their economies and took steps for the growth of this sector. Japan, India, Hong Kong and China have become leading producers of textile because of the availability of cheap labour which is a very important factor for this industry.

The future global market for textile and apparel is expected to expand in a significant way. The reasons for such expansion include growth of new consumption markets, global expansion of modern retail business, boom of air and sea shipments, growth of textile and related production in Europe, Russian, Turkey, South East Asia, India, China and United States of America.

8.2 Textile preparatory process

Desizing: Cotton fibres and cotton/synthetic fibre blends are sized, i.e. they are coated with a strengthening, adhesive like material (usually starch or a starch based material) to prevent damage during the weaving process. Size is usually applied to the warp thread, since this is particularly susceptible to mechanical strain during weaving. The material used for sizing is usually starch or a starch derivative. Starch size is most widely used (about 75% of textile industries use starch). The starch is usually obtained from potatoes, from maize and from rice. The size must be removed before a fabric can be bleached and dyed, since it affects the uniformity of wet processing. Various types of desizing methods are available. If the size is water soluble, an alkali wash with detergents may be used. Oxidative chemicals such as alkali or bromide and alkali may also be used (may be included in the bleaching process).

Enzymatic method: In the textile industry, amylases are used to remove starch-based size for improved and uniform wet processing. Amylase is a hydrolytic enzyme which catalyses the breakdown of dietary starch to short chain sugars, dextrin's and maltose. The advantage of these enzymes is that they are specific for starch, removing it without damaging to the support fabric (e.g. cotton and its blends). An amylase enzyme can be used for desizing processes at low-temperature. The optimum temperature is 30–60°C, optimum pH is 5.5–6.5.

Scouring: Scouring is the process of removal of natural and added impurities like fat, wax, sand or dust particles and oils. Conventionally, the scouring process is carried out by treating the fabric with caustic soda and sodium hydroxide at 70°C to 90°C.

The use of the traditional strongly alkaline process can have a detrimental effect on fabric weight and on the environment. Enzymatic scouring makes it possible to effectively scour fabric without negatively affecting the fabric or the environment. It also minimises health risks since operators are not exposed to aggressive chemicals.

Pectinase enzyme is effective and environmentally friendly for scouring. It breaks down the pectin in the cotton and thus assists in the removal of waxes, oils and other impurities. Novozymes offers an alkaline pectinase for bioscouring that gently but completely removes pectin and other impurities from cotton fibres and that can be used on a range of textile wet-processing machinery. The optimum temperature is 50–65°C and pH between 7.5–9.0.

Souring (bleach removal): In conventional method generally, bleaching of cellulose fibres with hydrogen peroxide is optimum at a pH of 10, 5 to 11, at a temperature between 80 and 100°C, and for a contact time between 45 minutes and 5–6 hours. After completion of the process, the bleached liquor is drained out (the bleaching chemical has to be removed before the dye is applied in order to prevent reaction between the bleach and dye) then the fabric is rinsed with water a number of times to remove the H_2O_2 from the bleached fabric. Alternatively, a mild reducing agent can be used to neutralise the bleach. In either case, large amounts of water (up to 40 litres per kg of fabric) are required for rinsing resulting in discharge of large volumes of wastewater.

An enzyme can be used to replace a chemical agent (such as thiosulphate) in bleach neutralisation, leading to reduction in water and energy consumption.

8.2.1   Dyeing process

Dyeing is the process of imparting colours to a textile material through a dye (colour). Dyes are obtained from flowers, nuts, berries and other forms of vegetables and plants as well as from animal and mineral sources. These are known as natural dyes. The other class of dyes is known as synthetic dyes.

These are based on a particular type of chemical composition. Some of these dyes are acid (anionic) dyes, basic (cationic) dyes, neutral-premetalised dyes, sulphur dyes, vat dyes, reactive dyes, pigment dyes, etc.

8.2.2   Dyeing methods

Colour is applied to fabric by different methods of dyeing for different types of fibre and at different stages of the textile production process. These methods include direct dyeing; stock dyeing; top dyeing; yarn dyeing; piece dyeing; solution pigmenting or dope dyeing; garment dyeing, etc. From these, direct dyeing and yarn dyeing methods are the most popular ones.

Drying is done after dewatering of fabric. In textile finishing unit, dryer is used for drying the knit, woven fabrics and dyed yarn. But the drying process and drying mechanism of yarn and fabrics is different from one to another. The main functions of a textile dryer is to dry the textile fabrics. Drying is defined as a process where the liquid portion of the solution is evaporated from the fabric.

Points for selecting a dryer: Following points should be considered during buying a dryer:

1. Heating methods: The textile fabrics may be heated by gas burner or steam.
2. Chamber: Number of chambers.
3. Burner: Number of burners.

8.2.3   Energy conservation in textile industries

The textile industry, in general, is not considered an energy-intensive industry. However, the textile industry comprises a large number of plants which all together consume a significant amount of energy. The textile industry uses large quantities of both electricity and fuels.

The share of electricity and fuels within the total final energy use of any one country's textile sector depends on the structure of the textile industry in that country.

8.2.4   Breakdown of energy used by end-use

There are significant losses of energy within textile plants. Around 36% of the energy input in textile industry is lost onsite. Motor driven systems have the highest share of onsite energy waste (13%) followed by distribution and boiler losses (7 to 8%). The share of losses could vary for the textile industry from country to country depending on the structure of the industry in those countries.

The motor driven systems are one of the major end-use energy in the textile industry. Material processing has the highest share of the energy used by motor driven systems (31%) followed by pumps, compressed air, and fan systems (19%, 15% and 14%, respectively). Again, these percentages in countries will highly depend on the structure of the textile industry in those countries. For instance, if the weaving industry in a country has a significantly higher share of air-jet weaving machines (which consume high amounts of compressed air) the share of total motor driven system energy consumed by compressed air energy systems would probably be higher.

8.2.5 Energy used by various textile processes

Energy used in the spinning process

Electricity is the major type of energy used in spinning plants, especially in cotton spinning systems. If the spinning plant just produces raw yarn in a cotton spinning system and does not dye or fix the produced yarn, the fuel may just be used to provide steam for the humidification system in the cold seasons for preheating the fibres before spinning them together. Therefore, the fuel used by a cotton spinning plant highly depends on the geographical location and climate in the area where the plant is located.

Energy used in wet-processing

Wet-processing is the major energy consumer in the textile industry because it uses a high amount of thermal energy in the forms of both steam and heat. The energy used in wet-processing depends on various factors such as the form of the product being processed (fibre, yarn, fabric, cloth), the machine type, the specific process type, the state of the final product, etc.

Thermal energy is used in a dyeing plant (with all dyeing processes included). It can be seen that a significant share of thermal energy in a dyeing plant is lost through wastewater loss, heat released from equipment, exhaust gas loss, idling, evaporation from liquid surfaces, unrecovered condensate, loss during condensate recovery, and during product drying (e.g. by over-drying).

Ring frames

Use of energy-efficient spindle oil: Synthetic-based spindle oils (energy-efficient grades) along with certain metal compatibility additives may result in higher energy savings, in the range of 5–7% depending upon viscosity.

Replacement of lighter spindles in place of conventional spindles in ring frames: Ring frames are the largest energy consumers in the ring spinning process. Within a ring frame, spindles rotation is the largest energy consumer. Thus, the weight of the spindles is directly related to the energy use of the machine. There are so-called high efficiency spindles on the market which are lighter than the conventional spindles and hence use less energy.

Windings, doubling and yarn finishing process

Installation of variable frequency drives on autoconer machines: Autoconer is the name of the machine which usually is used subsequent to ring frames in the yarn spinning process. The adoption of this measure in spinning plant resulted in electricity savings of 331.2 MWh/year.

Intermittent modes of the movement of empty bobbin conveyors in Autoconer/cone winding machines: The continuous movement of empty bobbin conveyor belts can be converted into an intermittent mode of movement. This measure results in not only substantial energy saving, but also results in maintenance cost savings and waste reduction.

Replacing electrical heating systems with steam heating systems for yarn polishing machines: Steam consumption can increase by about 31.7 T steam/year for each machine, while electricity use declined in average by about 19.5 MWh/year/machine.

Air conditioning and humidification system

Replacement of nozzles with energy-efficient mist nozzles in yarn conditioning rooms: In some textile plants, the yarn cones are put in a yarn conditioning room in which yarn is kept under a maintained temperature and humidity. In such rooms, usually water is sprayed in to the air to provide the required moisture for the yarn to improve its strength, the softness and quality and to increase its weight. The type of nozzles used for spraying the water can effectively influence the electricity use of the yarn conditioning system. Replacing jet nozzles with energy-efficient mist nozzles in a yarn conditioning room, will result in 31 MWh/year electricity savings.

Installation of VFD on humidification system pumps: In place of throttling valves, variable frequency drives can be installed for controlling relative humidity, and thereby the speed of the pumps can be reduced. This retrofit measure resulted in electricity savings equal to 35 MWh/year in a worsted spinning plant.

Energy-efficient control systems for humidification systems: Energy-efficient control systems consist of variable speed drives for supply to air fans, exhaust air fans and pumps in addition to control actuators for fresh air, recirculation and exhaust dampers. Energy savings in the range of 25% to 60% is possible by incorporating such control systems in the plants depending on the outside climate.

General energy-efficiency measures in spinning plants

Energy conservation measures in overhead travelling cleaner (OHTC): It is imperative for textile plants to have control over waste removal out of the processing area to ensure best yarn and fabric quality.

Energy-efficient blower fans for overhead travelling cleaners (OHTC): Existing blower fans of OHTCs can be replaced by energy-efficient fans with smaller diameters and less weight. An energy savings of about 20% is achievable with a quick payback period of less than 6 months.

Replacement of ordinary 'V–belts' with cogged 'V–belts' at various machines: In textile plants, many motors are connected to the rotating device with pulleys and belts. In many cases, a V-belt is used to transfer the motion. Ordinary V-belts can be replaced with cogged V-belts to reduce friction losses, thereby saving energy.

The implementation of such modification on 20 V-belt drives in a spinning plant resulted in electricity savings equal to 30 MWh/year.

8.2.6 Energy-efficiency technologies and measures in the weaving process

The list of measures/technologies for the weaving process are discussed below:

Evaluation and enhancement of the energy efficiency of compressed air systems in air-jet weaving plants: Air-jet weaving machines use compressed air for weft yarn insertion. The conversion efficiency to produce compressed air is fairly low (less than 15% efficiency without heat recovery). Most textile companies rely on compressed air in their production, and improving the use of compressed air will have significant economic benefit for plants.

General measures to save energy in weaving plants

Combine preparatory treatments in wet processing: Combining preparatory treatments such as the combined desizing, scouring and bleaching of a cotton fabric could lead to a process step reduction from the original eight-stage process to just two stages; this method would employ a steam purge and cold-pad-batch technique.

The elimination of three intermediate washings, one hot kier and a cold acid process could reduce energy requirements by as much as 80%.

Use of counter-flow currents for washing: By applying this counter-flow principle, it is possible to save both water and energy.

Installing automatic valves in continuous washing machines: Automatic stop valves which link the main drive systems of machines to water flows can save considerable amounts of energy and water by shutting off water flow as soon as a stoppage occurs.

Installing heat recovery equipment in continuous washing machines: Installing heat recovery equipment on a continuous washer is usually a simple but very effective measure since water inflow and effluent outflow are matched and this eliminates the need for holding tanks. Another measure is to install a simple plate heat exchanger with a pre-filter, which may have a higher initial cost, but which also has an efficiency that could be higher than 90%.

8.2.7   Dyeing and printing process

Installation of variable frequency drives on pump motors of Top dyeing machines: Top dyeing is a method for dyeing combed wool before spinning. The average electricity saving resulting from this retrofit was about 26.9 MWh/year/machine.

Heat insulation of high temperature/high pressure (HT/HP) dyeing machines: By reinforcement of the heat insulation of machines and steam systems in fabric wet-processing, plants can result in fuel savings of about 4 GJ/T fabric and electricity savings of 6.3 kWh/T fabric processed.

Installation of VFD on circulation pumps and colour tank stirrers: Circulation pumps are used to circulate chemicals in machine chambers in the dye house. In many plants, especially those with old equipment, the flow of chemicals is often controlled by closing ball valves. The implementation of these two retrofit measures in a plant resulted in 138 MWh/year electricity saving.

Reducing the process temperature in wet batch pressure-dyeing machines: A reduction in the process temperature may also be achieved in wet batch pressure-dyeing machines by introducing alternative processes. For example, under suitable circumstances, direct dyeing machines operated at 100–120°C may be replaced with reactive dyeing at 40–60°C, thus minimising water heating and radiation/convection losses.

Use of steam coils instead of direct steam heating in batch dyeing machines (winch and jigger): A steam coil submerged in the dye bath now allows for the recycling of the condensate, resulting in significant fuel savings.

Reducing the process time in wet batch pressure-dyeing machines: Processing times can sometimes be reduced simply by making modifications to the temperature profiles of certain dyeing cycles. This can result in energy savings and improved productivity.

Installation of covers or hoods in atmospheric wet batch machines: Using covers or hoods can reduce evaporative losses by approximately half.

Heat recovery of hot waste water in autoclaves: Autoclave (high temperature/high pressure) dyeing machines generate relatively high temperature waste water at 75°C, which in many plants is wasted away directly through drain disposal. On the other hand, fresh water at 13–25°C is heated to 130°C in the steam heater. A heat exchanger and surrounding equipment like water tanks and pumps for recovering heat from hot waste water as a heat source can be installed. A plate-type heat exchanger is usually recommended.

By implementation of this measure, a fuel saving of 554 MJ/batch in autoclave machines can be achieved.

Reducing the need for reprocessing in dyeing: Product quality and productivity can be improved while the use of dyes, chemicals, water and energy are optimised. Improved controls will typically lead to just 5% of the product requiring shading, which results in energy savings of around 10–12%.

Recovering heat from hot rinse water: In textile wet-processing, large amounts of hot (up to 80°C) water is used to rinse fabric or yarn. Plants may discharge a mass of rinse water up to thirty times the weight of the yarn/fabric that is rinsed. The heat from the rinse water can be captured and used for pre-heating the incoming water for the next hot rinse. This option provides the important ancillary benefit of reducing the temperature of the wastewater prior to treatment as well. Fuel savings of 1.4–7.5 GJ/T fabric rinsed by the implementation of this measure can be achieved.

Drying

Energy-efficiency improvements in cylinder dryer: Contact drying is mainly used for intermediate drying, rather than final drying (since there is no way of controlling fabric width), and for pre-drying prior to stentering. Fabric is passed around a series of cylinders, which are heated by steam supplied at pressures varying from 35 psi to 65 psi. Cylinders can be used to dry a wide range of fabrics.

Some of the energy-efficiency measures that can be implemented on cylinder dryers are explained below.

Introduce mechanical pre-drying: Mechanical pre-drying methods such as mangling, centrifugal drying, suction slot or air knife dewatering are used to reduce drying costs by removing some of the water from the fabric prior to contact drying in cylinder dryer. A slot is three times more energy intensive than a typical mangle, but consistently provides lower water retention rates over a range of fabric types.

End panel insulation in cylinder dryer: The insulation of end sections can reduce heat waste, thereby saving fuel. This measure, however, is more practical for cylinders with a diameter of one meter or more.

Operating cylinders at higher steam pressures in cylinder dryer: Cylinders can be operated at higher steam pressures and temperatures to reduce radiation and convection losses.

High-frequency reduced-pressure dryers for bobbin drying after the dyeing process: This equipment is a high-frequency reduced-pressure dryer employed in the bobbin drying after the bobbin dyeing process, and achieves 20%

electricity saving compared with conventional dry steam-type hot air dryers. A change in the method of temperature control in the drying process from fixed temperature controls to programmed temperature controls, and optimised control of the temperature of the drying vessel in accordance with the material and quantity permits a major reduction in electricity.

8.2.8   Finishing process

Energy-efficiency improvements in stenters: Stenters have an important role in the dyeing and finishing of fabrics. Stenters are mainly used in textile finishing for heat-setting, drying, thermosol processes and finishing. It can be roughly estimated that, in fabric finishing, the fabric is treated on average 2–3 times in a stenter.

Stenters can be heated in a variety of ways, such as direct gas firing and through the use of thermic fluid systems. Gas-fired stenters are highly controllable over a wide range of process temperatures. A typical energy breakdown for a stenter being used for hot-air drying. By far the greatest users of energy are the evaporation and air heating components. It is therefore necessary that the fabric moisture content is minimised before the fabric enters the stenter, and that exhaust airflow within the stenter is reduced.

Conversion of thermic fluid heating systems to direct gas firing systems in stenters and dryers: Often, thermic fluid heaters are used to provide the heating requirements of stenters and dryers. In this system, a fluid is heated up to 260°C and circulated in the plant through transmission lines. Heat is transferred from the hot fluid to the chambers using radiators. Substantial heat loss happens in the thermic fluid boilers, transmission lines and radiators. To reduce this heat loss, thermic fluid heating systems can be replaced with direct gas firing systems.

The direct gas firing system has several advantages over the thermic fluid heating system. First, there is a saving on fuel consumption with the reduced heat losses. The electricity required for pumping the thermic fluid and the risks involved in the circulation of the hot fluid are also eliminated.

A textile plant converted their thermic fluid system into a direct gas firing system can achieve 11000 GJ/year savings in fuel use (around 40% of the total fuel use) and 120 MWh/year savings in electricity use (around 90% of the total electricity use).

Stentering is an energy intensive process, so it is important to remove as much water as possible before the fabric enters the oven. This can be achieved using mechanical dewatering equipment or by using contact drying using heated cylinders. Up to 15% energy savings in the stenter (depending on the type of substrate) can be obtained if the moisture content of the fabric is reduced from 60% to 50% before it enters the stenter.

8.2.9   Electrical demand control

Demand control is a follow-up analysis that is normally conducted after the development of a demand/load profile by energy auditors. The main advantage of demand control and load management is the reduction of electricity costs.

Monitoring

Maintenance can be supported by monitoring by using proper instrumentation, including:

1. Pressure gauges on each receiver or main branch line and differential gauges across dryers, filters, etc.
2. Temperature gauges across the compressor and its cooling system to detect fouling and blockages.
3. Flow meters to measure the quantity of air used.
4. Dew point temperature gauges to monitor the effectiveness of air dryers.
5. kWh meters and hours run meters on the compressor drive.

Reduction of leaks (in pipes and equipment)

Leaks cause an increase in compressor energy and maintenance costs. The most common areas for leaks are couplings, hoses, tubes, fittings, pressure regulators, open condensate traps and shut-off valves, pipe joints, disconnects and thread sealants. Quick connect fittings always leak and should be avoided. In addition to increased energy consumption, leaks can make pneumatic systems/equipment less efficient and adversely affect production, shorten the life of equipment, lead to additional maintenance requirements and increased unscheduled downtime.

A simple way to detect large leaks is to apply soapy water to suspected areas. The best way is to use an ultrasonic acoustic detector, which can recognise the high frequency hissing sounds associated with air leaks.

Reduction of the inlet air temperature: Reducing the inlet air temperature reduces energy used by the compressor. In many plants, it is possible to reduce this inlet air temperature by taking suction from outside the building. As a rule of thumb, each 3°C reduction will save 1% compressor energy use.

Maximising allowable pressure dew point at air intake: Choose the dryer that has the maximum allowable pressure dew point, and best efficiency. A rule of thumb is that desiccant dryers consume 7–14% and refrigerated dryers consume 1–2% of the total energy of the compressor.

Optimising the compressor to match its load: The compressors consume more energy during part-load operation, this is something that should be avoided. In some cases, the pressure required is so low that the need can be met by a blower instead of a compressor which allows considerable energy savings, since a blower requires only a small fraction of the power needed by a compressor.

Proper pipe sizing: More than 85% of the electrical energy used by an industrial air compressor is converted into heat. A 150 hp compressor can reject as much heat as a 90 kW electric resistance heater or a 422 MJ/hour natural gas heater when operating. In many cases, a heat recovery unit can recover 50–90% of the available thermal energy for space heating, industrial process heating, water heating, makeup air heating, boiler makeup water preheating, industrial drying, industrial cleaning processes, heat pumps, laundries or preheating aspirated air for oil burners.

8.2.10   Use of renewable energy in the textile industry

There are various potentials for the use of renewable energy in the textile industry.

Installation of turbo ventilators that rotate using wind blowing over roofs: Few areas in the textile plant need to have a well-maintained standard temperature and humidity, for which HVAC systems are used. However, this is not the case for the whole plant. There are areas in many textile plants that do not need HVAC systems, such as most of the wet-processing plants and non-production areas. In these areas, usually, fans are used just to ventilate the air. Instead of using fans, turbo ventilators that rotate using natural wind can be installed on roofs. Energy savings and cost depend on the number of ventilators installed and the number of fans replaced.

Use of direct solar energy for fibre drying: In few textile plants, wet fibres need to pass through a drying process, such as in the drying of acrylic fibres after the dyeing process. If the plant is located in an area which gets sunshine for several months of the year, then there is the potential to use direct solar energy for fibre drying.

Use of solar energy for water heating in the textile industry: There is a high demand for low-temperature hot water in textile wet-processing which partially could be generated using solar energy. A feasibility study to understand the potential for solar energy utilisation in a textile plant can be conducted to find out whether or not it is economically feasible to use solar energy to provide low-temperature hot water, given the plant's geographical location and the climatic situation in the area.

8.3   Waste heat recovery in textile industries

By implementing the waste heat recovery methods we can conserve the energy in the textile industries. The improvements in the boiler blow down, condensate recovery, feed water management and waste water recovery will minimise the energy losses and improve the performance of the thermal systems in textile industries. As the industrial sector continues efforts to improve its energy efficiency, recovering waste heat losses provides an attractive opportunity for an emission free and less costly energy resource.

Textiles (dyeing and printing) are energy intensive industries. Steam is used as energy carrier for processing applications like dyeing and finishing in all textile industries. Hence boilers are the main fuel consumers in the textile industries. The main areas of waste heat recovery in textile dyeing are from boiler blow down flash steam, hot condensate flash steam and heat recovery from processed waste water.

8.3.1   Boiler blow down heat recovery

As water evaporates in the boiler steam drum, solids present in the feed water are left behind. The suspended solids form sludge or sediments in the boiler, which degrades heat transfer. Dissolved solids promote foaming and carryover of boiler water into the steam. To reduce the levels of suspended and total dissolved solids (TDS) to acceptable limits, water is periodically discharged or blown down from the boiler. Mud or bottom blow down is usually a manual procedure done for a few seconds on intervals of several hours. It is designed to remove suspended solids that settle out of the boiler water and form a heavy sludge. Surface or skimming blow down is designed to remove the dissolved solids that concentrate near the liquid surface. Surface blow down is often a continuous process.

Minimising blow down rate can substantially reduce energy losses, as the temperature of the blown-down liquid is the same as that of the steam generated in the boiler. Minimising blow down will also reduce makeup water and chemical treatment costs. Insufficient blow down may lead to carryover of boiler water into the steam, or the formation of deposits. Excessive blow down will waste energy, water, and chemicals. It is necessary to control the level of concentration of the solids and this is achieved by the process of 'blowing down', where a certain volume of water is blown off and is automatically replaced by feed water-thus maintaining the optimum level of total dissolved solids (TDS) in the boiler water. Blow down is necessary to protect the surfaces of the heat exchanger in the boiler. However, blow down can be a significant source of heat loss, if improperly carried out.

Concept of flash steam

Flash steam is vapour or secondary steam formed from hot condensate discharged into a lower pressure area. It is caused by excessive boiling of the condensate which contains more heat than it can hold at the lower pressure. Flash steam occupies many times the volume of water from which it forms.

For example, flash steam created by hot condensate flowing from 15 psig to an atmospheric pressure will have nearly 1600 times the volume of the high pressure hot water.

Heat content in the flash steam from boiler blow down and condensate can be recovered back to pre-heat the boiler feed water and flash steam produced due to excess boiler blow down also can reduced fuel consumption rate. It is found that Flash steam recovery from the boiler blow down also increases the efficiency of the boiler up to 2%. Heat recovery is used frequently to reduce energy losses that result from boiler water blow down.

8.3.2   Condensate heat recovery

Steam contains two types of energy: latent and sensible. When steam is supplied to a process application (heat exchanger, coil, etc.) the steam vapour releases the latent energy to the process fluid and condenses to a liquid condensate. The condensate retains the sensible energy the steam had. The condensate can have as much as 16% of the total energy in the steam vapour, depending on the pressure. Figure 8.1 illustrates a typical condensate recovery system. One of highest return on investments is to return condensate to the boiler.

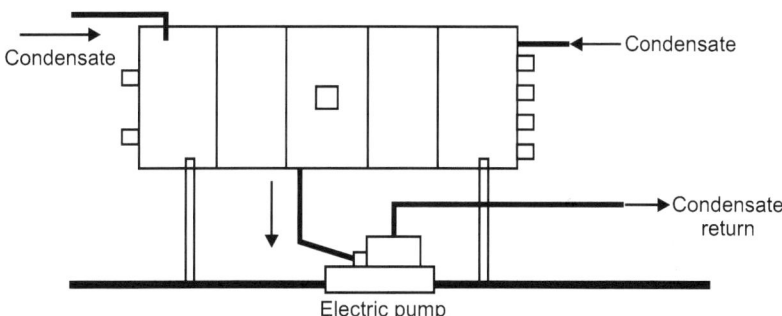

Figure 8.1: Pump and condensate receiver.

As fuel costs continue to rise, it's imperative to focus on recovering condensate in every industrial steam operation. Returning hot condensate to the boiler makes sense for several reasons. As more condensate is returned, less make-up water is required, saving fuel, makeup water, and chemicals and treatment costs. Less condensate discharged into a sewer system reduces disposal costs. Return of high purity condensate also reduces energy losses due to boiler blow down.

Flash steam recovery

Condensate is discharged through traps from a higher to a lower pressure. As a result of this drop in pressure, some of the condensate will then reevaporate into flash steam. The flash steam generated can contain up to half of the total energy of the condensate, hence flash steam recovery is an essential part of an energy efficient system.

It is clear that, flash steam recovery from hot condensate enhanced boiler efficiency and it will in turn reduce the fuel consumption rate.

The return of condensate represents huge potential for energy savings in the boiler house. Condensate has high heat content and approximately 1% less fuel is required for every 6°C temperature rise in the feed tank. The more the condensate recovery, the lesser will the condensate that is discharged into a sewer system be and the lower will the blow down be. This will reduce the sewer disposal costs.

Heat recovery from water

The first heat recovery option to consider is the reuse of the hot wastewater. In this way, water, residual chemicals as well as energy are recovered. In textile dyeing and finishing, operations involving acrylic fibres or wool where colourants are exhausted, wastewater reuse is possible. Similarly, wastewater from rinsing operations can make up new baths, for instance, for scouring. Dyeing and finishing specialists claim that wastewater from light shade operations can be reutilised up to 20 times.

Cooling water recovery

Cooling of baths is a common operation. The utilisation of cooling water, that is, of a stream of cold water to absorb heat from the hot bath, can also be considered as a heat recovery process. Subsequently, cooling water is collected and reutilised, thus, recovering heat and water. Under the most favourable conditions, cooling water recovery has been reported to have a payback period of 12 months.

Heat recovery from wastewater

Batch or non-continuous processing is common in textile dyeing plants. Thus, a large volume of wastewater is available intermittently from several machines at different locations in the plant. If wastewater can neither be reused nor can its heat be recovered locally, the feasibility of installing a centralised heat recovery system should be investigated. Figure 8.2 shows a typical setup for centralised wastewater recovery.

Equipment such as washing-, mercerising- and bleaching-machines often operate continuously for long hours, requiring a large volume of hot water and produce an equal volume of hot waste water simultaneously. This can be done by incorporation of heat exchangers on such textile machines with the purpose of heating up the incoming cold-water stream with hot wastewater

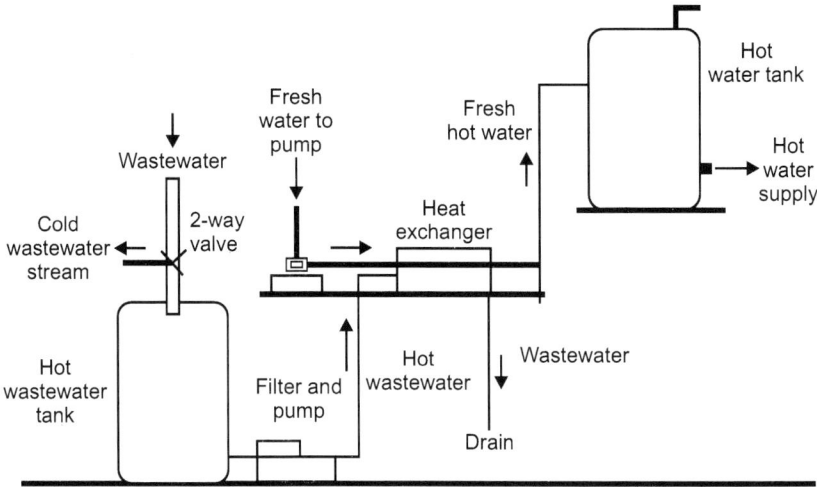

Figure 8.2: Centralised wastewater heat recovery.

leaving the machine. This water can be properly utilised for preheating the boiler feed water or dyeing purpose, it can save energy as well as water. Hot wastewater, produced in textile dyeing plants, can be a significant source of heat energy. In many instances, this valuable resource is discharged to wastewater treatment facilities without employing the heat it acquired during processing.

Most of the heat contained in the wastewater stream can be reclaimed and utilised, while providing significant cost reductions with attractive payback periods. Through utilisation of proper wastewater heat recovery system, reclaimed heat from the wastewater discharge can preheat incoming process water; thereby saving fuel costs, while enhancing the environment through the removal of thermal pollution.

Energy conservation in cement, ceramic and glass

9.1   Introduction

Cement is an inorganic, non-metallic substance with hydraulic binding properties, and is used as a bonding agent in building materials. It is a fine powder, usually gray in colour that consists of a mixture of the hydraulic cement minerals to which one or more forms of calcium sulphate have been added. Mixed with water it forms a paste, which hardens due to formation of cement mineral hydrates. Cement is the binding agent in concrete which is a combination of cement, mineral aggregates and water. Concrete is a key building material for a variety of applications.

9.2   Cement

The cement industry is made up of either Portland cement plants that produce clinker and grind it to make finished cement, or clinker-grinding plants that intergrind clinker obtained elsewhere, with various additives.

Clinker is produced through a controlled high-temperature burn in a kiln of a measured blend of calcareous rocks (usually limestone) and lesser quantities of siliceous, aluminous, and ferrous materials. The kiln feed blend (also called raw meal or raw mix) is adjusted depending on the chemical composition of the raw materials and the type of cement desired. Portland and masonry cements are the chief types produced all over the world.

Cement plants are typically constructed in areas with substantial raw materials deposits. Fuel costs are the single largest variable production cost at cement plants. Variable costs are typically about 50% of overall operating costs, so energy is frequently the single largest production cost, besides raw materials.

9.2.1   Cement production process

The most common raw materials used for cement production are limestone, chalk and clay. The major component of the raw materials, the limestone or chalk, is usually extracted from a quarry adjacent to or very close to the plant. Limestone provides the required calcium oxide and some of the other oxides, while clay, shale and other materials provide most of the silicon, aluminium and iron oxides required for the manufacture of Portland cement. The limestone is extracted from open-face quarries. The raw materials are selected, crushed, ground, and proportioned so that the resulting mixture has the desired fineness and chemical composition for delivery to the pyro-processing systems (Fig. 9.1). It is often necessary to raise the content of silicon oxides or iron oxides by adding quartz sand and iron ore, respectively. The quarried material is reduced in size by processing through a series of crushers. Normally primary size reduction is accomplished by a jaw or gyratory crusher, and followed by secondary size reduction with a roller or hammer mill. The crushed material is screened and stones are returned. More than 1.5 tons of raw materials are required to produce one ton of Portland cement.

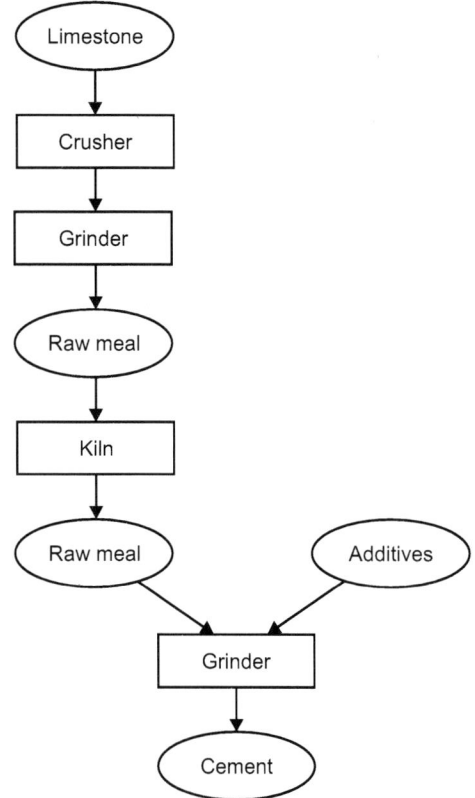

Limestone is the major process input. Other raw materials such as clay, shale, sand, quartz or iron ore may be added.

Figure 9.1: Simplified process flow diagram for cement making.

Raw material preparation: After primary and secondary size reduction, the raw materials are further reduced in size by grinding. The grinding differs with the pyroprocessing process used. In dry processing, the materials are ground into a flowable powder in horizontal ball mills or in vertical roller mills. In a ball (or tube) mill, steel-alloy balls (or tubes) are responsible for decreasing the size of the raw material pieces in a rotating cylinder, referred to as a rotary mill. Rollers on a round table fulfill this task of comminution in a roller mill. Utilising waste heat from the kiln exhaust, clinker cooler hood, or auxiliary heat from a stand-alone air heater before pyroprocessing may further dry the raw materials. The moisture content in the kiln feed of the dry kiln is typically around 0.5–0.7%.

When raw materials are very humid, as found in few countries and regions, wet processing can be preferable. In the wet process, raw materials are ground with the addition of water in a ball or tube mill to produce a slurry typically containing 36% water (range of 24–48%). Various degrees of wet processing exist, e.g. semi-wet (moisture content of 17–22%) to reduce the fuels consumption in the kiln.

Clinker production (pyro-processing): Clinker is produced by pyro-processing in large kilns. These kiln systems evaporate the inherent water in the raw meal, calcine the carbonate constituents (calcination), and form cement minerals (clinkerisation).

The main pyroprocessing kiln type used is the rotary kiln. In these rotary kilns, a tube with a diameter up to 25 feet is installed at a 3–4 degree angle that rotates 1–3 times per minute. The ground raw material, fed into the top of the kiln, moves down the tube countercurrent to the flow of gases and toward the flame-end of the rotary kiln, where the raw meal is dried, calcined, and enters into the sintering zone. In the sintering (or clinkering) zone, the combustion gas reaches a temperature of 3300–3600°F. While many different fuels can be used in the kiln, coal has been the primary fuel.

In a wet rotary kiln, the raw meal typically contains approximately 36% moisture. These kilns are developed as an upgrade of the original long dry kiln to improve the chemical uniformity in the raw meal. The water (due to the high moisture content of the raw meal) is first evaporated in the kiln in the low temperature zone. The evaporation step makes a long kiln necessary.

In a dry rotary kiln, feed material with much lower moisture content (0.5%) is used, thereby reducing the need for evaporation and reducing kiln length. The first development of the dry process is long dry kiln without preheating. Later developments have added multi-stage suspension preheaters (i.e. a cyclone) or shaft preheater. Pre-calciner technology was more recently developed in which a second combustion chamber has been added between the kiln and a conventional pre-heater that allows for further reduction of kiln fuel requirements.

Once the clinker is formed in the rotary kiln, it is cooled rapidly to minimise the formation of a glass phase and ensure the maximum yield of alite (tricalcium silicate) formation, an important component for the hardening properties of cement. The main cooling technologies are either the grate cooler or the tube or planetary cooler. In the grate cooler, the clinker is transported over a reciprocating grate through which air flows perpendicular to the flow of clinker. In the planetary cooler (a series of tubes surrounding the discharge end of the rotary kiln), the clinker is cooled in a countercurrent air stream. The cooling air is used as secondary combustion air for the kiln.

Finish grinding: After cooling, the clinker can be stored in the clinker dome, silos, bins, or outside. The material handling equipment used to transport clinker from the clinker coolers to storage and then to the finish mill is similar to that used to transport raw materials (e.g. belt conveyors, deep bucket conveyors and bucket elevators). To produce powdered cement, the nodules of cement clinker are ground to the consistency of face powder. Grinding of cement clinker, together with additions (3–5% gypsum to control the setting properties of the cement) can be done in ball mills, ball mills in combination with roller presses, roller mills, or roller presses.

While vertical roller mills are feasible, they have not found wide acceptance in the most of the countries. Coarse material is separated in a classifier that is recirculated and returned to the mill for additional grinding to ensure a uniform surface area of the final product.

Traditionally ball mills are used in finish grinding, while many plants use vertical roller mills. In ball or tube mills, the clinker and gypsum are fed into one end of a horizontal cylinder and partially ground cement exits from the other end. Modern state-of-the-art concepts utilise a high-pressure roller mill and the horizontal roller mill that are claimed to use 20–50% less energy than a ball mill. The roller press is a relatively new technology, and is more common in Western Europe than in North America.

Cement production energy use

Clinker production is the most energy-intensive stage in cement production, accounting for over 90% of total industry energy use, and virtually all of the fuel use. Fuel use for clinker production in a wet kiln can vary between 4.6 and 6.1 MBtu/short ton clinker. Typical fuel consumption of a dry kiln with 4 or 5-stage preheating can vary between 2.7 and 3.0 MBtu/short ton clinker, electricity use increases slightly due to the increased pressure drop across the system. A six stage preheater kiln can theoretically use as low as 2.5–2.6 MBtu/short ton clinker. The most efficient pre-heater, pre-calciner kilns use approximately 2.5 MBtu/short ton clinker. Alkali or Kiln Dust (KD) bypass systems may be required in kilns to remove alkalis, sulphates, and/or chlorides. Such systems lead to additional energy losses since sensible heat is removed with the bypass gas and dust. Power consumption for grinding depends on the surface area required for the final product and the additives used. Electricity use for raw meal and finish grinding depends strongly on the hardness of the material (limestone, clinker, pozzolana extenders) and the desired fineness of the cement as well as the amount of additives. Blast furnace slags are harder to grind and hence use more grinding power, between 45 and 64 kWh/short ton for a 3500 Blaine (expressed in cm^2/g). Modern ball mills may use between 29 and 34 kWh/short ton for cements with a Blaine of 3500.

9.2.2   Energy efficiency opportunities

Energy efficiency opportunities can fall into at least three primary categories:

1. O&M activities to ensure that the installed equipment is running efficiently.
2. Installation of high efficiency equipment/processes.
3. Control of the production process to ensure efficient use of inputs.

Operations and maintenance (O&M)

Operations and maintenance practices include elements such as motor and bearing lubrication, motor belt replacement, fan blade cleaning, fan wheel balancing and compressed air system maintenance including leak minimisation and filter replacement. Preventative maintenance is generally employed at the more efficient facilities but could be improved at other plants. Preventative maintenance includes training of plant staff to be attentive to energy consumption and efficiency. Energy savings of up to 2–3% are possible with the institution of a rigorous preventative maintenance programme.

High efficiency equipment/processes

In cement industry, as in other energy intensive process industries, the more generic measures, like high efficiency motors and lighting, are either already done or are so small that their impacts are 'below the radar'.

Significant energy savings projects typically involve major process and/or equipment modifications that are industry-specific and highly specialised. Often highly specialised expertise is necessary to identify and be able to quantify energy savings of technology improvements. Cement industry customers see their equipment vendors as 'business partners' because the vendors tend to have the specialised expertise and experience in their particular area (e.g. crushers/classifiers, kilns, conveyors).

Some of the energy efficiency equipment opportunities identified by customers, with a primary focus on electricity savings, include:

1. Efficient materials transport system: Most notably conversion of pneumatic conveyors to mechanical conveyors, with a savings of around 1% of total plant electricity use.

2 *Conversion of ball mills to roller mills:* For both raw materials and finish grinding the energy savings in raw materials preparation can be in the order of 5% of total electricity consumption, while installation of advanced finish grinding systems can save achieve savings in the 20% range.

3. *High efficiency classifiers:* These do a better job of separating out fine particles from coarse particles, which are returned to the mills. They prevent over-grinding of the fine particles that results in unnecessary power use in the mills. Savings can be around 8%.

4. *Conversion to more efficient kilns:* Such as vertical precalciner kilns, which will primarily improve the thermal efficiency of the kiln, saving on coal consumption.

5. *Variable speed drives*: For fans in the kilns, coolers, preheaters, separators, and mills, and for other drives associated with variable loads. A comprehensive conversion to VSDs could probably save about 5% of total plant electricity use.

6. *Compressed air system improvements:* This is one of the important part of a cement plant's total electricity use, there is often room for significant efficiency improvements in systems that have not been optimised.

In addition to the equipment-related opportunities listed above, there appears to be a good opportunity to recover waste heat from the clinker production process for the generation of electricity. There is significant waste heat from kilns even after it is used to the maximum possible degree to preheat incoming material. Pre-heater exhaust is often more than 700°F. Two of the studied facilities already have cogeneration plants, and several more have performed feasibility studies.

Process controls

Key opportunities for improved process controls involve clinker production and finish grinding, as well as operation of compressed air systems.

In clinker production, computerised controls can be used in a number of applications, such as:

1. Optimising the mix of raw materials entering the kilns to ensure proper chemical composition and provide for more steady kiln operation.

2. Optimising the combustion process and conditions in the kiln to improve product quality and grindability.

3. Improving heat recovery, material throughput, and emissions from the clinker cooler.

Grinding mill controls optimised the flow in the mill and classifiers to improve product quality and increase production. The increased production translates into energy savings per unit of output.

Overall, savings from advanced control systems are in the 2–5% range for plants that have not already installed such system.

9.2.3 Barriers to energy efficiency

Following are few key barriers identified in the process: (i) limited capital, (ii) production concerns, (iii) limited staff time, (iv) reliability concerns, (v) hassle, (vi) facility uncertainty, (vii) cost effectiveness and (viii) exit fees.

9.3 Ceramic

Ceramics comprise of all engineering materials or products (or portions thereof) that are chemically inorganic, except metals and alloys, and are usually rendered serviceable through high temperature processing. Ceramic materials are normally composed of both cationic and anionic species; their primary difference from other materials is in the nature of their chemical bonding. They are frequently termed ionic solids, i.e. possessing ionic bonding.

Modern ceramics encompass a wide range of materials ranging from single crystals and dense polycrystalline materials through glass-bonded aggregates to insulating foams and wholly vitreous substances. The need for high performance ceramic materials has increased steadily in the last ten to twenty years. As a class of materials, ceramics are better electrical and thermal insulators and more stable in chemical and thermal environments than metals, are appreciably stronger in compression than in tension, and exhibit greater rigidity, hardness, and temperature stability than polymers. Process flow diagram for ceramic manufacturing process is shown in Fig. 9.2.

Raw materials: The principal raw materials of the ceramic industry are clay (including shale and mudstone), silica, and feldspar. Other raw materials include a wide variety of rocks, minerals and synthetic compounds used to manufacture diverse products. The first step in the process is to weigh the raw materials required to manufacture a ceramic tile including all types of frit, feldspar and various clays. All the raw materials are accurately weighed, so that the quality of the product can be stabilised.

Fine grinding and milling: The basic beneficiation processes include crushing, grinding, and sizing or classification.

Primary crushing is used to reduce the size of coarse materials, such as clays, down to approximately 1–5 centimeters. Ball mills are the most commonly used piece of equipment for milling.

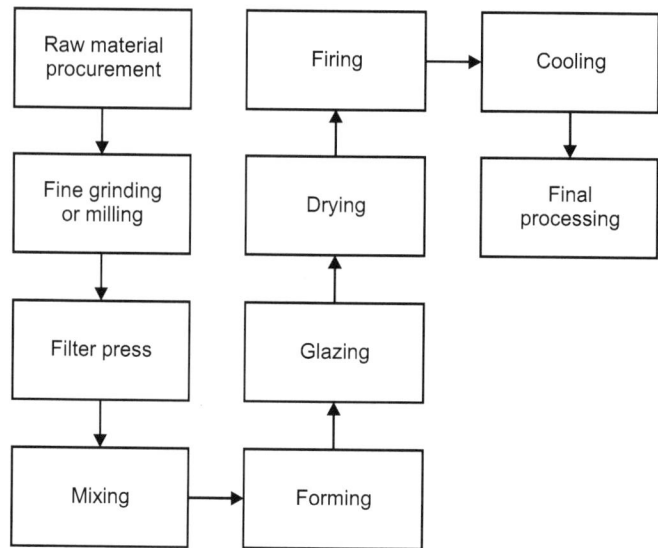

Figure 9.2: Process flow diagram of ceramic manufacturing process.

Filter press: During the process of making clay and ceramic slurries used for the manufacture of dinnerware, insulators, china clay, etc., the clay slurry goes through a dewatering step prior to further processing and molding into the desired form.

Mixing: The purpose of mixing or blunging is to combine the constituents of a ceramic powder to produce a more chemically and physically homogenous material for forming. Pug mills often are used for mixing ceramic materials. Mixing ensures a uniform distribution of clay in the solution. It also prevents the sedimentation of clay which is desirable for the process of ceramic formation. Pug mills are most commonly used for mixing in ceramic production.

Spray drying: Ceramic tiles are typically formed by dry pressing. Prior to pressing, many facilities granulate the ceramic mix to form a free-flowing powder, thereby improving handling and compaction. The most commonly used method of granulation is spray-drying.

Powder storage: The granules have to be kept in a storage bin for a few days so that its composition becomes even more homogeneous. This process makes the granules more pliable and less likely to stick to the mold.

Shaping: In the forming step, the ceramic mix is consolidated and molded to produce a cohesive body of the desired shape and size. Forming methods can be classified as either dry forming, plastic molding, or wet forming. Once the composition of the powder becomes homogenous, it is taken to the press where it is molded and squeezed under high pressure (of the order of hundreds of tons) to form a biscuit or greenware tile body.

Glazing: Glazing resemble glass in structure and texture. The purpose of glazing is to provide a smooth, shiny surface that seals the ceramic body. Not all ceramics are glazed. Those that are glazed can be glazed prior to firing, or can be glazed after firing, followed by refiring to set the glaze.

Car storage: After glazing, the biscuit is loaded into the stock car for storage, which is proceeded by the fully-automatic hydraulic controlled system.

Speed body drying: The drying process in the ceramic industry is the greatest energy consumer second to the firing process. Drying means loss of moisture from the surface of the substance by evaporation, and the drying speed depends on the temperature and humidity. When the substance is dried and moisture is lost, particles are put close to each other, resulting in shrinkage.

Firing: Firing is the process by which ceramics are thermally consolidated into a dense, cohesive body composed of fine, uniform grains. This process also is referred to as sintering or densification. Ceramics generally are fired at 50–75% of the absolute melting temperature of the material.

Ceramic products also are manufactured by pressure firing, which is similar to the forming process of dry pressing except that the pressing is conducted at the firing temperature.

The application of pressure enhances the densification of the ceramic during firing. Because of its higher costs, pressure firing is usually reserved for manufacturing ceramics that are difficult to fire to high density by conventional firing. In hot pressing, hydraulic presses and graphite dies commonly are used. In hot isostatic pressing, the pressing medium typically is a gas, such as argon or nitrogen.

Packing: The finished products are then packed and stored or shipped.

9.3.1 Energy performance

The primary energy use in ceramic manufacturing is for kiln. Natural gas, LNG and fuel oil are employed for most drying and firing operations. Nearly 30% of the energy consumed is used for drying. Over 60% of the energy consumed is used for firing. The percentage of the energy cost in the total ceramic production cost is between 5 and 20%, although it varies according to the product type and fuel price. Percentage share of electrical and thermal energy consumption in a typical ceramic industry varies from 15–20% and 75–80%, respectively. Typical thermal and electrical specific energy consumption range for different sub processes/kilns/type of firing in ceramic tile and sanitaryware manufacturing process.

Tunnel kilns and roller hearth kilns (roller kilns) are used for continuous firing. The primary advantages of tunnel kilns and roller kilns are lower energy consumption and the ease with which the ceramics can be transported through the firing process when compared to batch type kilns. A large number of tunnel and roller kilns use natural gas as fuel. Bell and shuttle kilns are used for batch type production of ceramic products. The main advantage of batch type kilns is that they can readily accommodate changes in firing temperature profile and cycle time to match the requirements of a wide variety of ceramic products.

9.3.2   Major energy consuming areas

Energy consumption in ceramic industry depends on payload of ceramic products, effectiveness and efficiency of various equipments. An indication of energy consumption of different processes in ceramic products manufacture is given below:

1. Kiln
2. Raw material preparation
3. Press machine
4. Glazing machine
5. Spray drier

9.3.3   Energy saving potential and major areas

The ceramic industry offers significant scope for energy efficiency improvements. Some of the common technological options applicable for ceramic industry are given below:

1. Tunnel kiln: Waste heat recovery, low thermal mass cars in sanitaryware units, use of hot air directly as combustion air, use of hot air from cooling zone to preheat input material.
2. Roller kiln: Maintain air-fuel ratio, improving insulation, preheating of combustion air using flue gases/hot air from cooling zone, energy efficient burner that can handle high temperature hot air.
3. Ball mill/blunger: Continuous multi-stage ball mill.
4. Spray dryer: Fuel switching to natural gas (NG).
5. Use of 'variable frequency drive' (VFD) in ball mills, blunger and agitation motors, presses and blowers.
6. Use of energy efficient motors in agitation systems and polishing line.
7. Improvement of kiln insulation.
8. Solar preheating of spray dryer input slurry.
9. Biomass/briquette firing in hot air generator.
10. Cogeneration system in NG based ceramic industries.

9.3.4   Possible energy efficiency measures for key processes/systems

The major energy efficiency measures in various processes as well as utilities in ceramic industries are provided below.

Kiln

1. Switching from intermittent type to continuous type kilns.
2. Auto interlock between brushing dust collection blowers and glazing lines.
3. Adopting best operating practices including optimising of excess air levels.

Spray dryer

1. Replacing LPG firing with diesel firing.
2. Arresting air infiltration in spray drier system.

Vertical dryer

1. Switch off chiller circuit when hydraulic press is not in operation.
2. Installing interlock to avoid idle operation of hydraulic press pump.

Apart from these, there are number of energy conservation options in utilities that vary from simple housekeeping measures to switching over to energy efficient equipment that can be adopted by ceramic industries.

9.4   Glass

Glass was formed naturally from common elements in the earth's crust long before anyone ever thought of experimenting with its composition, molding its shape, or putting it to the myriad of uses that it enjoys in the world today. Glass technology has evolved for six thousand years, and some of today's principles date back to early times. This includes what is today known about the structure of glass, its composition, properties, method of manufacture, and uses.

9.4.1   Manufacture of glass

Most glass articles are manufactured by a process in which raw materials are converted at high temperatures to a homogeneous melt that is then formed into the articles. The flow diagram 9.3 summarises the details of conventional glass manufacturing. The vapour deposition of SiO_2 from a flame fed with $SiCl_4$ and oxygen is the basis for manufacturing high-purity glass used for blanks that are redrawn into optical-waveguide fibres. Fused silica items that

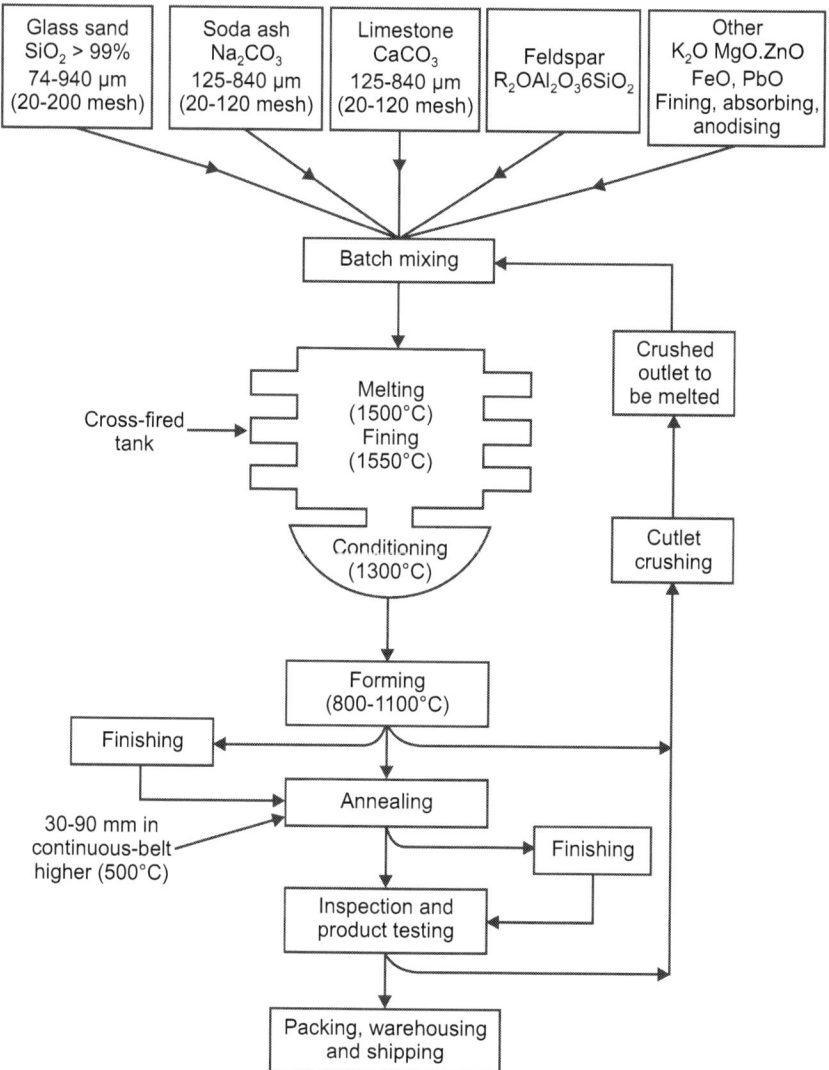

Figure 9.3: Glass manufacture. Temperatures are for common soda-lime glass. Other glasses may require appreciately different temperatures.

cannot be formed from viscous melts of SiO_2 or quartz are prepared by vapour deposition. Raw materials are selected according to purity, supply, pollution potential, ease of melting, and cost. Sand is the most common ingredient. Limestone is the source of calcium and magnesium. Powdered anthracite coal is a common reducing agent, and common colourants for glass include iron, chromium, cerium, cobalt and nickel.

Melting and fining depends on the batch materials interacting with each other at the proper time and in the proper order. Thus, extreme care must be taken to obtain materials of optimum grain size, to weigh them carefully, and mix them intimately. The efficiency of the melting operation and the uniformity and quality of the glass product are very often determined in the mix house. Batch handling systems vary widely through the industry, from manual to fully automatic. Melting units range from small pot furnaces for manual production to large and continuous tanks for rapid machine forming. The two common feeding designs used today are the screw feeder and the reciprocating pusher. Control devices have advanced from manual control to sophisticated computer-assisted operation. Radiation pyrometers in conjunction with thermocouples monitor furnace temperatures at several locations. Natural gas, oil, and electricity are the primary sources of energy; propane is used as a backup reserve in emergencies.

Molten glass is either molded, drawn, rolled, or quenched, depending on desired shape and use. Bottles, dishes, optical lenses, television picture tubes, etc., are formed by blowing, pressing, casting, and/or spinning the glass against a mold to cool it and to set its final shape. Window glass, tubing, rods, and fibre are formed by freely drawing the glass in air (or in a bath of molten tin as in the float process) until it sets up and can be cut to length. Art glass is usually hand-formed by freely blowing and shaping it while soft. Glass that is intended to be crushed into powder, called frit, is quenched between water-cooled rollers or ladled or poured directly into water and then dried (dri-gauged). Simultaneous forming of two glasses to produce a single article is possible if the two have similar viscosities in the appropriate temperature range.

Glass articles formed at high temperatures must be cooled in order to reduce the strain and associated stress caused by temperature gradients in the glass to a low level to prevent damage during finishing and subsequent use. The continuous solidification of glass on cooling and the methods of reducing strain are unique to glasses and do not apply to crystalline materials that undergo phase transitions. Lehring processes include annealing, tempering, densifying or compacting, and post heat treatments. The term annealing generally refers to removal of stress, and terms such as fine annealing or trimming of optical glass imply structural changes associated with lehring.

In secondary forming, a piece of preformed glass is reheated and reworked into the finished product. Secondary operations include mechanical finishing, chemical finishing, and cleaning.

Glass manufacturing is among the most energy-intensive industries.

The bulk of energy consumed in the glass manufacturing industry comes from natural gas combustion used to heat furnaces to melt raw materials to form glass. These furnaces are mainly natural gas-fired, but there are a small number of electrically-powered furnaces. Many glass furnaces also use electric boosting (supplementary electric heating systems) to increase throughput and quality. After the melting and refining process is complete, the glass is formed and finished to create the final product. Specific manufacturing processes depend on the intended product, and can include annealing (slow cooling), tempering, coating, and polishing, which require additional energy.

Different types of glass include:

1. Container glass for packaging foods, beverages, and cosmetics.
2. Flat glass for residential and commercial construction, automotive applications, tabletops and mirrors.
3. Fibreglass for insulation and roofing.
4. Specialty glass that is pressed and blown glass for tableware, cookware, lighting, laboratory equipment, and other uses.
5. Purchased glass serves as an input to produce glass containers, flat glass, glassware and high-value products such as some optical lenses and fibre optic cable.

There is substantial potential for energy efficiency improvements in glass manufacturing. Estimates range from 20–25%, mainly focused on the extremely energy-intensive melting and refining process.

Energy used in other key glass-making processes, such as forming, is highly variable depending on the product, so those processes don't have the same potential for efficiency gains.

9.4.2   Glass products and their characteristics

Glass products and their characteristics, as well as the production and processing routes involved in their production show large variations. However, batch preparation, melting and refining, conditioning and forming, and finishing (annealing, tempering, polishing or coating) steps are found in virtually all glass plants and are the most important from an energy point of view.

Melting furnaces–employing combustion-heating (with air-fuel or oxy-fuel burners), direct electrical heating, or a combination of the two (electric boosting)–are the major energy users. In general, the energy necessary for melting glass may account for over 75% of the total energy (in terms of final energy) requirements of glass manufacture, with an average of about 65% of the total energy input when considering all the sectors of the glass industry. Typically, melting furnaces operate with an overall efficiency of 50–60%, where structural and flue gas losses represent 20–25% and 25–35% of losses, respectively. Melting furnaces are also the most important area for efficiency improvement, followed by refining and conditioning.

9.4.3   Energy efficiency technologies

While the glass industry has made huge strides in improving its energy intensity, there are numerous technology opportunities available which can further improve energy efficiency and productivity. Barriers to implementation of these technologies are frequently cost and technical maturity.

Several potential opportunities are highlighted below:

Increasing cullet percentage: Glass from cullet requires less energy per ton to produce than glass from batch. Since substituting cullet for batch can be relatively easy to implement, industry will willingly utilise this technology.

Batch preheating: The use of process waste heat to preheat batch is clearly a winning way to conserve energy. The heat returned to the batch immediately lowers combustion demands. A number of means to carry out batch preheating have been tested at pilot scale and been installed in limited industrial use. Batch preheating is capital-intensive in most cases, sometimes requiring equipment of the same size scale as the melter itself.

Cullet preheating: When cullet can be used, cullet preheating is much more practical than batch preheating. Cullet can be heated to a higher temperature than batch before it softens, and cullet does not undergo decomposition reactions. For these reasons, cullet preheating is a promising means to reduce energy use in situations where capital costs warrant installation.

Oxy-fuel conversion: Conversion from air-gas to oxy-gas firing is the single most promising means to reduce energy use. Conversion to oxy-gas requires furnace rebuild and installation of various support equipment. Conversion will only be undertaken after careful economic analysis and at the end of a furnace campaign. Decreasing oxygen costs and increasing gas costs are making oxy-gas conversion more attractive.

More efficient burners: Combustion system providers regularly work to develop more efficient burners with lower emissions and tighter control capabilities.

Improved refractory: Refractory companies also work to produce products with superior thermal properties and longer life in the glass melter environment. Similar refractories are used throughout the industry, but variations are required based on glass chemistry. New refractories can only be employed at the time of furnace rebuild.

Improved control system: Control systems have improved dramatically over the last decade. Tighter control of the combustion and melting processes leads directly to energy savings. Control systems, however, are both costly and difficult to install on a working furnace. Although new control systems can be installed on a retrofit basis, they are almost always upgraded only at the time of furnace rebuild.

Alternatives to natural gas: Rising fuel prices have encouraged industry to consider other, less costly, fuel options. The amount of energy saved using alternative fuels is unknown, but the savings will be lower than savings from other techniques. Price would be the primary driving force in switching to alternatives, but the supply must be consistent and reliable before being seriously considered. Alternative fuels and combustion systems for them could be installed as a retrofit, but most companies would likely only consider at the time of furnace rebuild.

Exhaust gas heat recovery: Regenerators are used for exhaust gas heat recovery in air-fired furnaces. Technologies such as steam generation may be practical for air-fired melters, but cost constraints limit ability to recover much energy from the low temperature exhaust leaving the bottom of the regenerators. Oxy-gas furnaces exhaust only 30% of the volume of air-gas furnace exhaust, but no heat is currently recovered from oxy-gas melter exhaust. This high-level heat can potentially be used to generate steam, to preheat batch or cullet, to generate electricity by thermo-electrics, to generate needed oxygen, or to preheat oxygen or gas. Needed cost-effective technologies for heat recovery are not yet available, but rising fuel costs may spur development.

Convective melting: Glass is heated in a gas-fired melter predominantly by radiation and partially by convection. In convective melting, one or more burners are mounted on the crown and fired downward toward the melt surface. This combustion approach is purported to increase heat transfer and improve energy efficiency. Convective melting could also be installed at the time of rebuild.

9.4.4 Waste heat boilers

The temperature of the flue gases leaving the regenerator is usually between 300 and 600°C, and can be used to recover steam. Capturing the waste heat can be done before the flue gas cleaning (with subsequent cleaning) or after gas cleanup. The amount of heat that can be recovered is dictated by the outlet temperatures, which is limited to around 200°C in order to avoid condensation on boiler tubes.

Produced steam can be used to generate power (using steam turbines), drive blowers or compressors, and/or preheat and dry cullet. The applicability and economic feasibility of the technique is dictated by the overall efficiency that may be obtained (including effective use of the steam generated). In practice, waste heat boilers have only been considered to recover residual heat down stream from regenerator or recuperator systems.

In many cases, the quantity of recoverable energy is low for efficient power generation and supplementary firing may be needed to generate superheated steam to drive turbines. Recuperative furnaces with higher waste gas temperatures or installations where it is possible to group the waste gases from several furnaces offer more opportunities for power generation. Waste heat boilers are in industrial use on some container glass facilities but most applications are with float glass furnaces. All float furnaces in Germany and many in other member states have waste heat boilers.

Energy conservation in electrical system

10.1   Introduction

Electrical energy is universally accepted as an essential commodity for human beings. Energy is the prime mover of economic growth and is vital to the sustenance of a modern economy. Future economic growth crucially depends on the long-term availability of energy from sources.

Areas of application of energy conservation are power generating station, transmission and distribution system, consumers premises. Steps are to be taken to enhance the performance efficiency of generating stations. Energy conservation technology adopted in transmission and distribution system may reduce energy losses, to the tune of 35% of total losses in power system. Acceptance of energy conservation technology will enhances the performance efficiency of electrical apparatus used by end users. Implementation of energy conservation technology will lead to energy saving which means increasing generation of energy with available source.

Energy is the primary and the most universal measures of all kinds of work by human being and nature. Electrical energy is proved to be an ideal energy in all sorts of energy available in nature.

10.2   Energy conservation in electrical field

Energy Conservation (EC) means reduction in growth of energy consumption and is measured in physical terms. Energy conservation is the practice of decreasing the quantity of energy used while achieving a similar outcome of end use. This practice may result in increase of financial capital, environmental value, national security, personal security and human comfort. Energy conservation also means reduction or elimination of unnecessary energy used and wasted.

10.2.1   Area of application of energy conservation

Electrical system is a net work in which power is generated using non-renewable sources by conventional method and then transmitted over longer distances at high voltage levels to load centers where it is used for various energy conversion process. End user sector are identified as three major areas–power generating station, transmission and distribution systems, and energy consumers. Consumers are further classified as domestic, commercial and industrial consumers.

EC in power generating station: Power sector is an essential service and in the basis of industrialisation and agriculture. It plays a vital role in the socio-economic development. As the bulk of power generation, about 75%, is by thermal power stations, improvement in their performance would lead to increased availability and large scale energy conservation. Since the Plant Load Factor (PLF) has become a common yardstick for monitoring the availability of power stations, several efforts have been made to improve PLF. It has been estimated that one percentage point improvement in the overall PLF of thermal power sector will give additional generating capacity to the extent of 500 MW in a much shorter time and cost. However, the experience has shown that this alone has not been sufficient to bridge the gap between demand and supply. The PLF which over the years is being recognised as an index of plant performance, is not very appropriate as it itself depends upon the availability and besides other causes. The overall Availability Factor (OAF) will be a better index for comparing plant performance. Efforts are therefore required to secure operational efficiency of thermal power stations as well by identifying the various loss areas and taking appropriate actions, so as to maximise the power generation and loss make available the saved energy to the consumers. Therefore, improving efficiency of these thermal power stations in addition to increasing their PLF has become the need of the hour to bring the cost and maximise the generation levels.

Technical losses in T&D system: Power losses occurring in T&D sector due to imperfection in technical aspect which indirectly cause loss of investment in this sector, are technical losses. These technical losses are due to inadequate system planning, improper voltage and also due to poor power factor, etc.

Commercial losses: Commercial losses are those, which are directly responsible for wastage of money invested in transmission and distribution system. These losses are effects of inefficient management, improper maintenance, etc. Corruption is also the main reason contributing to the commercial losses. Metering losses include loss due to inadequate billings, faulty metering, overuse, because of meters not working properly and outright theft. Many of the domestic energy meters fail because of poor quality of the equipment.

10.3   Energy conservation techniques

10.3.1   EC techniques in transformers

1. Optimisation of loading of transformer:
    (a)   By proper location of transformer preferably close to the load center, considering other features like centralised control, operational flexibility, etc. This will bring down the distribution loss in cables.

(b) Maintaining maximum efficiency to occur at 38% loading [as recommended by Rural Electrification Corporation Limited (REC)], the overall efficiency of transformer can be increased and its losses can be reduced.

(c) Under fluctuating load condition, more than one transformer is used in parallel operation of transformers to share the load and can be operated close to the maximum efficiency range.

2. By improvisation in design and material of transformer:

(a) To reduce load losses in transformer, use thicker conductors so that resistance of conductor reduces and load loss also reduces.

(b) To reduce core losses use superior quality or improved grades of Cold Rolled Grain Oriented (CRGO) laminations.

3. Replacing by energy efficient transformers:

(a) By using energy efficient transformers, efficiency improves to 95–97%.

(b) By using amorphous transformers, efficiency improves to 97–98.5%.

(c) By using epoxy resin cast/encapsulated dry type transformer, efficiency improves to 93–97%.

Transformer losses can be divided into two main components: No-load losses and Load losses. These types of losses are common to all types of transformers, regardless of transformer application or power rating.

There are, however, two other types of losses; extra losses created by harmonics and losses which may apply particularly to larger transformers–cooling or auxiliary losses, caused by the use of cooling equipment like fans and pumps.

No-load losses

These losses occur in the transformer core whenever the transformer is energised (even when the secondary circuit is open). They are also called iron losses or core losses and are constant.

They are composed of:

1. Hysteresis losses, caused by the frictional movement of magnetic domains in the core laminations being magnetised and demagnetised by alternation of the magnetic field. These losses depend on the type of material used to build a core. Silicon steel has much lower hysteresis than normal steel but amorphous metal has much better performance than silicon steel. Nowadays hysteresis losses can be reduced by material processing such as cold rolling, laser treatment or grain orientation. Hysteresis losses are usually responsible for more than a half of total no-load losses (~50% to ~70%).

2. Eddy current losses, caused by varying magnetic fields inducing eddy currents in the laminations and thus generating heat. These losses can be reduced by building the core from thin laminated sheets insulated from each other by a thin varnish layer to reduce eddy currents. Eddy current losses nowadays usually account for 30–50% of total no-load losses. When assessing efforts in improving distribution transformer efficiency, the biggest progress has been achieved in reduction of these losses.

3. There are also marginal stray and dielectric losses which occur in the transformer core, accounting usually for no more than 1% of total no-load losses.

Load losses

These losses are commonly called copper losses or short circuit losses. Load losses vary according to the transformer loading.

They are composed of:

1. Ohmic heat loss: Ohmic heat loss, sometimes referred to as copper loss, since this resistive component of load loss dominates. This loss occurs in transformer windings and is caused by the resistance of the conductor. The magnitude of these losses increases with the square of the load current and is proportional to the resistance of the winding. It can be reduced by increasing the cross sectional area of conductor or by reducing the winding length. Using copper as the conductor maintains the balance between weight, size, cost and resistance; adding an additional amount to increase conductor diameter, consistent with other design constraints, reduces losses.

2. Conductor eddy current losses: Eddy currents, due to magnetic fields caused by alternating current, also occur in the windings. Reducing the cross-section of the conductor reduces eddy currents, so stranded conductors are used to achieve the required low resistance while controlling eddy current loss.

   Effectively, this means that the 'winding' is made up of a number of parallel windings. Since each of these windings would experience a slightly different flux, the voltage developed by each would be slightly different and connecting the ends would result in circulating currents which would contribute to loss.

   This is avoided by the use of Continuously Transposed Conductor (CTC), in which the strands are frequently transposed to average the flux differences and equalise the voltage.

Auxiliary losses

These losses are caused by using energy to run cooling fans or pumps which help to cool larger transformers.

Extra losses due to harmonics and reactive power

This category of losses includes those extra losses which are caused by reactive power and harmonics. The reactive component of the load current generates a real loss even though it makes no contribution to useful load power. Low power factor loads should be avoided to reduce losses related to reactive power. Power losses due to eddy currents depend on the square of frequency so the presence of harmonic frequencies which are higher than normal 50 Hz frequency cause extra losses in the core and winding.

Extra losses due to harmonics

Non-linear loads, such as power electronic devices, such as variable speed drives on motor systems, computers, UPS systems, TV sets and compact fluorescent lamps, cause harmonic currents on the network. Harmonic voltages are generated in the impedance of the network by the harmonic load currents. Harmonics increase both load and no-load losses due to increased skin effect, eddy current, stray and hysteresis losses. The most important of these losses is that due to eddy current losses in the winding; it can be very large and consequently most calculation models ignore the other harmonic induced losses.

The precise impact of a harmonic current on load loss depends on the harmonic frequency and the way the transformer is designed.

In general, the eddy current loss increases by the square of the frequency and the square of the load current. So, if the load current contained 20% fifth harmonic, the eddy current loss due to the harmonic current component would be $5 \times 5 \times 0.2 \times 0.2$ multiplied by the eddy current loss at the fundamental frequency–meaning that the eddy current loss would have doubled.

In a transformer that is heavily loaded with harmonic currents, the excess loss can cause high temperature at some locations in the windings. This can seriously reduce the life span of the transformer and even cause immediate damage and sometimes fire.

10.3.2   Energy efficient transformers

Most energy loss in dry-type transformers occurs through heat or vibration from the core. The new high-efficiency transformers minimise these losses. The conventional transformer is made up of a silicon alloyed iron (grain oriented) core. The iron loss of any transformer depends on the type of core used in the transformer. However the latest technology is to use amorphous material - a metallic glass alloy for the core. The expected reduction in energy loss over conventional (Si Fe core) transformers is roughly around 70%, which is quite significant. By using an amorphous core- with unique physical and magnetic properties- these new type of transformers have increased efficiencies even at low loads–98.5% efficiency at 35% load.

Electrical distribution transformers made with amorphous metal cores provide excellent opportunity to conserve energy right from the installation. Though these transformers are a little costlier than conventional iron core transformers, the overall benefit towards energy savings will compensate for the higher initial investment. At present amorphous metal core transformers are available up to 1600 kVA.

10.3.3   Electronic ballast

Role of ballast

In an electric circuit the ballast acts as a stabiliser. Fluorescent lamp is an electric discharge lamp. The two electrodes are separated inside a tube with no apparent connection between them. When sufficient voltage is impressed on these electrodes, electrons are driven from one electrode and attracted to the other. The current flow takes place through an atmosphere of low pressure mercury vapour.

Since the fluorescent lamps cannot produce light by direct connection to the power source, they need an ancillary circuit and device to get started and remain illuminated. The auxiliary circuit housed in a casing is known as ballast.

Conventional vs electronic ballasts

The conventional ballasts make use of the kick caused by sudden physical disruption of current in an inductive circuit to produce the high voltage required for starting the lamp and then rely on reactive voltage drop in the ballast to reduce the voltage applied across the lamp. On account of the mechanical switch (starter) and low resistance of filament when cold the uncontrolled filament current, generally tend to go beyond the limits specified by Indian standard specifications.

With high values of current and flux densities the operational losses and temperature rise are on the higher side in conventional choke.

The high frequency electronic ballast overcomes the above drawbacks. The basic functions of electronic ballast are:

1. To ignite the lamp.
2. To stabilise the gas discharge.
3. To supply the power to the lamp.

The electronic ballasts make use of modern power semi-conductor devices for their operation. The circuit components form a tuned circuit to deliver power to the lamp at a high resonant frequency (in the vicinity of 25 kHz) and voltage is regulated through an inbuilt feedback mechanism. It is now well established that the fluorescent lamp efficiency in the kHz range is higher than those attainable at low frequencies. At lower frequencies (50 or 60 Hz), the electron density in the lamp is proportional to the instantaneous value of the current because the ionisation state in the tube is able to follow the instantaneous variations in the current. At higher frequencies (kHz range), the ionisation state cannot follow the instantaneous variations of the current and hence the ionisation density is approximately constant, proportional to the RMS (Root Mean Square) value of the current. Another significant benefit resulting from this phenomenon is the absence of stroboscopic effect, thereby significantly improving the quality of light output. One of largest advantages of an electronic ballast is the enormous energy savings it provides.

This is achieved in two ways. The first is its amasingly low internal core loss, quite unlike old fashioned magnetic ballasts; and second is increased light output due to the excitation of the lamp phosphorus with high frequency. If the period of frequency of excitation is smaller than the light retention time constant for the gas in the lamp, the gas will stay ionised and, therefore, produce light continuously. This phenomenon along with continued persistence of the phosphorus at high frequency will improve light output from 8 to 12%. This is possible only with high frequency electronic ballast.

10.3.4 Adjustable-speed drive

Adjustable Speed Drive (ASD) or Variable Speed Drive (VSD) describes equipment used to control the speed of machinery. Many industrial processes such as assembly lines must operate at different speeds for different products. Where process conditions demand adjustment of flow from a pump or fan, varying the speed of the drive may save energy compared with other techniques for flow control.

Where speeds may be selected from several different pre-set ranges, usually the drive is said to be at adjustable speed. If the output speed can be changed without steps over a range, the drive is usually referred to as variable speed. Adjustable and variable speed drives may be purely mechanical (termed variators), electromechanical, hydraulic, or electronic.

Saving energy by using efficient adjustable speed drives: Some adjustable speed driven applications use less energy than fixed-speed operated loads, variable-torque centrifugal fan and pump loads are the world's most energy-intensive.

Since most of the energy used for such fan and pump loads is currently derived by fixed-speed machines, use of efficient adjustable speed drives for these loads in retrofitted or new applications offers the most future energy savings potential. For example, when a fan is driven directly by a fixed-speed motor, the airflow is invariably higher than it needs to be. Airflow can be regulated using a damper, but it is more efficient to directly regulate fan motor speed. According to affinity laws, motor-regulated reduction of fan speed to 50% of full speed can thus result in a power consumption drop to about 12.5% of full power.

10.4 Diesel generator (DG)

DG set is a combination of a diesel engine and an alternator. Diesel engine is the prime mover which drives an alternator to produce electrical energy. In the diesel engine, air is drawn into the cylinder and is compressed to a high ratio (14:1–25:1).

A metered quantity of diesel fuel is then injected into the cylinder which ignites spontaneously because of the high temperature. Hence, the diesel engine is also known as Compression Ignition (CI) engine. DG set can be classified according to cycle type as: two stroke and four stroke. However, the bulk of IC engines use the four stroke cycle. Types of fuel or energy used in DG sets are furnace oil and diesel.

10.4.1 Design and operation

A diesel generating set should be considered as a system since its successful operation depends on the well-matched performance of the components, namely:

1. The diesel engine and its accessories.
2. The AC generator.
3. The control systems and switchgear.
4. The foundation and power house civil works.
5. The connected load with its own components like heating, motor drives, lighting, etc.

It is necessary to select the components with highest efficiency and operate them at their optimum efficiency levels to conserve energy in this system. Various components of DG set are shown in Fig. 10.1. To make a decision on the type of engine, which is most suitable for a specific application, several factors need to be considered. The two most important factors are power and speed of the engine. The power requirement is determined by the maximum load. The engine power rating should be 10–20% more than the power demand by the end use. This prevents overloading the machine by absorbing extra

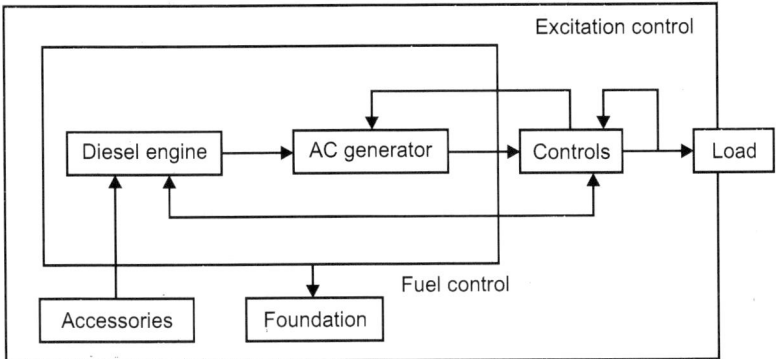

Figure 10.1: Diagram of DG set components.

load during starting of motors or switching of few types of lighting systems or when wear and tear on the equipment pushes up its power consumption.

An engine will operate over a range of speeds, with diesel engines typically running at lower speeds (1300–3000 rpm). Speed is measured at the output shaft and given in revolutions per minute (rpm). There will be an optimum speed at which fuel efficiency will be greatest. To determine the speed requirement of an engine, one has to again look at the requirement of the load. For some applications, the speed of the engine is not critical; but for other applications such as a generator, it is important to get a good speed match. If a good match can be obtained, direct coupling of engine and generator is possible; if not, then some form of gearing will be necessary - a gearbox or belt system, which will add to the cost and reduce the efficiency.

There are various other factors that have to be considered, when choosing a diesel engine for a given application. These include cooling system, abnormal environmental conditions (dust, dirt, etc.), fuel quality, speed governing (fixed or variable speed), poor maintenance, control system, starting equipment, drive type, ambient temperature, altitude, humidity, etc. Suppliers or manufacturers literature will specify the required information when purchasing an engine. The efficiency of an engine depends on various factors, for example, load factor (percentage of full load), engine size, and engine type. With the steady development of the diesel engine, the specific fuel consumption can come down. With the arrival of modern high efficiency turbochargers, it is possible to use an exhaust gas driven turbine generator to further increase the engine rated output.

The net result would be lower fuel consumption per kWh and further increase in overall thermal efficiency. The diesel engine is able to burn the poorest quality fuel oils, unlike gas turbine, which is able to do so with only costly fuel treatment equipment.

Diesel generator (DG) set selection and installation factors

1. If a DG set is required for 100% standby, then the entire connected load in HP/kVA should be added. After finding out the diversity factor (demand/connected load), the correct capacity of a DG set can be found out.

2. For an existing installation, record the current, voltage and power factor reading at the main bus-bar of the system at every half-an-hour interval for a period of 2–3 days; and during this period, the factory should be conducting its normal operations. The non-essential loads should be switched off to find the realistic current taken for running essential equipment. This will give a fair idea about the current taken from which the rating of the set can be calculated.

3. For a new installation, an approximate method of estimating the capacity of a DG set is to add full load currents of all the proposed loads to be run in DG set. Then, applying a diversity factor depending on the industry, process involved and guidelines obtained from other similar units, correct capacity can be arrived at.

Unbalanced load effects: It is always recommended to have the load as much balanced as possible, since unbalanced loads can cause heating of the alternator, which may result in unbalanced output voltages. The maximum unbalanced load between phases should not exceed 10% of the capacity of the DG sets.

Load pattern: In many cases, the load will not be constant throughout the day. If there is substantial variation in load, then consideration should be given for parallel operation of DG sets. In such a situation, additional DG sets are to be switched on when load increases. The typical case may be an establishment demanding substantially different powers in first, second and third shifts. By parallel operation, DG sets can be run at optimum operating points or near about, for optimum fuel consumption and additionally, flexibility is built into the system. This scheme can also be applied where loads can be segregated as critical and non-critical loads to provide standby power to critical load in the captive power system.

Energy performance assessment of DG sets

Routine energy efficiency assessment of DG sets involves following typical steps:

1. Ensure reliability of all instruments used for trial.
2. Collect technical literature, characteristics and specifications of the plant.
3. Conduct a 2 hour trial on the DG set, ensuring a steady load, wherein the following measurements are logged at 15 minutes intervals.
   (a) Fuel consumption (by dip level or by flow meter).
   (b) Amps, volts, PF, kW, kWh.

(c) Intake air temperature, Relative Humidity (RH).

(d) Intake cooling water temperature.

(e) Cylinder-wise exhaust temperature (as an indication of engine loading).

(f) Turbocharger rpm (as an indication of loading on engine).

(g) Charge air pressure (as an indication of engine loading).

(h) Cooling water temperature before and after charge air cooler (as an indication of cooler performance).

(i) Stack gas temperature before and after turbocharger (as an indication of turbocharger performance).

4. The fuel oil/diesel analysis is referred to from an oil company data.

Energy saving measures of DG sets

The following options will ensure that your diesel genset is operating at best efficiency and you can tap potential energy savings:

1. Ensure steady load conditions on the DG set, and provide cold, dust free air at intake (use of air washers for large sets, in case of dry, hot weather, can be considered).

2. Improve air filtration.

3. Ensure fuel oil storage, handling and preparation as per manufacturers' guidelines/oil company data.

4. Consider fuel oil additives in case they benefit fuel oil properties for DG set usage.

5. Calibrate fuel injection pumps frequently.

6. Ensure compliance with maintenance checklist.

7. Ensure steady load conditions, avoiding fluctuations, imbalance in phases, harmonic loads.

8. In case of a base load operation, consider waste heat recovery system adoption for steam generation or refrigeration chiller unit incorporation. Even the jacket cooling water is amenable for heat recovery.

9. In terms of fuel cost economy, consider partial use of biomass gas for generation. Ensure tar removal from the gas for improving availability of the engine in the long run.

10. Consider parallel operation among the DG sets for improved loading and fuel economy thereof.

11. Carry out regular field trials to monitor DG set performance and maintenance planning as per requirements.

10.4.2 Energy conservation in transmission line

1. To reduce line resistance 'R', solid conductors are replaced by stranded conductors (ACSR or AAC) and by bundled conductors in HT line.

2. High Voltage Direct Current (HVDC) is used to transmit large amount of power over long distances or for interconnections between asynchronous grids.

3. By transmitting energy at high voltage level reduces the fraction of energy lost due to Joule heating. (V $\alpha 1/I$ so I^2 R losses reduces).

4. As load on system increases terminal voltage decreases. Voltage level can be controlled by using voltage controllers and by using voltage stabiliser.

5. If required reactive power is transmitted through transmission lines, it causes more voltage drop in the line. To control receiving-end voltage, reactive power controllers or reactive power compensating equipments such as static VAR controllers are used.

10.4.3 Energy conservation in distribution line

1. Optimisation of distribution system: The optimum distribution system is the economical combination of primary line (HT), distribution transformer and secondary line (LT), to reduce this loss and improve voltage HT/LT line length ratio should be optimised.

2. Balancing of phase load: As a result of unequal loads on individual phase sequence, components causes over heating of transformers, cables, conductors, motors. Thus, increasing losses and resulting in the motor malfunctioning under unbalanced voltage conditions.

3. Harmonics: With increase in use of non-linear devices, distortion of the voltage and current waveforms occurs, known as harmonics. Due to presence of harmonic currents excessive voltage and current in transformers terminals, malfunctioning of control equipments and energy meter, over effect of power factor correction apparatus, interference with telephone circuits and broad casting occurs. Distribution Static Compensator (DASTACOM) and harmonic filters can reduce this harmonics.

4. Energy conservation by using power factor controller: Low power factor will lead to increased current and hence increase losses and will affect the voltage. We can use power factor controller or automatic power factor controller that can be located near receiving substations, load centers or near loads.

5. Energy conservation by demand side management control demand-side management is used to describe the actions of a utility, beyond the

customer's meter, with the objective of altering the end-use of electricity whether it be to increase demand, decrease it, shift it between high and low peak periods, or manage it when there are intermittent load demands in the overall interests of reducing utility costs. Nearly energy of 15,000 MW can be saved through end-use energy efficiency.

10.4.4 Energy conservation in lighting system

Good lighting is required to improve the quality of work, to reduce humans/ workers fatigue, to reduce accidents, to protect his eyes and nervous system. In industry it improves production, and quality of products/work. To view economy of lighting system, cost of initial installation cost, running cost, and effect on production/work are to be considered as main parameters. The power consumption by the industrial lighting is nearly 2–10% of total power consumption, depending on type of industries.

1. Optimum use of natural light: Whenever the orientation of a building permits, day lighting has to be used in combination with electric lighting. The maxim use of sunlight can be transmitted by means of transparent roof sheets, north light roof, etc.

2. Replacing incandescent lamps by Compact Fluorescent Lamps (CFL's): CFL's are highly suitable for places such as living rooms, hotel lounges, bars, restaurants, pathways, building entrances, corridors, etc.

3. Replacing conventional fluorescent lamp by energy efficient fluorescent lamp: Energy efficient lamps are based on the highly sophisticated technology. They offer excellent colour rendering properties in addition to the very high luminous efficacy.

4. Replacement of mercury/sodium vapour lamp by Halides lamp: Mercury Halide Lamp (MHL) provides high colour rendering index and offer efficient white light. Hence for critical applications where higher illumination levels are required, these lamps are used. They are highly suitable for applications such as assembly line, inspection area, painting shops, etc.

5. Replacing HPMV lamps by High Pressure Sodium Vapour Lamp (HPSV): Where colour rendering is not critical for such applications, e.g. street lighting, yard lighting because CRI of HPSV is low but offer more efficiency.

6. Replacing filament lamps on panels by LED: LED lamps consumes less power (1 W lamp), withstand high voltage fluctuation in the power supply, longer operating life (>100,000 hrs). Hence nowadays they are also used in street lighting, signalling, advertising boards, even as replacement for tube light or CFL.

7. Replacement of conventional ballast by electronic ballast: Installation of high frequency (28–32 Mhz) electronic ballast in place of conventional ballasts helps to reduce power consumption up to 35%.

8. Installation of separate transformer for lighting: In most of the industries, the net lighting load varies between 2 and 10%. If power load and lighting load fed by same transformer, switching operation and load variation causes voltage fluctuations. This also affects the performance of neighbouring power load apparatus, lighting load equipments and also reduces lamps. Hence, the lighting equipment has to be isolated from the power feeders. This will reduce the voltage related problems, which in turn provides a better voltage regulation for the lighting, this also increases the efficiency of the lighting system.

9. Installation of servo stabiliser for lighting feeder: Wherever, installation of separate transformer for lighting is not economically attractive, then servo stabiliser can be installed for the lighting feeders.

10. Control over energy consumption pattern: Occupancy sensors, daylight linked control are commonly used in commercial buildings, malls, offices, where more number of lights are to be controlled as per operational hours microprocessor based light control circuits are used. As a single control unit it can be programmed to switch on/off as per the month wise, year wise and even season wise working schedule.

11. Periodic survey and adequate maintenance programme: Illumination level reduces due to accumulation of dirt on lamps and luminaries. By carrying periodic maintenance, i.e. cleaning, dusting of lamps and luminaries will improve the light output/luminance. As part of maintenance programme, periodic surveys of installation, lightning system with respect lamp positioning and illumination levels, proper operation of control gears should be conducted to take advantage of energy conservation opportunities as user requirements changes.

Energy conservation in motors: Considering all industrial applications 70% of total electrical energy consumed by only electric motors driven equipments.

1. Improving power supply quality: Maintaining the voltage level within the BIS standards, i.e. with tolerance of +/–6% and frequency with tolerance of +/–3% motor performance improves and also life.

2. Optimum loading: Proper selection of the rating of the motor will reduce the power consumption. If the motor is operating at less than 50% of loading ($\eta < 50\%$) significant power saving can be obtained by replacing with properly sized high efficiency motors. If the motor is operating at loads below 40% of its capacity, an inexpensive and effective measure might be to operate in star mode.

3. Improving transmission efficiency: Proper selection of power transmission means (belts, gears) will reduces transmission losses.

4. Stopping idle or redundant running of motors or lights will save 100% power.

5. By use of soft starter: Soft starters are essentially stator voltage controllers, helps to overcome above problem. It helps to restrict starting current and also provide smooth start and stop operation.

6. By improving power factor (PF): For improving PF, connect the capacitor bank, which will improve the PF of the system from installation to generating station. Maximum improvement in overall system efficiency is achieved, which also reduces maximum demand of the system and that will reflect in energy bill.

7. Use of high efficiency or energy efficient motors.

The energy efficient motors can reduce losses through improved design, better materials and improved manufacturing techniques. Generally motor life doubles for each 10°C reduction in operating temperature. While selecting EEM, select with 1.15 service factor, design for operation at 85% of rated load.

Energy conservation in thermal power plants

11.1 Introduction

Electricity is a convenient form of energy. Thermal power plants convert the energy in coal to electricity. For example, consider a coal with a high heating value of 20,000 kJ/kg. Theoretically this is equivalent to 5.56 kwhr of electrical energy. The first process of energy conversion is the combustion where the potential energy in coal is converted to heat energy. The efficiency of this conversion is around 90%.

1. Due to practical limitations in heat transfer, all the heat produced by combustion is not transferred to the water; some is lost to the atmosphere as hot gases.

2. The coal contains moisture. Also coal contains a small per cent of hydrogen, which also gets converted to moisture during combustion. In the furnace, moisture vapourises taking latent heat from the combustion heat and exits the boiler along with the hot gases.

3. Improper combustion of coal, hot ash discharged from the boiler and radiation are some of the other losses.

The second stage of conversion is the thermodynamic stage. The heat from combustion is transferred to the water to produce steam. The energy of the steam is converted to mechanical rotation of the turbine. The steam is then condensed to water and pumped back into the boiler for reuse. This stage works on the principle of the Rankine cycle. For plants operating with steam at subcritical pressures (less than 221 bar) and steam temperatures of 570°C, the Rankine cycle efficiency is around 43%. For the state of the art plants running at greater than supercritical pressure and steam temperatures near to 600°C, the efficiency is around 47%.

1. The steam is condensed for reuse. During this process, the latent heat of condensation is lost to the cooling water. This is the major loss and is almost 40% of the energy input.

2. Losses in the turbine blades and exit losses at turbine end are some of the other losses.

3. The Rankine cycle efficiency is dictated by the maximum temperature of steam that can be admitted into the turbine. Due to metallurgical constraints steam temperatures are at present limited to slightly more than 600°C.

The third stage converts the mechanical rotation to Electricity in a generator. Copper, magnetic and mechanical losses account for 5% loss in the generator. Another 3% is lost in the step-up transformer which makes the power ready for transmission to the consumer. To operate the power plant it is required to run various auxiliary equipment like pulverises, fans, pumps and precipitators. The power to operate these auxiliaries has to come from the power plant itself. For large power plants around 6% of the generator output is used for internal consumption. This brings the overall efficiency of the power plant to around 33.5%. This means we get only 1.9 kwhr of electrical energy from one kg of coal instead of the 5.56 kwhr that is theoretically available in the coal. The efficient use of primary energy in the power-generation process makes a decisive contribution towards lowering CO_2 emissions and is a key issue in power-plant fleet. The energy process is an organised approach to identify energy waste in a facility, determining how this waste can be eliminated at a reasonable cost with a suitable time frame. Energy efficiency is widely used and many have different meaning depending on energy service companies. Encrgy audit of a building can range from a short a walk through of the facility to a detailed analysis. It not only serves to identify energy use among the various services and to identify opportunities for energy conservation but it is also a crucial first step in establishing an energy management programme. The efficiency will produce the data on which such a programme is based.

The study should reveal to the owner, manager, or management team of the building the options available for reducing energy waste, the costs involved, and the benefits achievable from implementing those energy conserving opportunities (ECOs). The energy analysis contains valuable information such as energy consumption patterns of the factory and the identification of high energy intense equipments, possible energy saving measures and cost benefit analysis of energy saving measures.

11.2   Energy management cell (EMC)

11.2.1   Objectives of EMC

1. To operate the power plant at highest efficiency and optimum cost.
2. To create awareness about energy conservation amongst all the stake-holders.
3. EMC can achieve objective of 'high energy efficiency and at optimum cost' through following steps:
   (a)   Regular internal energy efficiency.
   (b)   Documentation for energy management activity.
   (c)   Regular energy efficiencies through accredited energy efficiency firms.

(d) Regular filling of energy returns to state level designated agency.

(e) Energy conservation projects-identification, evaluation and implementation.

(f) Establishing energy efficiency test procedures and schedules for all equipments and systems.

4. EMC can achieve objective of 'awareness drive' through following initiatives:

(a) Display of posters and slogans in the plant area.

(b) Celebration of energy conservation week.

(c) Film show.

Energy management cell structure is shown in Fig. 11.1.

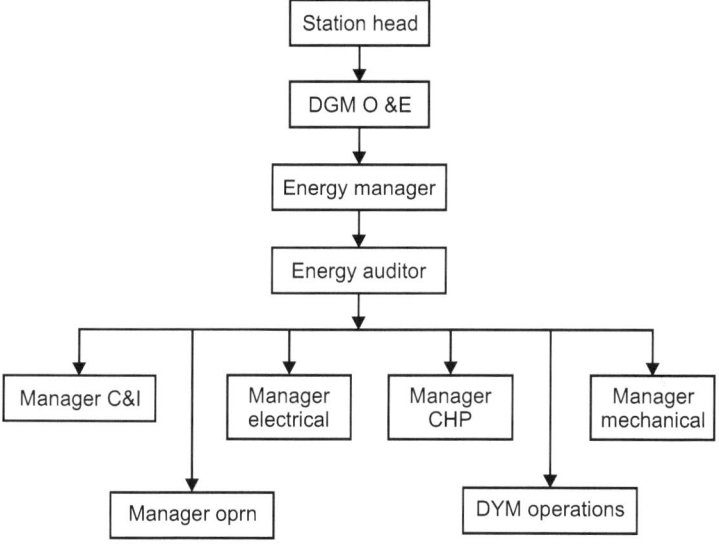

Figure 11.1: EMC structure.

11.3   Auxiliary power consumption (APC) in thermal power plant

The auxiliary power consumption plays a major role in enriching the energy efficiency of the thermal power plant. As per the norms APC should well within the 10%. Since Thermal power plant also falls under energy intensive consumer category like railways, metal industries, port trust, etc. according to Electricity Act features. It is paramount importance to analyse the consumption pattern of the plant and work on various areas so as to boost up the efficiency of cycles and sub-cycle.

11.3.1 Factors affecting the APC

1. Plant load factor is high.
2. Operational efficiency of the equipment is moderate.
3. Startup and shutdown is low.
4. Age of the plant is high.
5. Coal quality is moderate to high.

11.4 Tips for major energy saving potential areas in thermal power plant

Most thermal power plant uses 30–40% of energy value of primary fuels. The remaining 60–70% is lost during generation, transmission and distribution of which major loss is in the form of heat. Thermal power consist of various sub cycles/systems like air and flue gas cycle, main steam, feed water and condensate cycle, fuel and ash cycle, equipment cooling water (ECW), auxiliary cooling water (ACW) system, compressed air system, electrical auxiliary power and lighting system, HVAC system, etc. There is tremendous scope of energy saving potential in each system/cycle which is given below.

11.4.1 Air and flue gas cycle

Optimising excess air ratio

It reduces FD fan and ID fan loading.

Replacement of oversize FD and PA fan: Many thermal power plants have oversize fan causing huge difference between design and operating point leads to lower efficiency. Hence fan efficiency can be improved by replacing correct size of fan. If replacement is not possible, use of HT VFD for PA and ID fan can be the solution.

Attending the air and flue gas leakages: Leakages in air and flue gas path increases fan loading. Use of thermo vision monitoring can be adopted to identify leakages in flue gas path. Air Preheater (APH) performance is one crucial factor in leakage contribution. If APH leakage exceeds design value then it requires corrective action.

11.4.2 Steam, feed water and condensate cycle

BFP scoop operation in three element mode instead of DP mode: In three element mode throttling losses across Financial Reporting System (FRS) valve reduces leads to reduction in Boiler Feed Pump (BFP) power.

Optimisation of level set point in LP and HP heater: Heater drip level affects Terminal Temperature Difference (TTD) and Drain Cooler Approach

(DCA) of heater which finally affect feed water O/L temp. Hence it requires setting of drip level set point correctly.

Charging of APRDS from CRH line instead of MS line: Air Pressures Reducing Desuperheating (APRDS) charging from Cold Reheat (CRH) is always more beneficial than from MS line charging.

Isolation of steam line which is not in use: It is not advisable to keep steam line charged unnecessarily if steam is not being utilised, since the energy loss will occur due to radiation. For example, deareator extraction can be charged from turbine extraction/CRH or from APRDS. In normal running, APRDS extraction is not used so same can be kept isolated.

Replacement of BFP cartridge: BFP draws more current if cartridge is worn out, causing short circuit of feed water flow inside the pump. It affects pump performance. Hence, cartridge replacement is necessary.

Attending passing recirculation valve of BFP: BFP power consumption increases due to passing of R/C valve. It requires corrective action.

Installation of HT VFD for CEP: Country Environmental Profile (CEP) capacity is underutilised and also the pressure loss occurs across deareator level control valve. There is large scope of energy saving which can be accomplished by use of HT VFD for CEP or impeller trimming.

11.4.3   Fuel and ash cycle

Optimised ball loading in ball tube mill: Excessive ball loading increases mill power. Hence, ball loading is to be optimised depending upon coal fineness report.

Use of wash coal or blending with A-grade coal: F-grade coal has high ash content. Overall performance can be improved by using wash coal or blending of F-grade coal with A-grade coal instead of only using F-grade coal.

Use of dry ash evacuation instead of WET deashing system: Dry deashing system consumes less power and also minimises waste reduction.

Optimise mill maintenance: Mill corrective/preventive maintenance is to be optimised depending parameters like running hours, mill fineness, bottom ash unburnt particle, degree of reject pipe chocking, etc.

11.4.4   Electrical and lighting system

Optimising voltage level of distribution transformer: It is found that operating voltage level is on higher side than required, causing more losses. It is required to reduce the voltage level by tap changing.

Use of auto star/delta/star converter for under loaded motor lighting: The methods which can be used are, use of electronic chock instead of conventionally used copper chock, use of CFL, replacement of mercury vapour lamp by metal halide lamp, use of timer for area lighting. Lighting has tremendous potential of saving.

11.4.5   ECW and ACW system

Isolating ECW supply of standby auxiliaries: Many times standby coolers are kept charged from ECW side. Also, standby equipment's auxiliaries like lube oil system are kept running for reliability. We can isolate standby cooler from ECW system and switching off standby auxiliaries, doing trade-off between return and reliability.

Improving condenser performance by condenser tube cleaning and use of highly efficient debris filter: Tube cleaning by bullet shot method increases condenser performance; condenser tube cleaning is necessary which is to be carried out in overhaul. Also, highly advanced debris filter contribute to condenser performance.

Application of special coating on CW pump impeller: It improves pump impeller profile condition, increasing pump performance.

11.4.6   Compressed air system

Optimising discharge air pressure by tuning loading/unloading cycle: It will reduce power consumption.

Use of heat of compression air dryer instead of electrically heated air dryer: Heat of compression air dryer use heat generated in compression cycle, thus reduces power consumption.

Use of screw compressor instead reciprocating compressor: Power consumption of screw compressor is less than reciprocating air compressor leads to reduce auxiliary power consumption.

11.4.7   HVAC system

1. Cooling tower performance improvement.
2. Installing absorption refrigeration system instead of vapour compression system.
3. Use of wind turbo ventilators instead of conventional motor driven exhauster.

11.4.8   Tips to improve energy efficiency in power plants

1. Detailed measurements to identify thermal losses.
2. Mills and combustion (e.g. online coal-quality control).
3. Fixing leakages in flue-gas ducts.

4. Air heaters: Overhaul/upgrade.
5. Turbine: New blading.
6. Condenser: Overhaul/upgrade.
7. Cooling tower: New packings.
8. Raising the steam temperature.
9. Lowering internal consumption by optimising operations or exchanging electrical consumers.

11.5   Waste heat to power systems

Waste heat to power (WHP) is the process of capturing heat discarded by an existing industrial process and using that heat to generate power (Fig. 11.2). Energy intensive industrial processes—such as those occurring at refineries, steel mills, glass furnaces, and cement kilns—all release hot exhaust gases and waste streams which can be harnessed with well-established technologies to generate electricity (see Appendix). The recovery of industrial waste heat for power is a largely untapped type of combined heat and power (CHP), which is the use of a single fuel source to generate both thermal energy (heating or cooling) and electricity.

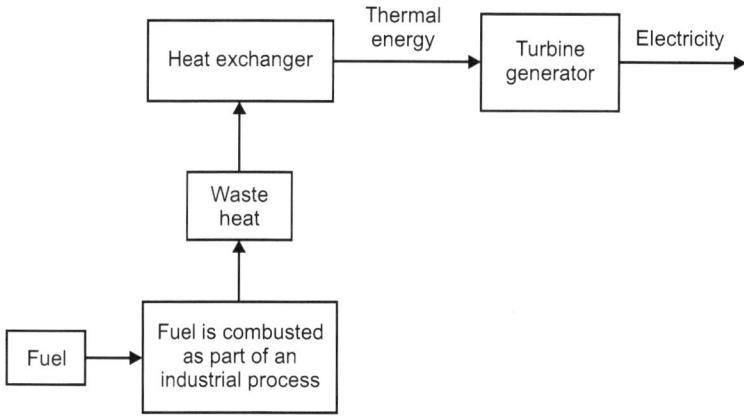

Figure 11.2: Waste heat to power diagram.

CHP generally consists of a prime mover, a generator, a heat recovery system, and electrical interconnection equipment configured into an integrated system. CHP is a form of distributed generation, which, unlike central station generation, is located at or near the energy-consuming facility. CHP's inherent higher efficiency and its ability to avoid transmission losses in the delivery of electricity from the central station power plant to the user results in reduced primary energy use and lower greenhouse gas (GHG) emissions.

The most common CHP configuration is known as a topping cycle, where fuel is first used in a heat engine to generate power, and the waste heat from the power generation equipment is then recovered to provide useful thermal energy. As an example, a gas turbine or reciprocating engine generates electricity by burning fuel and then uses a heat recovery unit to capture useful thermal energy from the prime mover's exhaust stream and cooling system. Alternatively, steam turbines generate electricity using high-pressure steam from a fired boiler before sending lower pressure steam to an industrial process or district heating system.

Waste heat streams can be used to generate power in what is called bottoming cycle CHP—another term for WHP. In this configuration, fuel is first used to provide thermal energy in an industrial process, such as a furnace, and the waste heat from that process is then used to generate power. The key advantage of WHP systems is that they utilise heat from existing thermal processes, which would otherwise be wasted, to produce electricity or mechanical power, as opposed to directly consuming additional fuel for this purpose.

11.6   Opportunity for waste heat power (WHP)

Industrial energy use represents the largest potential source of WHP generation. Roughly one-third of the energy consumed by industry is discharged as thermal losses directly to the atmosphere or to cooling systems. These discharges are the result of process inefficiencies and the inability of the existing process to recover and use the excess energy streams. Most of this waste energy, however, is of low quality (i.e. available in waste streams with temperatures below 300°F or dissipated as radiation heat loss) and is typically not practical or economical to recover with current technology.

The efficiency of generating power from waste heat recovery is heavily dependent on the temperature of the waste heat source. In general, economically feasible power generation from waste heat has been limited primarily to medium- to high-temperature waste heat sources (i.e. > 500°F). Emerging technologies, such as organic Rankine cycles, are beginning to lower this limit, and further advances in alternative power cycles will enable economic feasibility of generation at even lower temperatures over time.

Estimates of the amount of industrial waste heat available at a temperature high enough for power generation with today's technologies (i.e. > 500°F) are in the range of 0.6–0.8 Quads (or 6000–8000 megawatts [MW] of electric generating capacity) on a national basis. Nonindustrial applications, such as exhaust from natural gas pipeline compressor drives and landfill gas engines, represent an additional 1000–2000 MW of power capacity for a total of seven to ten gigawatts. At the project level, a number of factors in addition to the temperature of the waste heat must be considered to determine the economic feasibility of power generation from waste heat sources:

1. Is the waste heat source a gas or a liquid stream?
2. What is the availability of the waste heat—is it continuous, cyclic, or intermittent?
3. What is the load factor of the waste heat source—are the annual operating hours sufficient to amortise the capital costs of the WHP system?
4. Does the temperature of the waste stream vary over time?
5. What is the flow rate of the waste stream, and does it vary?
6. Is the waste stream at a positive or negative pressure and does this vary?
7. What is the composition of the waste stream?
8. Are there contaminants that may corrode or erode the heat recovery equipment?

The answers to these questions will determine system design and, ultimately, the economic viability of a WHP project. Many high-temperature waste heat sources are straightforward to capture and use with existing technologies.

Other sources must be cleaned prior to use. The cleaning process is typically expensive, and removing the contaminants often removes the heat at the same time. Other waste heat sources are difficult to recover because of equipment configuration or operational issues.

11.7   Applicable technologies

Steam Rankine cycle (SRC): The most commonly used system for power generation from waste heat involves using the heat to generate steam in a waste heat boiler, which then drives a steam turbine. Steam turbines are one of the oldest and most versatile prime mover technologies. Heat recovery boiler/steam turbine systems operate thermodynamically as a Rankine cycle, as shown in Fig. 11.3. In the steam Rankine cycle, the working fluid–water–is first pumped to elevated pressure before entering a heat recovery boiler. The pressurised water is vapourised by the hot exhaust and then expanded to lower temperature and pressure in a turbine, generating mechanical power that can drive an electric generator. The low-pressure steam is then exhausted to a condenser at vacuum conditions, where heat is removed by condensing the vapour back into a liquid. The condensate from the condenser is then returned to the pump and the cycle continues.

Organic Rankine cycles (ORC): Other working fluids, with better efficiencies at lower heat source temperatures, are used in ORC heat engines. ORCs use an organic working fluid that has a lower boiling point, higher vapour pressure, higher molecular mass, and higher mass flow compared to

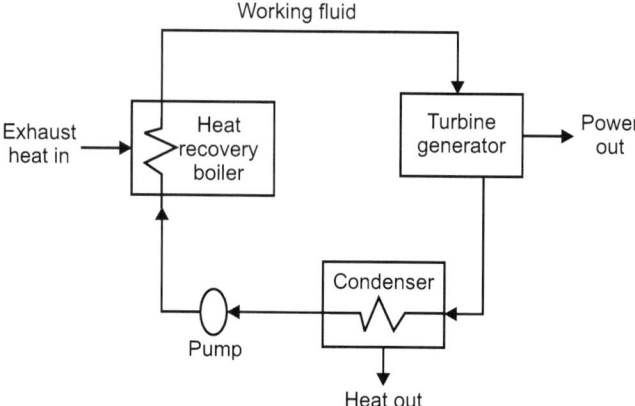

Figure 11.3: Rankine cycle heat engine.

water. Together, these features enable higher turbine efficiencies than in an SRC. ORC systems can be utilised for waste heat sources as low as 300°F, whereas steam systems are limited to heat sources greater than 500°F. ORCs have commonly been used to generate power in geothermal power plants, and more recently, in pipeline compressor heat recovery applications.

Kalina cycle: Kalina cycle is another Rankine cycle, using a mixture of water and ammonia as the working fluid, which allows for a more efficient energy extraction from the heat source. The Kalina cycle has an operating temperature range that can accept waste heat at temperatures of 200°F–1000°F and is 15 to 25% more efficient than ORCs at the same temperature level. Kalina cycle systems are becoming increasingly popular overseas in geothermal power plants, where the hot fluid is very often below 300°F. The three types of Rankine power cycles discussed above overlap to a certain degree.

There are advantages to each, however:

1. SRCs are the most familiar to industry and are generally economically preferable where the source heat temperature exceeds 800°F.

2. For lower temperatures, ORC or Kalina cycle systems are used. They can be applied at temperatures lower than for steam turbines, and they are more efficient in moderate temperature ranges.

3. Kalina systems have the highest theoretical efficiencies. Their complexity makes them generally suitable for large power systems of several megawatts or greater.

4. ORC systems can be economically sized in small, sub-megawatt packages and they are also well suited for using air-cooled condensers, making them appropriate for applications such as pipeline compressor stations that do not have access to water.

In addition to Rankine cycle systems, there are a number of advanced technologies in the research and development stage that can generate electricity directly from heat and that could in the future provide additional options for power generation from waste heat sources. These technologies include thermo-electric, piezoelectric, thermionic, and thermo-photovoltaic (thermo-PV) devices. Several of these have undergone prototype testing in automotive applications and are under development for industrial heat recovery.

11.7.1 Applications of waste heat power

Economically feasible WHP applications are generally based on recovering waste heat from combustion exhaust streams with temperatures above 500°F. Industrial processes which produce these temperatures include calcining operations (cement, lime, alumina, and petroleum coke), metal melting, glass melting, petroleum fluid heaters, thermal oxidisers, and exothermic synthesis processes. Key WHP opportunities within these operations are provided below:

1. Primary metals: Primary metals manufacturing involves a large number of high-temperature processes from which waste heat can be recovered. Steel mills, for example, have various high-temperature heat-recovery opportunities. In integrated mills, waste heat can be recovered from coke ovens, Blast furnaces for iron production, and basic oxygen furnaces for steel production. There are also opportunities to recover waste heat from electric arc furnaces. In the aluminum industry there is energy recovery potential from the exhaust of the Hall Héroult cells and secondary melting processes. Metal foundries have a variety of waste heat sources, such as melting furnace exhaust, ladle preheating, core baking, pouring, shot-blasting, castings cooling, heat treating, and quenching.

2. Nonmetallic mineral product manufacturing: There are a number of strong opportunities for WHP in this sector. Calcining in rotary kilns is a high-temperature process that is used in the manufacture of cement, gypsum, alumina, soda ash, lime, and kaolin clay. The glass industry uses raw material melting furnaces, annealing ovens, and tempering furnaces, all operated at high temperatures.

3. Petroleum refining: Basic processes used in petroleum refineries include distillation (fractionation), thermal cracking, catalytic and treatment. These processes use large amounts of energy, and many involve exothermic reactions that also produce heat. Modern refineries are highly integrated systems that recover heat from one process to use in other processes. However, many operations still release high-quality waste heat that could be recovered for power production. An example is the exhaust from petroleum coke calciners. Petroleum coke is heated to 2400°F, and the exhaust is typically 900 to 1000°F leaving the calciner.

4. Chemical industry: There are several major segments of the industry, including petrochemicals, industrial gases, alkalies and chlorine, cyclic crudes and intermediates (e.g. ethylene, propylene and benzene/toluene/xylene), plastics materials, synthetic rubber, synthetic organic fibres, and agricultural chemicals (fertilisers and pesticides), in which high-temperature exhaust is released that could be recovered for power generation.

5. Fabricated metals: Processes generating waste heat include metal pre-heating, heat treatment, cleaning, drying, and furnace heating.

6. Landfill gas energy systems: Landfills that use reciprocating internal combustion engines or turbines to produce power could generate additional power with ORC systems using exhaust gases. Other landfills could install ORC systems to generate power from the waste heat from flaring.

7. Oil and gas production: There are a number of flared energy sources in oil and gas production that could utilise WHP.

Besides above there are natural gas compressor stations using exhaust heat of gas turbine-driven compressors.

11.8   Cogeneration and combined heat and power

Cogeneration through combined heat and power (CHP) is the simultaneous production of electricity with the recovery and utilisation heat. Cogeneration is a highly efficient form of energy conversion and it can achieve primary energy savings of approximately 40% by compared to the separate purchase of electricity from the national electricity grid and a gas boiler for onsite heating (Fig. 11.4). Combined heat and power plants are typically embedded close to the end user and therefore help reduce transportation and distribution losses, improving the overall performance of the electricity transmission and distribution network.

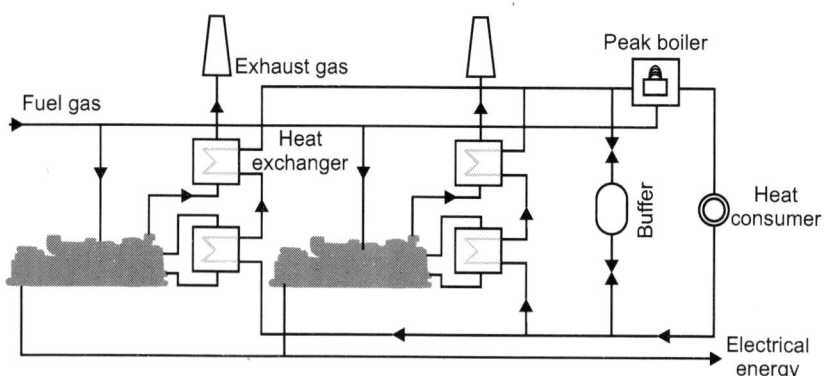

Figure 11.4: Cogeneration and combined heat and power.

For power users where security of supply is an important factor for their selection of power production equipment and gas is abundant, gas-based cogeneration systems are ideally suited as captive power plants (i.e. power plants located at site of use).

11.8.1   Benefits of gas engine CHP

The high efficiency of a CHP plant compared with conventional bought in electricity and site-produced heat provides a number of benefits including:

1. On site production of power.
2. Reduced energy costs.
3. Reduction in emissions compared to conventional electrical generators and onsite boilers.

11.8.2   Heat sources from a gas engine

The heat from the generator is available in from 5 key areas:

1. Engine jacket cooling water.
2. Engine lubrication oil cooling.
3. First stage air intake intercooler.
4. Engine exhaust gases.
5. Engine generator radiated heat, second stage intercooler.

1, 2 and 3 are recoverable in the form of hot water, typically on a 70/90°C flow return basis and can be interfaced with the site at a plate heat exchanger. The engine exhaust gases typically leave the engine at between 400 and 500°C. This can be used directly for drying, in a waste heat boiler to generate steam, or via an exhaust gas heat exchanger combining with the heat from the cooling circuits. The heat from the second stage intercooler is also available for recovery as a lower grade heat. Alternatively new technologies are available for the conversion of heat to further electricity, such as the organic rankine cycle engine.

CHP applications

A variety of different fuels can be used to facilitate cogeneration. In gas engine applications CHP equipment is typically applied to natural gas (commercial, residential and industrial applications), biogas and coal gas applications.

CHP system efficiency

Gas engine combined heat and power systems are measured based upon the efficiency of conversion of the fuel gas to useful outputs. Figure 11.5 illustrates this concept. Firstly the energy in the fuel gas input is converted into mechanical energy via the combustion of the gas in the engine's cylinders and their resulting

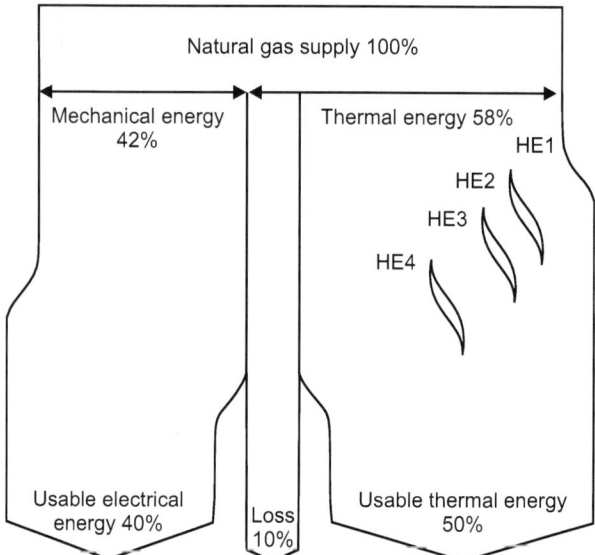

Figure 11.5: Concept showing usable electrical, thermal and mechanical energy. (Diagram key: HE1–Mixture intercooler, HE2–Oil exchange heater, HE3–Engine jacket water heat exchanger, HE4–Exhaust gas heat exchanger.

action in the turning of the engine's crankshaft. This mechanical energy is in turn used to turn the engine's alternator in order to produce electricity. There is a small amount of inherent loss in this process and in this example the electrical efficiency of the engine is 40% (in reality GE Jenbacher gas engines are typically between 40–48.7% electrically efficient).

11.8.3   Achieving greater thermal plant efficiency

To achieve greater thermal plant efficiency, a broad array of issues must be addressed, including:

1. Power plant efficiency, while not difficult to measure, is not routinely measured in many plants owing to the lack of test equipment/ instrumentation and training.
2. Little financial incentive exists for many utilities to improve their thermal efficiency, especially when fuel costs can be passed on to the consumer.
3. Some efficiency improvements can be perceived as risky to plant operators or electricity regulators.

A few of the guidelines being addressed are as follows:

1. Identify, implement, and train workers in best practices, and employing dedicated plant efficiency engineers.

2. Optimise processes using advanced computational tools and software.
3. Conduct on-line, real-time performance monitoring of efficiency.
4. Reduce air, water, steam, and flue gas leakage.
5. Optimise coal pulveriser performance and balance coal and air flows to the burners.
6. Upgrade steam turbines.
7. Use variable speed motors.
8. Lower stack temperature.

Understanding and comparing generating efficiency across power generation assets is a very challenging problem due to differences in technology, operation, fuel, size, and age. Even what in concept would seem very straightforward, such as monitoring fleet generating efficiency to produce realistic improvement targets, is a more difficult task than one might think at first glance. The reasons are many and include factors such as the simple fact that it is not easily measured, asset diversity and normal asset degradation. From the perspective of producing improvement targets, even a comparison of a unit to its own historical or design performance is a challenge due to changes in operation, modifications made to address environmental regulations, normal degradation, changes in fuel quality/sourcing or equipment upgrades.

One important factor to consider when comparing generating assets is to ensure that all metrics of efficiency are compared on an equal footing.

The primary metric of unit efficiency used in the industry is the heat rate of the unit, which is a ratio of the energy required to produce a unit of electricity– such as how many Btu/hr of fossil fuel are required to produce 1 kW of electricity at the generator terminal. Design heat rates vary significantly based on plant type. Actual heat rates may vary by as much as 10–15% from factors including normal degradation, fuel source, how well it's operated, etc. As an example, the heat rate of a large coal plant may be reduced by 3–4% by switching from a bituminous fuel to a low sulphur sub-bituminous coal. This efficiency loss, when coupled with the expected increases in unit auxiliary power which are coincident with using a lower-quality coal, can result in a reduction in the plant efficiency of 5% or more. A combustion turbine based plant burning natural gas may perform 1–3% better than the same plant design burning oil.

Another factor affecting many coal plants is the addition of emissions equipment such as Selective Catalytic Reduction (SCR) systems and flue gas scrubbers which may hurt performance by as much as 1–2% in exchange for reducing SO_2 emissions. One of the large hurdles to CO_2 sequestration is the huge impact on unit performance of as much as 50%. Steam turbines are at the heart of coal-fired power plants. As shown in the simplified schematic of a pulverised coal plant in Fig. 11.6, a steam electric power plant consists of a number of basic components. Coal is crushed and fed into a boiler where it is burned to heat water into steam. The steam is injected under pressure into a turbine which turns a generator (where essentially a magnet turns in a coil of wire causing electrons to flow thus creating an electric current).

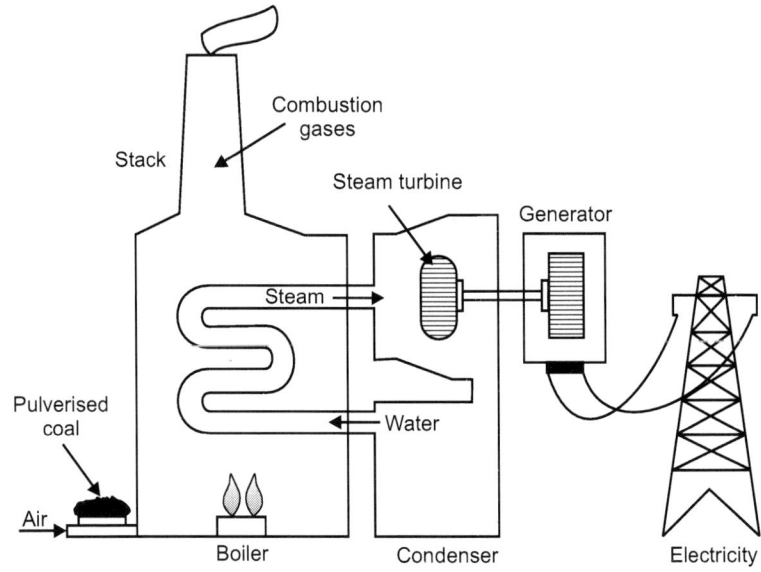

Figure 11.6: Electric power generation.

Steam returning from the turbine is then cooled in a condenser, and the water is fed back by a feedwater pump to the boiler to continue the process. The expansion of water into steam vapour (and condensation back into liquid water) in this manner is called a Rankine cycle, and is the basis for most electric power generation in the United States and other developing countries.

11.8.4   Economics of waste heat power

The total cost to install WHP systems include the costs associated with the waste heat recovery equipment (boiler or evaporator), the power generation equipment (steam, ORC or Kalina cycle), power conditioning and inter-connection equipment. It would also include the soft costs associated with designing, permitting and constructing the system. The installed costs of Rankine cycle power systems (steam, ORC or Kalina) are fairly similar, differing more as a function of project size and the complexity of site integration than type of system. Operation and maintenance (O&M) cost estimates can vary widely. Rankine cycle power systems themselves have relatively low maintenance costs. However, maintenance requirements of the heat recovery boilers and balance of plant must also be included and these can vary by technology and by site conditions. As an example, steam systems may require on-site boiler operators while ORCs can often run unattended. There are no fuel costs for true waste heat to power projects (i.e. no supplemental fuel use).

11.8.5 Current market status

Other options are entering the market that can be used at lower temperatures and smaller sizes, including ORCs, ammonia-water systems (e.g. Kalina cycles), and thermo-electric generators (still in development) that use solid state systems that require no moving parts and sit directly in the waste stream. Utilising liquid streams below 200°F and gas streams below 500°F typically remains economically impractical with today's technologies, however. Conversion to electricity is less efficient with all these technologies compared to traditional electric generators, and project costs currently run high for a variety of reasons, including the cost of the equipment and the cost of integrating the waste heat recovery system with the waste heat source. WHP is generally considered only when the waste heat cannot be used directly within the process, or other recovery methods are not practical within the facility. While the costs of these systems currently remain high, and commercial demonstration is limited, the technologies continue to evolve rapidly.

11.8.6 Market barriers

Technical barriers

The principal hurdle for WHP systems is the heat recovery itself. While the power generation equipment is commercially established and relatively standardised, each heat recovery situation presents unique challenges.

Some of the project-specific technical issues that affect project economics include:

1. The waste heat sources at a plant are dispersed and difficult to reach or consolidate, or are from noncontinuous or batch processes.
2. Seasonal operations and low-volume operations reduce the economic benefits of WHP.
3. Waste heat sources often contain chemical and/or mechanical contaminants that impact the complexity, cost and efficiency of the heat recovery process.
4. There may be added cost and complexity for integrating the WHP system controls with existing process controls.
5. Space limitations and equipment configurations make WHP systems difficult or impossible to site economically.

Appendix: Waste heat streams classified by temperature.

Temperature classification	Waste heat source	Characteristics	Commercial waste heat to power technologies
High (>1200°F)	Furnaces • Steel electric arc • Steel heating • Basic oxygen • Aluminum reverberatory • Copper reverberatory • Nickel refining • Copper refining • Glass melting Iron cupolas Coke ovens Fume incinerators Hydrogen plants	High quality heat High heat transfer High power-generation efficiencies Chemical and mechanical contaminants	Waste heat boilers and steam turbines
Medium (500–1200°F)	Prime mover exhaust streams • Gas turbine	Medium power-generation efficiencies	Waste heat boilers and steam turbines (>500°F)
	• Reciprocating engine Heat-treating furnaces	Chemical and mechanical contaminants (some streams such as cement kilns)	Organic Rankine cycle (<800°F)
	Ovens • Drying • Baking • Curing Cement kilns		Kalina cycle (<1000°F)
Low (< 500°F)	Boilers Ethylene furnaces	Energy contained in numerous small sources	Organic Rankine cycle (>300°F gaseous streams.
	Steam condensate	Low power-generation efficiencies	>175°F liquid streams)
	Cooling water		Kalina cycle (>200°F)

(Cont'd...)

Temperature classification	Waste heat source	Characteristics	Commercial waste heat to power technologies
	• Furnace doors	Recovery of combustion	
	• Annealing furnaces	streams limited due to acid	
	• Air compressors	concentration if temperatures	
	• IC engines	reduced below 250°F	
	• Refrigeration condensers		
	Low-temperature ovens		
	Hot process liquids or solids		

Energy efficiency in motors, fans and compressors

12.1 Introduction

Electric motor systems account for about 60% of global industrial electricity consumption. Electric motors drive both core industrial processes, like presses or rolls, and auxiliary systems, like compressed air generation, ventilation or water pumping. They are utilised throughout all industrial branches, though the main applications vary. Studies showed a high potential for energy efficiency improvement in motor systems in developing as well as in developed countries.

Particularly, system optimisation approaches that consider the whole motor system's efficiency show great potential. Many of the energy efficiency investments show payback times of only a few years. Still, market failures and barriers like lack of capital, higher initial costs, lack of attention by plant managers and principal agent dilemmas hamper the investment in energy efficient motor systems.

12.2 Monitoring motor failures

Present day continuous process plants demand high reliability of rotating equipments, which are mostly driven by electrical motors. Forced outages of motors could lead to production loss besides loss of energy during start up and shutdown, motor repairs and man hours lost.

Motors are designed and manufactured as per the standards laid down by various institutes and manufacturers to give trouble free operation. Series of inspections and tests are carried out during every stage of manufacture before the motors are certified and released for use. A study of more than 50,000 LT induction motor failures show that the actual life achieved is short and motor failures are more due to insulation and bearing failures. Other causes for failure are hot running of motors, over loading and jamming, corrosive fluid and water entry, single phasing. If these are attended to in time, a number of motor failures and unforeseen stoppages could be avoided.

12.2.1 Voltage variations

Insulation failure could be due to either unbalanced terminal voltage, wide variation in supply voltage compared to permissible limits given by the manufacturer or insufficient cooling. There could be unbalanced currents in the motor accompanied by a decrease in available torque and efficiency, and increase in slip, vibration, noise and motor heat losses. In the phase with the highest current, the percentage increase in temperature rise will be approximately two times the square of the percentage voltage inbalance. International specification of induction motor in USA are IEEE 112-B and Germany IEC 34-2 and Indian specification are IS: 13529. These specifications give the effect of unbalanced voltage on the performance of 3 phase induction motors. In case the inbalance voltage is high, motors should be derated to prevent the failure.

12.2.2   Poor insulation resistance

The insulation resistance of the windings should be tested periodically during service and maintained as per IS: 900, and BIS specification code of practice for installation and maintenance of induction motors. Insulation resistance should also be checked for standby motors before starting, during the rainy season. In case of weak insulation, resistance becomes a regular feature, the windings should be given a coat of good insulating varnish after the machine has been dried out. Standby motors should be run alternatively to avoid unexpected stoppages.

Generally, standard rated outputs of motors are based on temperature rise (75°C) for insulation class B, where the temperature rise is more, old motors should be replaced with higher class °F (95°C).

12.2.3   Misalignment and vibrations

The principal causes for bearing failure could be misalignment of motor with driven equipment and subsequent vibrations, cumulative errors in tolerances, wrong size or type of bearing, ingress of foreign matter, inadequate and contaminated lubrication, etc. It accounts for 20–25% of failures. The abnormal bearing noise because of misalignment and/or vibration, serves as an early warning for further trouble, e.g. bearing seizure. Hence, alignment should be checked and confirmed by a senior person before starting the equipment, which had been taken out for maintenance.

12.2.4   Corrosive fluids/water entry

High corrosive atmosphere in a chemical plant causes rapid deterioration leading to early failure. This accounts for 12–15% of breakdowns of motors. Leakages of chemicals/corrosive materials from the driven equipment are not only safety hazards for plant operations but potential problems for the motors as well. Hence, all such leakages from driven equipments and pipelines should be arrested immediately. To avoid such failures, motors could be provided with expeller on the shaft to prevent entry of any fluid in to the motor.

12.2.5   Higher temperature/overloading

15–20% of failures of LT induction motors take place because of higher temperature, overloading, electrical load inbalance, incorrect voltage and frequency, motor winding with loose connections, etc. Motors are intended for operation in maximum ambient temperature of 45°C. The motors near dryers, furnaces driving hot air blowers, fans, etc., in foundry, and textile plants, which are operated at higher temperatures, should be derated as below to avoid premature failure. Figure 12.1 shows the location of motor failing frequently due to higher temperature because of steam leakage from the inspection door.

Environment temp °C →	45	50	55	60	65	75	85
Permitted output → as % of rated output	100	96.5	93.0	90.0	86.5	79.0	70.0

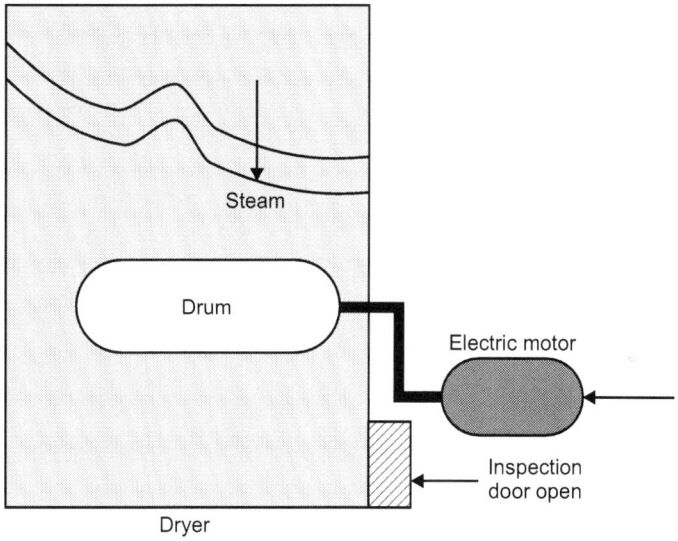

Figure 12.1: Showing wrong location of the motor.

Overloading and jamming of motors is mainly due to change in the process parameters. Frequent load variations for more than 15 seconds, due to process upsets or jamming because of mechanical problems, damage the insulation which results in motor break down. Hence, overloading due to change in process parameters should be identified and rectified. Many faults develop because of frequent motor starts. It causes the motor to carry starting loads and currents for a long time leading to overheating. In such cases, direct on line start current can be reduced with soft start circuits. Electronic timed over load protection can be also be used.

12.2.6   Inadequate protection

In a three phase motor blowing of a fuse, by burning out or mechanical interruption of one of the supply leads, by a faulty switch contact or circuit breaker could cause single phasing. When the motor runs on single phase, it gradually shows signs of distress, increase in noise, decrease in speed and overheating occurs. If adequate protection is not provided, the motors would burn out. The burn out is caused by an abnormal increase in the current flowing in the motor winding.

12.2.7   Operating parameter

Monitoring the operating parameters of electrical motors is essential to prevent forced plant outages due to any of the above reasons. This would give a timely and reliable evaluation of the motor condition. It could help plan maintenance and prevent failure, thereby avoid expensive repairs.

The tests and inspections could be carried out at regular intervals during plant overhauls. The motor operating conditions could be recorded and analysed for failure of each motor and rectified to avoid re-replacing them by energy efficient ones can save energy. Energy efficient motors are manufactured as per IEEMA 19–2000 and cost 20–25% more. Their cost can be recovered with 1500–2000 running hours, i.e. less than a year. Most of the motors in the plants are running around 60–70% loads, which results in higher losses, whereas energy efficient motors have constant efficiency between 50–100% of full load. This means savings in energy at lower loads also. Hence actual load at operating parameters should be checked with efficiency curves given by the suppliers.

12.3   Energy-efficiency improvement opportunities in electric motors

When considering energy-efficiency improvements to a facility's motor systems, a systems approach incorporating pumps, compressors, and fans must be used in order to attain optimal savings and performance. In the following, considerations with respect to energy use and energy saving opportunities for a motor system are presented. Pumping, fan and compressed air systems are discussed in addition to the electric motors.

12.3.1   Motor management plan

A motor management plan is an essential part of a plant's energy management strategy. Having a motor management plan in place can help companies realise long-term motor system energy savings and will ensure that motor failures are handled in a quick and cost effective manner.

The following are the key elements for a sound motor management plan:

1. Creation of a motor survey and tracking programme.
2. Development of guidelines for proactive repair/replace decisions.
3. Preparation for motor failure by creating a spares inventory.
4. Development of a purchasing specification.
5. Development of a repair specification.
6. Development and implementation of a predictive and preventive maintenance programme.

Maintenance

The purposes of motor maintenance are to prolong motor life and to foresee a motor failure. Motor maintenance measures can therefore be categorised as either preventative or predictive. Preventative measures, include voltage imbalance minimisation, load consideration, motor alignment, lubrication and motor ventilation.

Some of these measures are further discussed below.

Note that some of them aim to prevent increased motor temperature which leads to increased winding resistance, shortened motor life, and increased energy consumption. The purpose of predictive motor maintenance is to observe ongoing motor temperature, vibration, and other operating data to identify when it becomes necessary to overhaul or replace a motor before failure occurs. The savings associated with an ongoing motor maintenance programme could range from 2% to 30% of total motor system energy use.

12.3.2 Energy-efficient motors

Energy-efficient motors reduce energy losses through improved design, better materials, tighter tolerances, and improved manufacturing techniques. With proper installation, energy-efficient motors can also stay cooler, may help reduce facility heating loads, and have higher service factors, longer bearing life, longer insulation life, and less vibration. The choice of installing a premium efficiency motor strongly depends on motor operating conditions and the life cycle costs associated with the investment. In general, premium efficiency motors are most economically attractive when replacing motors with annual operation exceeding 2000 hours/year. Sometimes, even replacing an operating motor with a premium efficiency model may have a low payback period.

Rewinding of motors

In some cases, it may be cost-effective to rewind an existing energy-efficient motor, instead of purchasing a new motor. As a rule of thumb, when rewinding costs exceed 60% of the costs of a new motor, purchasing the new motor may be a better choice. When repairing or rewinding a motor, it is important to choose a motor service center that follows best practice motor rewinding standards in order to minimise potential efficiency losses. When best rewinding practices are implemented, efficiency losses are typically less than 1%. Software tools such as Motor Master can help identify attractive applications of premium efficiency motors based on the specific conditions at a given plant.

Proper motor sizing

It is a persistent myth that oversized motors, especially motors operating below 50% of rated load, are not efficient and should be immediately replaced with appropriately sized energy-efficient units. In actuality, several pieces of information are required to complete an accurate assessment of energy savings. They are the load on the motor, the operating efficiency of the motor at that load point, the full-load speed [in revolutions per minute (rpm)] of the motor to be replaced, and the full-load speed of the downsized replacement motor.

The efficiency of both standard and energy-efficient motors typically peaks near 75% of full load and is relatively flat down to the 50% load point. Motors in the larger size ranges can operate with reasonably high efficiency at loads down to 25% of rated load. There are two additional trends: larger motors exhibit both higher full- and partial-load efficiency values, and the efficiency decline below the 50% load point occurs more rapidly for the smaller size motors. Software packages such as Motor Master can aid in proper motor selection.

Adjustable speed drives (ASDs)

Adjustable-speed drives better match speed to load requirements for motor operations, and therefore ensure that motor energy use is optimised to a given application. As the energy use of motors is approximately proportional to the cube of the flow rate, relatively small reductions in flow, which are proportional to pump speed, already yield significant energy savings.

Power factor correction

Power factor is the ratio of working power to apparent power. It measures how effectively electrical power is being used. A high power factor signals efficient utilisation of electrical power, while a low power factor indicates poor utilisation of electrical power. Inductive loads like transformers, electric motors, and HID lighting may cause a low power factor. The power factor can be corrected by minimising idling of electric motors (a motor that is turned off consumes no energy), replacing motors with premium-efficient motors, and installing capacitors in the AC circuit to reduce the magnitude of reactive power in the system.

Minimising voltage unbalances

A voltage unbalance degrades the performance and shortens the life of three-phase motors. A voltage unbalance causes a current unbalance, which will result in torque pulsations, increased vibration and mechanical stress, increased losses, and motor overheating, which can reduce the life of a motor's winding insulation. Voltage unbalances may be caused by faulty operation of power factor correction equipment, an unbalanced transformer bank, or an open circuit. A rule of thumb is that the voltage unbalance at the motor terminals should not exceed 1% although even a 1% unbalance will reduce motor efficiency at part load operation. A 2.5% unbalance will reduce motor efficiency at full load operation. By regularly monitoring the voltages at the motor terminal and through regular thermographic inspections of motors, voltage unbalances may be identified. It is also recommended to verify that single-phase loads are uniformly distributed and to install ground fault indicators as required. Another indicator for voltage unbalance is a 120 Hz vibration, which should prompt an immediate check of voltage balance. The typical payback period for voltage controller installation on lightly loaded motors is 3 years.

12.3.3   Energy savings

Energy savings can be achieved by either using energy efficient motors or reducing motor failures and losses, which is possible by:

1. Running the motors at rated load.
2. Reducing copper losses by increasing the volume of copper wire.
3. Reducing iron losses by using superior grade of stampings and/or longer core length.

Failure of insulation decreases the properties of materials because of heating. This decreases the efficiency by about 0.5%, every time re-insulation is done. Observations have shown that the contractors do re-insulation of motors with the materials available with them. No efficiency tests are carried out in their shop after the repairs. Hence the motors, which are repaired 5–6 times, have more losses and lower efficiency. Therefore, monitoring of motor failures and regular checks and upkeep of motors as recommended by the manufacturers and applicable standards. Regular monitoring of the working conditions and general inspection can discover faults like leakage of lubricants burning smell, higher temperature, noise, and looseness of bolts, etc. which with timely rectification could avoid motor failures in future and save energy.

12.4   Fans and blowers

Fans and blowers provide air for ventilation and industrial process requirements. Fans generate a pressure to move air (or gases) against a resistance caused by ducts, dampers, or other components in a fan system. The fan rotor receives energy from a rotating shaft and transmits it to the air. Fans fall into two general categories: centrifugal flow and axial flow. In centrifugal flow, airflow changes direction twice - once when entering and second when leaving (forward curved, backward curved or inclined, radial). In axial flow, air enters and leaves the fan with no change in direction (propeller, tubeaxial, vaneaxial).

Blower is a rotary, positive displacement type of machine used to move gas and air and is used in a variety of methods. Rotary blowers and positive displacement blowers are just two of several types that are available.

Because fan systems often directly support production processes, many fans operate continuously. These long run times translate into significant energy consumption and substantial annual operating costs.

The operating costs of large fans are often high enough that improving fan system efficiency can offer a quick payback.

1. Inspite of this, facility personnel often do not know the annual operating costs of an industrial fan, or how much money could be saved by improving fan system performance.

2. Operating costs of fan systems primarily include electricity and maintenance costs. Of these two components, electricity costs can be determined with simple measurements. In contrast, maintenance costs are highly dependent on service conditions and need to be evaluated case-by-case. A particularly useful method of estimating these costs is to review the maintenance histories of similar equipment in similar applications.

3. The cost of operating a fan system is affected by the amount of time and the percentage of full capacity (load factor) at which the fan motor operates. Because the fan system does not usually operate at rated full load all the time, an estimate of its average load factor must be made.

12.4.1 Determination of the load factor

The load factor of a fan system can be determined by listing the number of operating hours at each level of output over a typical plant cycle like one week. By multiplying the number of hours by the level of output, adding the results and dividing by the total number of hours in the entire period, one obtains the average load factor of the fan system.

12.4.2 Calculating electricity consumption

Electricity consumption can be determined by several methods, including:

1. Using motor nameplate data.
2. Using direct electrical measurements.
3. Using fan performance curve data.

In systems with widely varying operating conditions, simply taking data once will probably not provide a true indication of fan energy consumption. It is better to use data for several operating points and use a weighted average based on hours of operation at each point.

The motors used on most fans have a 1.15 continuous service factor. This means that a motor with a nominal nameplate rating of 150 brake horse power (bhp) may be operated continuously up to 172.5 bhp, although motor efficiency drops slightly above the rated load. Using nameplate data to calculate energy costs on motors that operate above their rated loads will understate actual costs. A more accurate way to determine electricity consumption requires taking electrical measurements of motor current, voltage and power factor over a range of operating conditions.

12.4.3   Matching fans/blowers to motor

Fans are usually selected to match the maximum pressure and flow requirement of the system. Although the system may not be required to operate at the maximum conditions all the time, the fan-motor combination must be capable of delivering the required air flow when needed.

1. The motor selected for the drive system must be capable of supplying the fan with the required driving power.
2. Fans are typically driven by alternating current (AC) motors. In industrial fan applications, the most common motor type is the squirrel-cage induction motor, selected for its durability, low cost, reliability and low maintenance requirements. These motors are commonly available with 2 or 4 poles which, on a 60- hertz system, translate to nominal operating speeds of 3600 revolutions per minute (rpm) and 1800 rpm, respectively.
3. Although motors with 6 poles or more are used in slower fan systems, they are relatively expensive. Motors can operate safely over long periods of time above the rated output or with a 'service factor' above 1. Service factors range from 1.1 to 1.15, meaning that the motors can safely operate at loads between 110 and 115% of their output power rating.

12.4.4   Energy saving opportunities

Minimising demand on the fan

1. Minimising excess air level in combustion systems to reduce FD fan and ID fan load.
2. Minimising air in-leaks in hot flue gas path to reduce ID fan load, especially in case of kilns, boiler plants, furnaces, etc. Cold air in-leaks increase ID fan load tremendously, due to density increase of flue gases

and in-fact choke up the capacity of fan, resulting as a bottleneck for boiler/furnace itself.

3. In-leaks/out-leaks in air conditioning systems also have a major impact on energy efficiency and fan power consumption and need to be minimised.

The findings of performance assessment trials will automatically indicate potential areas for improvement, which could be one or a more of the following:

1. Change of impeller by a high efficiency impeller along with cone.

2. Change of fan assembly as a whole by a higher efficiency fan.

3. Impeller derating (by a smaller dia impeller).

4. Change of metallic/glass reinforced plastic (GRP) impeller by the more energy efficient hollow FRP impeller with aerofoil design, in case of axial flow fans, where significant savings have been reported.

5. Fan speed reduction by pulley dia modifications for derating.

6. Option of two speed motors or variable speed drives for variable duty conditions.

7. Option of energy efficient flat belts, or, cogged raw edged V belts, in place of conventional V belt systems, for reducing transmission losses.

8. Adopting inlet guide vanes in place of discharge damper control.

9. Minimising system resistance and pressure drops by improvements in duct system.

12.4.5   Energy-efficiency in compressed air systems

Air compressor is a compressor that takes in air at atmospheric pressure and delivers it at a higher pressure. More than 85% of the electrical energy input to an air compressor is lost as waste heat, leaving less than 15% of the electrical energy consumed to be converted to pneumatic compressed air energy. This makes compressed air an expensive energy carrier compared to other energy carriers. Many opportunities exist to reduce energy use of compressed air systems. For optimal savings and performance, it is recommended that a systems approach is used. In the following, energy saving opportunities for compressed air systems are presented. Also, Energy Assessment for Compressed Air Systems (ASME) has published a standard that covers the assessment of compressed air systems that are defined as a group of subsystems of integrated sets of components for consistent, reliable, and efficient use of energy. In this standard the procedure of conducting a detailed energy assessment of the compressed air system as well as the energy efficiency opportunities are described.

1. Reduction of demand.

2. Maintenance.

3. Monitoring.

 4. Reduction of leaks (in pipes and equipment).
 5. Electronic condensate drain traps (ECDTs).
 6. Reduction of the inlet air temperature.
 7. Maximising allowable pressure dew point at air intake.
 8. Optimising the compressor to match its load.
 9. Proper pipe sizing.
10. Heat recovery.
11. Installing adjustable speed drives (ASDs).

Reduction of demand: Because of the relatively expensive operating costs of compressed air systems, the minimum quantity of compressed air should be used for the shortest possible time, constantly monitored and reweighed against alternatives.

Maintenance: Inadequate maintenance can lower compression efficiency, increase air leakage or pressure variability and lead to increased operating temperatures, poor moisture control and excessive contamination. Better maintenance will reduce these problems and save energy.

Monitoring: Maintenance can be supported by monitoring using proper instrumentation, including following:

 1. Pressure gauges on each receiver or main branch line and differential gauges across dryers, filters, etc.
 2. Temperature gauges across the compressor and its cooling system to detect fouling and blockages.
 3. Flow meters to measure the quantity of air used.
 4. Dew point temperature gauges to monitor the effectiveness of air dryers.
 5. kWh meters and hours run meters on the compressor drive.

Reduction of leaks (in pipes and equipment): Leaks cause an increase in compressor energy and maintenance costs.

The most common areas for leaks are:

1. Couplings.
2. Hoses.
3. Tubes.
4. Fittings.
5. Pressure regulators.
6. Open condensate traps and shut-off valves.
7. Pipe joints.
8. Disconnects.
9. Thread sealants.

Quick connect fittings always leak and should be avoided: In addition to increased energy consumption, leaks can make pneumatic systems/equipment less efficient and adversely affect production, shorten the life of equipment, lead to additional maintenance requirements and increased unscheduled downtime. A typical plant that has not been well-maintained could have a leak rate between 20 and 50% of total compressed air production capacity. Leak repair and maintenance can sometimes reduce this number to less than 10%. Overall, a 20% reduction of annual energy consumption in compressed air systems is projected for fixing leaks.

A simple way to detect large leaks is to apply soapy water to suspect areas. The best way is to use an ultrasonic acoustic detector, which can recognise the high frequency hissing sounds associated with air leaks.

Electronic condensate drain traps (ECDTs): Due to the necessity to remove condensate from the system, continuous bleeding, achieved by forcing a receiver drain valve to open, often becomes the normal operating practice, but is extremely wasteful and costly in terms of air leakage. ECDTs offer improved reliability and are very efficient as virtually no air is wasted when the condensate is rejected. The payback period depends on the amount of leakage reduced, and is determined by the pressure, operating hours, the physical size of the leak and electricity costs.

Reduction of the inlet air temperature: Reducing the inlet air temperature reduces energy used by the compressor. In many plants, it is possible to reduce this inlet air temperature by taking suction from outside the building. Importing fresh air has paybacks of up to 5 years, depending on the location of the compressor air inlet.

As a rule of thumb, each 3°C reduction will save 1% compressor energy use.

Maximising allowable pressure dew point at air intake: Choose the dryer that has the maximum allowable pressure dew point, and best efficiency. A rule of thumb is that desiccant dryers consume 7–14% and refrigerated dryers consume 1–2% of the total energy of the compressor.

Consider using a dryer with a floating dew point. Note that where pneumatic lines are exposed to freezing conditions, refrigerated dryers are not an option.

Optimising the compressor to match its load: Plant personnel have a tendency to purchase larger equipment than needed driven by safety margins or anticipated additional future capacity. Given the fact that compressors consume more energy during part-load operation, this is something that should be avoided. Some plants have installed modular systems with several smaller compressors to match compressed air needs in a modular way.

In some cases, the pressure required is so low that the need can be met by a blower instead of a compressor which allows considerable energy savings, since a blower requires only a small fraction of the power needed by a compressor.

Proper pipe sizing: Pipes must be sized correctly for optimal performance or resized to fit the compressor system. Inadequate pipe sizing can cause pressure losses, increase leaks and increase generating costs. Increasing pipe diameter typically reduces annual compressor energy consumption by 3%.

Heat recovery: As already mentioned, more than 85% of the electrical energy used by an industrial air compressor is converted into heat. A 150 hp compressor can reject as much heat as a 90 kW electric resistance heater or a 422 MJ/hour natural gas heater when operating.

In many cases, a heat recovery unit can recover 50–90% of the available thermal energy for space heating, industrial process heating, water heating, makeup air heating, boiler makeup water preheating, industrial drying, industrial cleaning processes, heat pumps, laundries or preheating aspirated air for oil burners. With large water-cooled compressors, recovery efficiencies of 50–60% are typical. When used for space heating, the recovered heat amount to 20% of the energy used in compressed air systems annually.

'Paybacks are typically less than one year'.

In some cases, compressed air is cooled considerably below its dew point in refrigerated dryers to condense and remove the water vapour in the air. The waste heat from these after coolers can be regenerated and used for space heating, feed water heating or process-related heating.

Installing adjustable speed drives (ASDs): When there are strong variations in load and/or ambient temperatures there will be large swings in compressor load and efficiency. In those cases, installing an ASD may result in attractive payback periods.

'Implementing adjustable speed drives in rotary compressor systems has saved 15% of the annual compressed air system energy consumption'.

Energy efficiency in boilers

13.1   Introduction

A boiler is an enclosed pressure vessel that provides means for combustion heat to be transferred into water until it becomes steam. The steam under pressure is then usable for providing heat for an industrial process. When water is boiled into steam, its volume increases about 1600 times, producing a force that is almost as explosive as gunpowder.

This makes a boiler an extremely dangerous piece of equipment that must be treated with utmost care. A boiler system comprises three parts:

1. A feed water system.
2. A steam system.
3. A fuel system.

The feed water system provides water to the boiler and regulates it automatically to meet the steam demand. Various valves provide access for maintenance and repair. The steam system collects and controls the steam produced in the boiler. Steam is directed through a piping system to the point of use. Throughout the system, steam pressure is regulated using valves and checked with steam pressure gauges. The fuel system includes all the equipment used to provide fuel to generate the necessary heat. The equipment required in the fuel system depends on the type of fuel used by the system.

13.2   Heating surfaces in a boiler

The amount of heating surface of a boiler is expressed in square meters. Any part of the boiler metal that actually contributes to making steam is a heating surface. The larger the heating surface a boiler has, the higher will be its capacity to raise steam.

Heating surfaces can be classified into several types:

1. Radiant heating surfaces (direct or primary) include all water-backed surfaces that are directly exposed to the radiant heat of the combustion flame.
2. Convection heating surfaces (indirect or secondary) include all those water-backed surfaces exposed only to hot combustion gases.
3. Extended heating surfaces include economisers and super heaters used in certain types of water tube boilers.

13.3 Boiler types and classification

Broadly, boilers can be classified into four types: fire tube boilers, water tube boilers, packaged boilers and fluidised bed combustion boilers.

13.3.1 Fire tube boilers

Fire tube boilers contain long steel tubes through which the hot gases from a furnace pass and around which the water to be converted to steam circulates. It is used for small steam capacities (up to 12000 kg/hr and 17.5 kg/cm^2). The advantages of fire tube boilers include their low capital cost and fuel efficiency (over 80%). They are easy to operate, accept wide load fluctuations and because they can handle large volumes of water, produce less variation in steam pressure. Flow diagram of fire tube boiler is shown in Fig. 13.1.

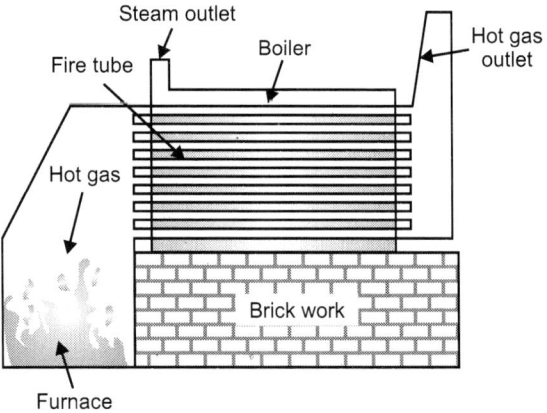

Figure 13.1: Fire tube boilers.

13.3.2 Water tube boilers

In water tube boilers, water passes through the tubes and the hot gasses pass outside the tubes. These boilers can be of single- or multiple-drum type. They can be built to handle larger steam capacities and higher pressures, and have higher efficiencies than fire tube boilers. They are found in power plants whose steam capacities range from 4.5–120 T/hr, and are characterised by high capital cost. These boilers are used when high pressure high-capacity steam production is demanded. They require more controls and very stringent water quality standards. Flow diagram of water tube boiler is shown in Fig. 13.2.

13.3.3 Packaged boilers

The packaged boiler is so called because it comes as a complete package. Once delivered to a site, it requires only steam, water pipe work, fuel supply,

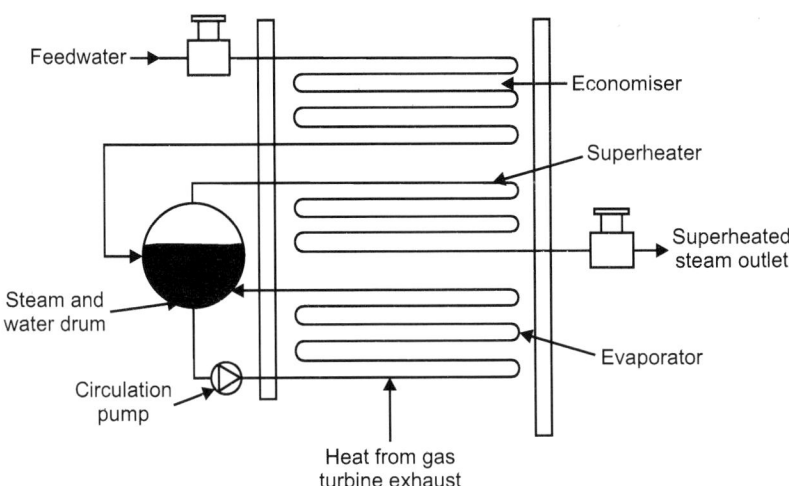

Figure 13.2: Water tube boilers.

and electrical connections in order to become operational. Package boilers are generally of shell type with fire tube design so as to achieve high heat transfer rates by both radiation and convection. These boilers are classified on the basis of the number of passes (the number of times the hot combustion gases pass through the boiler). The combustion chamber is taken as the first pass, after which there may be one, two, or three sets of fire tubes. The most common boiler of this class is a three-pass unit with two sets of fire tubes and with the exhaust gases exiting through the rear of the boiler.

13.3.4   Fluidised bed combustion (FBC) boilers

In fluidised bed boilers, fuel burning takes place on a floating (fluidised) bed in suspension. When an evenly distributed air or gas is passed upward through a finely divided bed of solid particles such as sand supported on a fine mesh, the particles are undisturbed at low velocity. As air velocity is gradually increased, a stage is reached when the individual particles are suspended in the air stream. A further increase in velocity gives rise to bubble formation, vigorous turbulence, and rapid mixing, and the bed is said to be fluidised. Fluidised bed boilers offer advantages of lower emissions, good efficiency, and adaptability for use of low calorific-value fuels like biomass, municipal waste, etc.

13.4   Performance evaluation of boilers

The performance of a boiler, which include thermal efficiency and evaporation ratio (or steam to fuel ratio), deteriorates over time for reasons that include poor combustion, fouling of heat transfer area, and inadequacies in operation and maintenance. Even for a new boiler, deteriorating fuel quality and water quality can result in poor boiler performance. Boiler efficiency tests help us to calculate deviations of boiler efficiency from the design value and identify areas for improvement.

13.4.1 Thermal efficiency

Thermal efficiency of a boiler is defined as the percentage of heat input that is effectively utilised to generate steam. There are two methods of assessing boiler efficiency – direct and indirect. In the direct method, the ratio of heat output (heat gain by water to become steam) to heat input (energy content of fuel) is calculated. In the indirect method, all the heat losses of a boiler are measured and its efficiency computed by subtracting the losses from the maximum of 100.

13.4.2 Evaporation ratio

Evaporation ratio, or steam to fuel ratio, is another simple, conventional parameter to track performance of boilers on-day-to-day basis. For small capacity boilers, direct method can be attempted, but it is preferable to conduct indirect efficiency evaluation, since an indirect method permits assessment of all losses and can be a tool for loss minimisation. In the direct method, steam quality measurement poses uncertainties. Standards can be referred to for computations and methodology of evaluation.

Example of direct efficiency calculation:

Calculate the efficiency of the boiler from the following data:

Type of boiler	:	Coal-fired
Quantity of steam (dry) generated	:	8 TPH
Steam pressure (gauge)/temp	:	10 Kg/cm^2 (g)/180°C
Quantity of coal consumed	:	1.8 TPH
Feed water temperature	:	85°C
GCV of coal	:	3200 kcal/kg
Enthalpy of steam at 10 kg/cm^2 (g) pressure	:	665 kcal/kg (saturated)
Enthalpy of inlet fed water	:	85 kcal/kg

$$\text{Boiler efficiency } (\eta) = \frac{8 \text{ TPH} \times 1000 \text{ Kg} \times (665 - 85) \times 100}{1.8 \text{ TPH} \times 1000 \text{ Kg} \times 3200}$$

$$= 80.0 \%$$

Evaporation ratio = 8 TPH of steam/1.8 TPH of coal = 4.4

13.5   Boiler water treatment

Boiler water treatment is an important area for attention since water quality has a major influence on the efficiency of a boiler as well as on its safe operation. The higher the pressure rating, the more stringent the water quality requirements become. Boiler water quality is continuously monitored for buildup of total dissolved solids (TDS) and hardness, and blow down is carried out (involving heat loss) to limit the same.

Boiler water treatment methods are dependent upon quality limits specified for TDS and hardness by the manufacturers, the operating pressure of the boiler, the extent of make-up water used, and the quality of raw water at the site. For small-capacity and low-pressure boilers, water treatment is carried out by adding chemicals to the boiler to prevent the formation of scale, and by converting the scale-forming compounds to free-flowing sludge, which can be removed by blow down.

Limitations: Treatment is applicable to boilers where feed water is low in hardness salts, where low pressure–high TDS content in boiler water is tolerated, and where only small quantities of water need to be treated. If these conditions are not met, then high rates of blow down are required to dispose of the sludge, and treatment become uneconomical based on heat and water loss considerations.

Chemicals used: Sodium carbonate, sodium aluminate, sodium phosphate, sodium sulphite, and compounds of vegetable or inorganic origin are used for treatment. Internal treatment alone is not recommended.

13.6   Parameters for selection of boilers

Steam boiler is a very important equipment for all process industries. There are many codes in use for design of boilers internationally. All these codes mainly take care of safety aspects of boilers from angle of mechanical strength. Some codes stipulate norms for furnace sizing on thermal input basis. Many users who have limited knowledge of boilers tend to believe that any two boilers designed as per same design code are technically at par. This is far from the truth. In today's modern world, mechanical strength is only one of the many criteria, which decides the superiority of any boiler. There are many other more important aspects like efficiency, availability round the clock, ease in maintenance, environmental compliance, etc. This section provides guidelines for any boiler user to evaluate various brands of boilers and quantify the strengths/weaknesses. The evaluation criteria and its importance are explained in brief as under.

13.6.1 Safety and reliability

Apart from mechanical strength, it is the control logic and instrumentation, which decides safety and reliability of any modern boiler. Few of the important aspects are discussed below.

Number of boiler water level controllers

Keeping proper water level in the boiler is of paramount importance from boiler safety point of view. This instrument not only maintains necessary operating water level by controlling the water inflow, but also ensures burner stoppage in case of the level falling below safe limit. It is advisable to have two instruments, considering the criticality of the function.

Number of fusible plugs

Fusible plug avoids dry running of a boiler by sparging high-pressure water in the furnace when water level goes below the topmost area of radiation heat transfer zone. This is the ultimate safety device which can save furnace from collapse and rupture due to dry running. It generally consists of three parts where the innermost and outermost parts are held together with 'low melting point alloy metal'. In case of dry running, this part melts creating an opening through which water in the boiler can enter the furnace extinguishing flame. It is advisable to have two fusible plugs.

Tube overheat controller

This works as an overriding control in case the water level controller does not function and the burner keeps operating inspite of very low water level. It senses the temperature of flue in the topmost row of tube. When the level drops down, this row gets exposed and flue gas temperature in these tubes rises much higher than the bulk temperature in such eventuality, this controller sounds an alarm and can also stop the burner depending on the logic.

High stack alarm controller

The stack temperature is an indicator of fouling of heat transfer surfaces in the boiler from flue and waterside. This not only results into higher fuel consumption but also overheating of tubes and furnace (in case of waterside fouling). This instrument sounds an alarm in such conditions, cautioning the operator to clean the surfaces.

Sinking time calculation

Sinking time is the time required to lower the water level in the boiler from normal working level to the furnace crown when the feed water pump fails and burner keeps firing at high flame due to failure of all safety devices. The furnace is subjected to very high temperature flame and hence is the most critical component of boiler. In case of dry running the furnaces become the first failure points. Boilers with bottom furnace type design have much higher sinking time than those having furnaces on one side. This gives more time for corrective action in a crisis, thereby avoiding damage to the furnace and possibility of an accident.

Fuel pressure monitoring system

Most of modern oil fired boilers use pressure jet burners. It is necessary to maintain fuel pressure above the minimum desired limit to ensure atomisation of fuel and complete combustion. Fuel pressure sensing system should be provided for tripping the burner in case the fuel pressure falls below the safe limit.

Fuel temperature monitoring system

For heavy oil fired boilers, the fuel needs to be heated to reduce viscosity and improve atomisation. Low fuel temperature can result in incomplete combustion, unstable flame and backfiring. Fuel temperature monitoring system should stop the burner firing below safe temperature.

Combustion air pressure monitoring system

This will ensure availability of air for combustion. Unavailability/shortage of air results in similar situations mentioned above. The burner should trip automatically in case air is not sufficiently available.

Steam pressure modulation

Steam pressure tends to change due to fluctuations in demand from plant. Immediate correction in fuel firing rate is necessary to maintain steady fuel pressure. Stepless or continuous modulation adjusts the fuel input constantly by checking steam pressure feedback. High-low or step modulation adjusts the fuel in stages.

Stepless modulation can maintain steam pressure on the boiler within a tolerance of $0.1–0.2$ kg/cm^2. With step or high-low type of modulation, you can expect variation of $1.0–1.5$ kg/cm^2. Above is subject to steam demand being lower than boiler capacity at any given time.

Steam pressure limit switch

If the steam demand drops to a very low level, the steam pressure rises inspite of burner firing at minimum possible level. Steam pressure limit switch cuts off the burner and eliminates possibility of safety valve popping up, saving precious fuel.

Safety valves

Safety valves release steam without any need for electronic signal from instruments. This is a very important device and is a must as per all codes. The release capacity should be more than that of the steam generation capacity of boiler.

Automatic blow down/continuous blow down

Salts in feed water does not evaporate with steam and hence the concentration in boiler keeps on increasing. It is essential to drain the highly concentrated boiler water and add some extra feed water, which has comparatively much lower Total Dissolved Solids (TDS). Conventionally, the operator used to give blow done a periodic intervals in full day.

This is done 3 to 6 times a day. In this process, the TDS in boiler keeps fluctuating. If the operator delays the blow down, the TDS increases beyond acceptable limits resulting in salts getting deposited on tubes and the furnace. In automatic blow down steam, the TDS in boiler is sensed constantly and the opening of blow down valve is adjusted to maintain the TDS of boiler water below desired limits. Anyhow, proper care has to be taken while selecting the automatic blow down system since there are very few systems which have performed in the field without problems. The automatic systems would be expensive for smaller sizes of boiler. One can always have a continuous blow down system where blow down valve can be adjusted by calculating the per cent blow down required based on boiler load and feed water TDS. It can be adjusted considering full capacity of boiler. A heat recovery system will save the heat going out from boiler due to blow down.

Boiler water TDS meter/conductivity meter

Considering importance of the TDS level in the boiler, the boiler water needs to be checked for TDS periodically. Necessary meters should be provided to the operator for this job.

Furnace water TDS meter

TDS of water in the tubes of water walled furnace (mainly for solid fuel fired boiler) is always much higher than the drum. On line separate TDS sensing arrangement can be very useful for such boilers.

Furnace draft alarm for solid fuel boilers

The furnaces of solid fuel fired boilers are kept at a slight negative pressure (on flue side) to avoid flame, hot flue gas coming out from firing doors and the fuel feeding system. Alarm can be provided to provide warning of higher pressure than desired.

Pilot flame ignition

For gaseous fuels, pilot flame is essential to ensure flame stability during ignition. It is more so in case of lean gases like biogas. In case of liquid fuels, burner with rotary, steam and air atomisation are generally provided with pilot flame ignition.

Electrical panel with fuses, O/L relays, ELCBs, earthings, etc.

The electricals are equally important. Panel should be equipped with fuses and O/L relays for all motors. Miniature Circuit Breaker (MCB) is necessary for the safety of electrical components. Earth Leakage Circuit Breaker (ELCB) provides safety to the operating staff.

Instrumentation for measurement of parameters related to safety of a boiler

Few parameters are very important for safety of a boiler. Instruments need to be provided for measurement of these parameters. On-line instruments are preferred since they provide continuous data. Records connected to these instruments create record of data round the clock which can be very useful in trouble shooting. Off-line instruments have to be used by operating staff for periodic monitoring of these parameters.

Operator discipline is very crucial for these instruments to be effectively used. Parameters such as feed water TDS, boiler water TDS, boiler water level, steam pressure, stack temperature, and tube overheat temperature are very important.

13.6.2 Environmental compliance

Environmental aspects are becoming more important day by day. Generally local pollution control boards have limits specified for polluting elements in flue gases. Constituents such as CO, NO_x, particulate matters, SO_2/SO_3, hydrocarbons, in flue gases should be measured and must be below the limits specified by the pollution control boards.

Following aspects are equally important even though many of them do not get covered under any statutory requirements:

1. Boiler water TDS and treatment before disposal.
2. No fuel oil spillage/proper spillage recovery (for oil fired boilers).
3. Noise level in boiler house.
4. Normal boiler house ambient.
5. Proper soot disposal system while tube cleaning and after tube cleaning.
6. Ash disposal system (for solid fuel boilers).
7. Lighting and illumination in boiler house.

8. Fire extinguishers in boiler house.
9. First aid kit in boiler house.
10. Space for operator movement.

13.6.3 Availability to user for maintenance without stoppage

Companies without a standby boiler need to look at this aspect in detail. Facilities for maintenance of components without stoppage of boiler can save the investment of a standby boiler. It is necessary to have facility to carry maintenance work on components like water pump, fuel pump, fuel oil heater, filters for fuel and water, and instruments (viz., pressure gauge and temperature gauge), without stopping the boiler operation.

Mechanical and manual cleaning convenience

Convenience for cleaning saves time required for preventive maintenance shutdown. Convenience can be categorised in three areas, viz, time, manpower, and efforts. These should be evaluated for both 'water side' and 'flue gas side' of any boiler.

Chemicals cleaning convenience

On some occasions chemical cleaning is required to be done. This eliminates opening of boiler and saves time. Hence provisions for such cleaning should be provided.

Repair convenience

In some designs, this aspect is completely neglected. This is very important criterion. If proper care is not taken while designing, repairs can be very expensive and unaffordable. Pressure and non-pressure parts, tubes and furnace should be studied carefully from this point of view.

Operating convenience

If it is inconvenient to carry out certain function for the operator, there is tendency to skip that particular operation, which can result in accidents or inefficient operation. Visibility of instruments and ease of access for observation are the two important factors aiding operating convenience.

Instrumentation reliability

Providing lot of instrumentation can be counter effective if the instruments are not reliable.

The following factors can ensure reliability of instruments:

1. Instrument manufacturers certified by instrument societies.

2. Calibration certificate available.
3. Repair convenience.
4. Replacement convenience.
5. Control panel reliability:
   (a) Margins on power ratings of instruments.
   (b) Dust proof enclosure.
   (c) Control panel architecture.
   (d) Maintenance convenience.

13.6.4 Trouble shooting logic diagnostics and support

Trouble shooting can be nightmare if lot of interlocks are provided without visual indications on panel.

It can become very easy with:

1. On-line logic analyser.
2. Data acquisition and control systems.
3. Indication for all parameter status.
4. Audio/visual alarms.

Life expectancy

Every purchaser can use his own yardstick for this aspect. Depending on the industry, market conditions and many other factors this can change. But nevertheless it is a very important aspect which needs to be deliberated upon before making a final decision.

Space/dimensions/weight

The cost of installation does not involve only boiler. The civil and steel structural requirements, cost of land occupied must be considered while evaluating commercially.

Site start-up time

The investment in boiler can start paying only after the same is commissioned. Amount of site work decides the time required for starting after the boiler reaches site.

Transportability

Transport costs can escalate appreciably if the shape and size of each individual component is such that it is difficult to transport. This is very significant if user is far away from the manufacturer.

Aesthetics

Even though this does not help in day to day functioning directly, good aesthetics can have positive psychological effect on operating staff.

External facility dependence

A system design which demands many external facilities result in high initial and running cost. Typically with proper designing of boiler system, one can work without such extra facilities like fuel ring main.

Water treatment system simplicity

Different designs of boilers have varying requirements of feed water. A boiler design, which does not demand for very stringent water quality norms, saves initial and running cost.

Thus, steam boiler selection can be done after evaluating the technical merits on various aspects. Proper selection after detailed study can avoid problems during use of boiler. Boiler being a capital equipment is not procured on routine basis. Hence analysis of all minor and major aspects mentioned above can provide necessary inputs for selection.

13.7　Energy conservation opportunities

The various energy efficiency opportunities in boiler system can be related to combustion, heat transfer, avoidable losses, high auxiliary power consumption, water quality and blow down.

Examining the following factors can indicate if a boiler is being run to maximise its efficiency:

Stack temperature: The stack temperature should be as low as possible. However, it should not be so low that water vapour in the exhaust condenses on the stack walls. This is important in fuels containing significant sulphur as low temperature can lead to sulphur dew point corrosion. Stack temperatures greater than 200°C indicates potential for recovery of waste heat. It also indicate the scaling of heat transfer/recovery equipment and hence the urgency of taking an early shut down for water/flue side cleaning.

Feed water preheating using economiser: Typically, the flue gases leaving a modern 3-pass shell boiler are at temperatures of 200–300°C. Thus, there is a potential to recover heat from these gases. The flue gas exit temperature from a boiler is usually maintained at a minimum of 200°C, so that the sulphur oxides in the flue gas do not condense and cause corrosion in heat transfer surfaces. When a clean fuel such as natural gas, LPG or gas oil is used, the economy of heat recovery must be worked out, as the flue gas temperature may be well below 200°C.

The potential for energy saving depends on the type of boiler installed and the fuel used. For a typically older model shell boiler, with a flue gas exit temperature of 260°C, an economiser could be used to reduce it to 200°C, increasing the feed water temperature by 15°C. Increase in overall thermal efficiency would be in the order of 3%. For a modern 3-pass shell boiler firing natural gas with a flue gas exit temperature of 140°C a condensing economiser would reduce the exit temperature to 65°C increasing thermal efficiency by 5%.

Combustion air preheat: Combustion air preheating is an alternative to feed water heating. In order to improve thermal efficiency by 1%, the combustion air temperature must be raised by 20°C. Most gas and oil burners used in a boiler plant are not designed for high air preheat temperatures. Modern burners can withstand much higher combustion air preheat, so it is possible to consider such units as heat exchangers in the exit flue as an alternative to an economiser, when either space or a high feed water return temperature make it viable.

Incomplete combustion: Incomplete combustion can arise from a shortage of air or surplus of fuel or poor distribution of fuel. It is usually obvious from the colour or smoke, and must be corrected immediately.

In the case of oil and gas fired systems, CO or smoke (for oil fired systems only) with normal or high excess air indicates burner system problems. A more frequent cause of incomplete combustion is the poor mixing of fuel and air at the burner. Poor oil fires can result from improper viscosity, worn tips, carbonisation on tips and deterioration of diffusers or spinner plates.

With coal firing, unburned carbon can comprise a big loss. It occurs as grit carry-over or carbon-in-ash and may amount to more than 2% of the heat supplied to the boiler. Non uniform fuel size could be one of the reasons for incomplete combustion. In chain grate stokers, large lumps will not burn out completely, while small pieces and fines may block the air passage, thus causing poor air distribution. In sprinkler stokers, stoker grate condition, fuel distributors, wind box air regulation and over-fire systems can affect carbon loss. Increase in the fines in pulverised coal also increases carbon loss.

Excess air control: Table 13.1 gives the theoretical amount of air required for combustion of various types of fuel. Excess air is required in all practical cases to ensure complete combustion, to allow for the normal variations in combustion and to ensure satisfactory stack conditions for some fuels. The optimum excess air level for maximum boiler efficiency occurs when the sum of the losses due to incomplete combustion and loss due to heat in flue gases is minimum.

This level varies with furnace design, type of burner, fuel and process variables. It can be determined by conducting tests with different air fuel ratios. Typical values of excess air supplied for various fuels are given in Table 13.2.

Table 13.1: Theoretical combustion data–common boiler fuels.

Fuel	Kg of air req./kg of fuel	Kg of flue gas/kg of fuel	m³ of flue/kg of fuel	Theoretical CO₂% in dry flue gas	CO₂% in flue gas achieved in practice
Solid fuels					
Bagasse	3.2	3.43	2.61	20.65	10–12
Coal (bituminous)	10.8	11.7	9.4	18.7	10–13
Lignite	8.4	9.1	6.97	19.4	9–13
Paddy husk	4.6	5.63	4.58	19.8	14–15
Wood	5.8	6.4	4.79	20.3	11.13
Liquid fuels					
Furnace oil	13.9	14.3	11.5	15	9–14
LSHS	14.04	14.63	10.79	15.5	9–14

Table 13.2: Excess air levels for different fuels.

Fuel	Type of furnace or burners	Excess air (% by wt)
Pulverised coal	Completely water-cooled furnace for slag-tap or dry-ash removal	15–20
	Partially water-cooled furnace for dry-ash removal	15–40
Coal	Spreader stoker	30–60
	Water-cooler vibrating-grate stokers	30–60
	Chain-grate and travelling-gate stokers	15–50
	Underfeed stoker	20–50
Fuel oil	Oil burners, register type	15–20
	Multi-fuel burners and flat-flame	20–30
Natural gas	High pressure burner	5–7
Wood	Dutch over (10–23% through grates) and Hofft type	20–25
Bagasse	All furnaces	25–35
Black liquor	Recovery furnaces for draft and soda-pulping processes	30–40

Controlling excess air to an optimum level always results in reduction in flue gas losses; for every 1% reduction in excess air there is approximately 0.6% rise in efficiency.

Various methods are available to control the excess air:

1. Portable oxygen analysers and draft gauges can be used to make periodic readings to guide the operator to manually adjust the flow of air for optimum operation. Excess air reduction up to 20% is feasible.

2. The most common method is the continuous oxygen analyser with a local readout mounted draft gauge, by which the operator can adjust air

flow. A further reduction of 10–15% can be achieved over the previous system.

3. The same continuous oxygen analyser can have a remote controlled pneumatic damper positioner, by which the readouts are available in a control room. This enables an operator to remotely control a number of firing systems simultaneously.

The most sophisticated system is the automatic stack damper control, whose cost is really justified only for large systems.

Radiation and convection heat loss: The external surfaces of a shell boiler are hotter than the surroundings. The surfaces thus lose heat to the surroundings depending on the surface area and the difference in temperature between the surface and the surroundings. The heat loss from the boiler shell is normally a fixed energy loss, irrespective of the boiler output. With modern boiler designs, this may represent only 1.5% on the gross calorific value at full rating, but will increase to around 6%, if the boiler operates at only 25% output. Repairing or augmenting insulation can reduce heat loss through boiler walls and piping.

Automatic blow down control: Uncontrolled continuous blow down is very wasteful. Automatic blow down controls can be installed that sense and respond to boiler water conductivity and pH. A 10% blow down in a 15 kg/cm^2 boiler results in 3% efficiency loss.

Reduction of scaling and soot losses: In oil and coal-fired boilers, soot buildup on tubes acts as an insulator against heat transfer. Any such deposits should be removed on a regular basis. Elevated stack temperatures may indicate excessive soot buildup. Also same result will occur due to scaling on the water side.

High exit gas temperatures at normal excess air indicate poor heat transfer performance. This condition can result from a gradual build-up of gas-side or waterside deposits. Waterside deposits require a review of water treatment procedures and tube cleaning to remove deposits. An estimated 1% efficiency loss occurs with every 22°C increase in stack temperature.

Stack temperature should be checked and recorded regularly as an indicator of soot deposits. When the flue gas temperature rises about 20°C above the temperature for a newly cleaned boiler, it is time to remove the soot deposits. It is, therefore, recommended to install a dial type thermometer at the base of the stack to monitor the exhaust flue gas temperature. It is estimated that 3 mm of soot can cause an increase in fuel consumption by 2.5% due to increased flue gas temperatures. Periodic off-line cleaning of radiant furnace surfaces, boiler tube banks, economisers and air heaters may be necessary to remove stubborn deposits.

Reduction of boiler steam pressure: This is an effective means of reducing fuel consumption, if permissible, by as much as 1 to 2%. Lower steam pressure gives a lower saturated steam temperature and without stack heat recovery, a similar reduction in the temperature of the flue gas temperature results. Steam is generated at pressures normally dictated by the highest pressure/temperature requirements for a particular process. In some cases, the process does not operate all the time, and there are periods when the boiler pressure could be reduced. The energy manager should consider pressure reduction carefully, before recommending it. Adverse effects, such as an increase in water carryover from the boiler owing to pressure reduction, may negate any potential saving. Pressure should be reduced in stages, and no more than a 20% reduction should be considered.

Variable speed control for fans, blowers and pumps: Variable speed control is an important means of achieving energy savings. Generally, combustion air control is effected by throttling dampers fitted at forced and induced draft fans. Though dampers are simple means of control, they lack accuracy, giving poor control characteristics at the top and bottom of the operating range. In general, if the load characteristic of the boiler is variable, the possibility of replacing the dampers by a VSD should be evaluated.

Effect of boiler loading on efficiency: The maximum efficiency of the boiler does not occur at full load, but at about two-thirds of the full load. If the load on the boiler decreases further, efficiency also tends to decrease. At zero output, the efficiency of the boiler is zero, and any fuel fired is used only to supply the losses.

The factors affecting boiler efficiency are:

1. As the load falls, so does the value of the mass flow rate of the flue gases through the tubes. This reduction in flow rate for the same heat transfer area, reduced the exit flue gas temperatures by a small extent, reducing the sensible heat loss.
2. Below half load, most combustion appliances need more excess air to burn the fuel completely. This increases the sensible heat loss.

In general, efficiency of the boiler reduces significantly below 25% of the rated load and as far as possible, operation of boilers below this level should be avoided.

Proper boiler scheduling: Since the optimum efficiency of boilers occurs at 65–85% of full load, it is usually more efficient, on the whole, to operate a fewer number of boilers at higher loads, than to operate a large number at low loads.

Boiler replacement: The potential savings from replacing a boiler depends on the anticipated change in overall efficiency.

A change in a boiler can be financially attractive if the existing boiler is:
1. Old and inefficient.
2. Not capable of firing cheaper substitution fuel.
3. Over or undersized for present requirements.
4. Not designed for ideal loading conditions.

The feasibility study should examine all implications of long-term fuel availability and company growth plans. All financial and engineering factors should be considered. Since boiler plants traditionally have a useful life of well over 25 years, replacement must be carefully studied.

Heating, ventilation and air conditioning systems

14.1 Introduction

Heating, Ventilating, and Air Conditioning (HVAC) equipment perform heating and/or cooling for residential, commercial or industrial buildings. The HVAC system may also be responsible for providing fresh outdoor air to dilute interior airborne contaminants such as odours from occupants, volatile organic compounds (VOC's) emitted from interior furnishings, chemicals used for cleaning, etc. A properly designed system will provide a comfortable indoor environment year round when properly maintained. Whilst energy efficiency optimisation is becoming an increasingly important business strategy for managing costs and supporting environmental compliance, the way in which Heating, Ventilation and Air Conditioning (HVAC) systems are used could be thwarting companies' best intentions to save energy and money.

HVACs are subjected to more misuse than any other type of equipment in business sector. Poor maintenance, lack of knowledge on how to use them efficiently, overuse, and the large number of old and inefficient systems at work in the sector, make HVACs a significant contributor to the country's demand for energy. By virtue of their energy intensiveness, HVACs account for a large percentage of companies' energy costs; continued misuse can dramatically impact the effectiveness of companies' energy-saving projects and initiatives, which are mostly driven by the objective to reduce operating costs. Replacing energy inefficient systems with energy-efficient alternatives, and using HVACs sparingly and smartly, offer huge opportunities for companies to save electricity in many instances, these opportunities reduce operating costs and remain untapped.

14.2 Tackling costs

By targeting HVAC systems as part of an overall energy management plan, electricity smart companies can ensure continued comfort and safety of their workers without incurring dramatic increases in energy costs. Using the advice given below will help to optimise the energy efficiency of HVAC systems in building by:

1. Making regular maintenance a priority.
2. Doing continuous monitoring.

3. Ensuring electricity smart operating.
4. Adopting energy-saving behaviours and practices.
5. Switching to energy-efficient technologies.

14.3 Unpacking HVAC

Heating, Ventilation and Air Conditioning (HVAC) systems control temperature, humidity and quality of air in a building to a set of chosen or preferred conditions. To achieve this, systems need to transfer heat and moisture into and out of the air and control the level of air pollutants, either by directly removing it or by diluting it to acceptable levels.

1. Heating is needed to increase the temperature in a space to compensate for heat loss.
2. Ventilation is needed to supply air to a space and extract polluted air from it.
3. Cooling is needed to lower temperature in a space where heat gains are caused by the sun, activity of people and the function of equipment.

HVAC systems vary widely in terms of size, functions they perform and the amount of energy they consume. Factors that influence energy usage include:

1. The design, layout and operation of a building, affects how the external environment impacts on internal temperature and humidity levels.
2. HVAC systems will use more energy when the required indoor temperature and air quality in extreme temperatures or in the case of operations where greater precision or more refined air quality is required.
3. The heat generated internally by lighting, equipment and people - all have an impact on how warm the building is, and the load on the HVAC system.
4. The design and efficiency of older HVAC system tend to be less energy-efficient.
5. How, when and for how long the HVAC system is operated every day.
6. How well the HVAC system is monitored and maintained.

14.4 Effective maintenance

Proper maintenance of system components keep HVACs operating at peak efficiency implement a maintenance programme to ensure that all components including motors, pumps, fans, compressors, ducting and filters are intact and working effectively. This not only conserves energy but also helps to extend equipment life and prevent costly breakdowns.

Maintenance should be performed continuously on a regular scheduled basis - simple maintenance procedures can be undertaken in-house whilst more comprehensive maintenance interventions should be carried out by a qualified service provider - simple maintenance procedures include:

1. Cleaning heat exchange surfaces.
2. Inspecting ductwork for air leakages - seal all leaks by taping or caulking.
3. Inspecting ductwork insulation - repair or replace as necessary.
4. Inspecting damper blades and linkages - adjust on a regular basis and clean the oil.
5. Cleaning or replacing air filters.
6. Inspecting and cleaning coils.
7. Inspecting coils and casings for leakage - seal all leaks.
8. Inspecting all room air outlets and inlets (diffusers, registers and grilles) - these should be kept clean and free of dirt and obstructions.
9. Lubricating motor and drive bearings.
10. Checking for over-voltage or low-voltage conditions on motors.
11. Checking excessive noise and vibration.
12. Keeping fan blades clean.
13. Inspecting piping for leakage at joints - repair as and when necessary.
14. Inspecting strainers and cleaning regularly.
15. Inspecting vents and remove all clogs - clogged vents retard efficient air elimination and reduce system efficiency.
16. Replacing leaking dampers on ventilation systems.
17. Ensuring condensing and evaporating devices are clean and well maintained - check that condensers are not obstructed by equipment or vegetation, for example.
18. Ensuring that the cooling plant is regularly maintained to avoid reduced levels of operational efficiency.
19. Replacing insulation on refrigerant pipework and paying specific attention to pipework located outside a building - insulation in poor conditions will affect the temperature of the refrigerant flowing through the system and therefore, consumes more energy in maintaining the required temperature.

14.5   Smart operating of electricity

Controlling how, when and why HVAC systems are operated can save significant amounts of electricity.

14.5.1 Minimise the temperature difference

Air conditioning systems will use more electricity when needed to maintain an internal temperature that is lower than the outside temperature - keep the thermostat within a 10-degree range of the outside temperature.

In summer, set the average building temperature to 23°C. In winter, maintain it in the 'golden zone' between 18 and 22°C.

14.5.2 Reduce the cooling load

The amount of electricity air conditioning systems use also depends on the cooling load - the amount of heat the system has to remove.

Reduce the cooling load on systems by:

1. Installing variable speed drives on HVAC fans and pumps: This allows motor-driven loads (such as fans and pumps) to operate in response to varying load requirements instead of simply operating in 'on/off' mode.

2. Insulating the cooled space: Implementing various measures such as ceiling insulation, window glazing, blinds, awnings and door sweeps will contribute to creating a thermally-efficient shell that can dramatically reduce the cooling load on HVAC systems whilst ensuring that comfortable internal temperatures are maintained.

3. Reducing warm air filtration into the cooled space: Keep windows and doors closed when HVAC systems are in use.

4. Minimising the use of appliances and lighting: Emitting heat, lights and equipment that are not required at any particular time should be switched off to help reduce the cooling load.

14.5.3 Adjust control set points

Proper maintenance control is essential for optimal HVAC system operation. Situations may prevail where the existing controls are not appropriate or are not capable of controlling the systems properly - a symptom of this may be as simple as a thermostat that fails to effectively control comfort levels in an occupied space. Choose HVAC units with thermostat controls and programmable timers - programmable thermostats can also be installed in place of standard thermostats.

14.5.4 Raise evaporator temperature (suction pressure)

The amount of electricity demanded by an air conditioning compressor, is also determined by the difference between the evaporator and the condenser temperature (or pressure) - if the system can tolerate a little increase in temperature at the evaporator, an opportunity exists to reduce compressor power. Consult an air-conditioner specialist to check if your system will tolerate an adjustment to the evaporator temperature.

14.5.5 Lower condensing temperature (discharge pressure)

An opportunity to lower compressor power exists through lowering the condensing temperature. However, since compressors are fine tuned systems, be cautious when considering adjustments to operating conditions. Consult an air-conditioner specialist to check if your system will tolerate an adjustment to the condensing temperature.

14.5.6 Keep units cool

Providing cooler air to condensers can help compressors operate more efficiently. Rooftop cooling units containing compressors and condensers generally draw air near the rooftop; cooler air may be available just a bit higher - at 1.2 to 1.5 meters off the roof.

Keep units cool by:

1. Installing condenser units on the southern side of a building.
2. Positioning units to avoid direct sunlight during the day.
3. Ensuring sufficient natural cross-ventilation over the condenser unit to remove any hot air more effectively.
4. Avoiding installing condenser units in closed or confined areas where hot air might accumulate.
5. Providing enough vents for hot air to escape.
6. Installing whirley birds to remove hot air efficiently.

14.5.7 Set time controls to match occupancy

Check that controls are appropriately set and displaying the correct time and date - adjust if necessary to ensure heating and cooling only operate when and where required. Optimal start- and stop-controls can be used to minimise after hours operation.

Don't allow heating and cooling at the same time

Set controls to give a wide temperature gap at which heating and cooling systems turn on - a gap of around 4 to 5°C between the heating and cooling thermostat set points will create a comfortable 'dead band'. This will help to keep occupants comfortable and increase cost savings. Unless this measure is implemented, both systems may operate simultaneously and waste electricity and money.

Myth: Turning air conditioning thermostats down as low as possible, cools the building more quickly.

Reality: Temperature drops at the same rate but then 'overshoots', making it uncomfortable for staff and wasting electricity at the same time. If controls are not coordinated, the temperature could even go low enough for the heating system to be switched on.

Remedy: Set thermostats correctly and educate staff to dispel this myth. Where possible, as a last resort, protect thermostats and prevent tampering.

14.6 Reducing dependency on HVAC systems

By harnessing natural 'free energy' to heat, cool and ventilate a building, companies can reap substantial energy cost savings - it is not always necessary for HVAC systems to operate all the time. Making the most of natural ventilation is a simple and cost effective way to achieve substantial energy cost savings. This requires taking control of heat from the sun with ventilation so that nature provides a majority of fresh air and manages temperature levels. Natural ventilation relies on air flow through openings into a room or building, preferably from opposite sides - this also applies to rising hot air being replaced with cooler air sucked in through windows or vents from a lower level.

When cooling is required in a building - if it is cooler outside than inside - simply open doors, vents and windows. This will increase air flow, reduce heat and, possibly, provide all the ventilation that is needed.

Air-conditioner fans can also be used to draw in and circulate, cool air from outside during early morning hours. Some buildings are equipped with mixed mode systems, which use a combination of both natural and mechanical systems - mechanical systems kick-in only when needed.

14.6.1 Reduce overheating

Before installing cooling equipment, always identify where excess heat is coming from - sunlight, office equipment, lighting and refrigeration are often the main sources. Consider shading windows on the outside or replacing window panes with special heat reflective glass to prevent heat buildup. Alternatively, internal daylight blinds enable natural light to enter offices by redirecting it onto the ceiling, thereby alleviating any discomfort felt by the occupants from direct daylight - many daylight blinds also have perforated blades to not obstruct employees' view of the outside.

14.6.2 Night cooling

This is usually done in places with low humidity, an established technique where cool night air is passed through a building to remove heat that has accumulated during the day. Although more heat will be absorbed the following day, a cooled building fabric will produce lower internal temperatures. The movement of cool night air may be natural or fan assisted a technique that leads to a reduction in HVAC use and lower energy costs.

14.6.3   Consider zoning your building and save

Many buildings are serviced by an overall HVAC system, while having problematic areas with different time and temperature requirements. A solution is to 'zone' your building by installing separate time and temperature controls for individual areas - zoned areas can provide better conditions as occupants will have greater control over their respective environments. It is also an effective energy cost saving measure as HVACs can then be turned down or off in unused or unoccupied zones.

14.6.4   Use air-conditioners on demand

Another effective energy cost saving measure is to only use air-conditioners in occupied rooms. The best way to ensure that air-conditioners are not left on unnecessarily in unoccupied rooms is by installing occupancy or movement sensors. These detect when someone enters a room and then activate the system. Once the room has been left vacant for a certain period of time, the air-conditioner shuts down.

As with lights, air-conditioners are often left on unnecessarily when people leave offices and meeting rooms - everyone expects that someone else will do the switching off but, more often than not, nobody does.

With incidences of power outages late afternoons just before closing time, people leave work for home forgetting that air-conditioners were running before the power outage. This causes air-conditioners to spring back to life once power is restored and, therefore, consume electricity unnecessarily throughout the night.

14.7   Saving energy through behaviour change

Good housekeeping, understanding the thermal needs of staff in your building and educating the people who are responsible for HVAC on how to optimise the system's operation will contribute to a reduction in your energy costs.

14.7.1   Adjust your system

At times, it makes sense to use the outside temperature to adjust the conditions inside your building. However, when the HVAC system is on, it is possible to save up to one third on energy costs by reducing the amount of outside air that enters a building. It is always better to adjust the system rather than to open a door or window and let heated or cooled air out - when it is too hot, people tend to open windows or doors to make their space more comfortable. Ensure that staff understands the implications of opening windows and doors when air-conditioners are in use - try adjusting the thermostats instead.

14.7.2 Turn off and power down

Electrical equipment and lights emit heat and can contribute to the heating and cooling load on HVAC systems. Try limiting the time that equipment is switched on and, where possible, use 'power-down' facilities on copiers, faxes, printers and computers during the day. Dim or switch off lighting if there is sufficient daylight and use as little light at night as possible.

14.7.3 Train staff on how to operate air conditioning units and heating controls

Staff should receive guidance on recommended operating temperatures and how to set heating or cooling units correctly. Louvres - movable slats to guide cool or heated air - are a feature on most air conditioning systems; staff should be able to operate these to maintain comfortable temperature levels. Display instructions on individual units and ensure that remote controls have accessible and obvious storage spaces.

14.7.4 Ensure controls are in place and HVAC system operation reflects demand

HVAC loads vary at different times and in different parts of a building throughout the day. Well-set time controls should ensure that systems only operate when and where required during core business hours. It is also crucial to regularly check settings - many systems are set incorrectly because 'someone has made a short-term adjustment and then forgotten about it'. Controlling how and when HVACs are operated can save significant amounts of power - follow these guidelines and encourage all employees to do the same:

1. Close windows while the air-conditioner is running.
2. Only switch on air-conditioners in rooms that are occupied.
3. Switch off units 30 minutes before leaving the office.
4. Adjust blinds and curtains in rooms that receive direct sunlight rather than powering up the air-conditioner.

14.8   Switch to energy-efficient technologies

Consider replacing old HVAC systems - new systems offer enhanced control functionality and can use up to 50% less electricity than energy-intensive systems.

Savings on the day-to-day running costs of electricity saving systems can quickly recoup the investment in energy-efficient HVAC technology solutions.

14.8.1 Upgrading and refurbishing

When putting in a new HVAC system, always choose the most energy-efficient technology solution you can afford and one that fits your building's requirements - avoid simply exchanging like with like, in the belief that it will minimise disruptions to your business.

When replacing inefficient components also avoid simply exchanging like with like - ensure that the replacement is of the highest possible efficiency.

Consider

1. Replacing conventional boilers with condensing boilers.
2. Replacing standard motors with high efficiency motors.
3. Investing in Variable Speed Drives (VSDs) for motors to match speed with output demand.
4. Investing in direct drive pumps and fans, which are more efficient than those that are belt driven.
5. Opportunities for heat recovery and recirculation in your building.
6. Installing a Building Energy Management System (BMS or BEMS), which offers close control and monitoring of building services performance, including heating, ventilation and air conditioning displayed on a computer screen in real time and allowing system performance to be monitored and settings to be changed quickly and easily, BEMS can reduce energy costs by up to 10%.

14.8.2 Seek advice

Always seek advice from HVAC system specialists before you upgrade or refurbish. HVAC system action checklist for reducing energy costs is given below:

1. Reduce the need: Switch off unnecessary equipment during the day and, especially, after hours to reduce heat buildup in your building (unless your building is 'night-cooling').
2. Consider installing automatic controls to ensure equipment stays off.
3. Set higher switch-on temperatures for cooling and lower switch-on temperatures for heating: A temperature control gap or 'dead band' between heating and air conditioning of about 5°C will improve occupants' comfort, cut operating costs and reduce wear and tear on both systems.

4. Look into areas that appear too hot or too cold and consider localised thermostatic controls.

5. Check for draughts, especially around poor fitting windows and doors - install draught-proofing to reduce heat loss and increase staff comfort.

6. Check insulation levels and increase it wherever practical to reduce the need for heating.

7. Walk around your building at different times of the day and during different seasons to see how and when systems are on: Check time and temperature settings.

8. Take advantage of free-cooling: Where outside temperatures are colder than the required internal temperature you can ventilate the building with fresh air ('night cooling' is especially efficient in summer).

14.9 Types of systems

14.9.1 Individual systems

The design, installation and control of heating, ventilation and air conditioning functions are usually integrated into one or more HVAC systems. For very small buildings, contractors usually select the systems and equipment to be installed. For larger buildings, building service designers, mechanical engineers or building services engineers specify the systems according to buildings' requirements. Building permits and code compliance inspections of installations are normally required for all sizes of buildings.

14.9.2 District networks

Although HVACs operate 'individually' in buildings, the equipment involved is sometimes an extension of a larger District Heating (DH) or District Cooling (DC) network, or a combined DHC network.

For example, at a given time one building may be utilising chilled water for air conditioning and the warm water it returns may be used in another building for heating, or for the overall heating portion of the DHC network.

14.9.3 Heaters

Heaters are appliances used solely for warming a building. A central heating system provides warmth to the whole interior of a building (or portion of a building) from one point. It contains a boiler, furnace or heat pump to heat water, steam or air at a central location such as a furnace room.

When combined with other systems to control a building's climate, a central heating system may form part of a fully integrated HVAC (heating, ventilation and air conditioning) system.

14.9.4   Hot water heating systems

In hot water heating systems, piping transports heat to a building's rooms. Most modern hot water heating systems have a circulator - which is a pump to move hot water through the distribution system. The heat can be transferred to the surrounding air using radiators, hot water coils (hydro-air) or other heat exchangers. The heated water can also supply an auxiliary heat exchanger to supply hot water for bathing and washing.

14.9.5   Warm air systems

Warm air systems distribute heated air and return air through metal or fibre glass ducts. Many systems use the same ducts to distribute air cooled by an evaporator coil for air conditioning. The air supply is typically filtered through air cleaners to remove dust and pollen particles.

14.9.6   Dual duct

These systems mix the outside air with return air through dampers, which is then split into hot and cold ducts. The air moving through the hot duct is heated by heating coils and the air moving through the cold duct is cooled through conditioning coils. In each portion of the building, the air is controlled by mixing it from the two ducts, which is regulated in the mixing box controlled by a thermostat.

14.9.7   Single duct

Just like with the dual duct, return air and outside air are blown through a single cold air duct by a supply fan. The cooled air is then moved through a heating unit - typically a hot water coil - before it enters the room.

14.9.8   Air-conditioners

An air conditioning system, or a stand-alone air-conditioner, provides cooling and humidity control for all or part of a building by removing heat through radiation, convection or conduction. A vent generally draws fresh air from outside into an indoor heat exchanger section, creating positive air pressure. The percentage of fresh air can usually be manipulated by adjusting the opening of the vent.

14.9.9   Free cooling systems

Free cooling systems can have very high efficiencies and are sometimes combined with seasonal thermal energy storage so that the cold of winter can be used for summer air conditioning. Common storage mediums are deep aquifers or a natural underground rock mass accessed via a cluster of small diameter, heat exchanger equipped boreholes. Some systems with small storages are hybrids, using free cooling early in the cooling season and later employing a heat pump to chill the circulation coming from energy storage.

14.9.10 Split systems

These systems, although most often seen in residential applications, are gaining popularity in small commercial buildings. The evaporator coil is connected to a remote condenser unit using refrigerant piping between an indoor and outdoor unit instead of ducting air directly from the outdoor unit. Indoor units with directional vents mount onto walls, suspend from ceilings or fit into ceilings. Other indoor units mount inside the ceiling cavity so that short lengths of duct handle air from the indoor unit to vents or diffusers around buildings' rooms.

14.9.11 Dehumidifiers

A dehumidifier is a device like an air-conditioner that controls the humidity of a room or building. It is often employed in basements, which have a relatively higher humidity because of their lower temperature (and propensity for damp floors and walls). In food retail establishments, large open chiller cabinets are highly effective at dehumidifying internal air.

Dehumidifiers draw excess moisture from the air – helping to combat condensation, prevent mold growth and reduce damp on walls.

There are two main types of dehumidifier to choose from – refrigerant (also known as compressor) and desiccant. They work in different ways and claim to be better suited to different environments.

1. Refrigerant (or compressor) dehumidifiers: These draw in air through a filter and over cold coils. Water then condenses on the coils and drips into the water tank. It's often claimed that they work better at higher temperatures and higher humidities, and so are the better choice for most occupied homes in Britain.

2 Desiccant dehumidifiers: These use an adsorbent material to extract water from the air, and the material is then heated so that the moisture drips into the water tank. Desiccant dehumidifiers are designed to work more effectively in lower temperatures – the sort of environment you might expect in your typical garage or conservatory. It's often claimed that desiccant dehumidifiers tend to use more energy than refrigerant dehumidifiers because of the way they use heat to warm the adsorbent material.

Energy conservation in pumps

15.1 Introduction

Pumping systems account for a significant percentage of energy consumption of the total industrial energy usage. Although pumps are typically purchased as individual units, they provide service only when operating as part of a system. The energy and materials used by a system depends on the design of the pump, installation and the way the system is operated. These factors are interdependent and must be carefully matched to each other, throughout their life time. The initial purchase price is only a small component of the life cycle cost for high usage pumps. While operating requirements may sometimes override energy cost considerations, an appropriate solution is still possible if the system is analysed suitably as per operating practices. Typically, in several industrial sectors, centrifugal pumps are the ones widely employed for any given duty accounting for bulk of industry use. Reciprocating pumps or hydraulic pumps are also employed for special duty applications. In industries, a dip in pump efficiency would be attributable to several reasons. But choosing the right solution plays a vital role in implementing energy conservation proposals. When any plant person encounters a pump operating at low efficiency, the immediate tendency is to replace it with a new pump without actually analysing the real reasons behind the inefficient operation. At times in industries, even a new pump installed in the system operates inefficiently due to improper matching of the system-head characteristics, inadequate pipe sizing, etc.

15.2 Methodology adopted for performance evaluation of pumping system

The prime objective of pump performance evaluation is to estimate the actual efficiency of the pump. The efficiency of the pump is directly proportional to the flow and head. Pump performance analysis can be carried out by knowing the liquid flow rate, liquid pressure, motor input power and pump rotating speed. Table 15.1 gives brief information about different operating parameters and typical measuring instruments required in pumping sets.

By knowing the liquid flow, operating head, and motor input power, pump efficiency can be evaluated by the following expression:

$$\eta_{pump} = (Q \times H \times \rho)/(102 \times kW \times \eta_{motor})$$

where,

η_{pump} = Operating efficiency of pump, in %

Q = Capacity, in m^3/s

H = Head, in m

P = Density of liquid handled, in kg/m

kW = Power consumed by motor, in kW

η_{motor} = Efficiency of motor, in %

Table 15.1: Brief information about different operating parameters and typical measuring instruments required for pumping sets.

Parameters	Typical measurement instruments
Liquid flow rate: Comparing liquid rates in different parts of the system can help pinpoint leaks, high pipe friction, and real time pumping requirements.	Differential pressure devices, such as orifice meters and venturi meters.
	Velocity flow meters, such as pitot tubes
	Open flow meters
	Positive displacement meters
	Ultrasonic flow meters
Liquid pressure: Monitoring water pressure can help find leaks, reduce unnecessary pumping, and maintain constant service.	Bourdon tubes
	Bellows
	Diaphragm gauges
	Piezo-resistive transducers
Motor input power: Input power readings can help determine if a motor is operating at its optimal loading (efficiency).	Ammeters
	Voltmeter
	Power factor meters
Pump rotating speed: Rotating speed data can help determine if a motor is operating at its optimal loading (efficiency)	Stroboscope

15.2.3 Possible energy conservation options in pumping systems

1. Selection of low/high head pumps.
2. Capacity regulation by opting for Variable Speed Drives (VSD) in place of throttling control.
3. Avoiding recirculation/bypass by opting for two way valves and VSD.
4. Adequate sizing of suction and discharge pipe.
5. Parallel operation of pumps.
6. Trimming of impellers.
7. Reduction of frictional losses by selection of low friction pipes.

8. Reduction of frictional losses by adequate sizing of pipes.

9. Downsizing of pumps with low capacity pumps.

10. Providing sufficient net positive suction head (NPSH).

11. Installation of overhead tanks.

12. Matching of correct sized motor for the pumps.

Selection of low/high head pumps: In process industries, pumping systems are designed to deliver through the shortest possible route by minimising the number of bends, valves, etc. However, when new pumps are required for the system, an adhoc approach is taken and pumps with a higher head (by about 20%) are installed. This leads to operation of pumps away from the best efficiency point. Being continuous duty pumps, when they are operated inefficiently, huge amount of energy is lost considering the life cycle of the pumps. The buyer or plant personnel should always install a pump as close to the system head requirements as possible to avoid inefficient operations. Same is the case for high head applications too.

Sometimes when system head requirements are higher, plant people install a low head pump for given duty. This results in low output of the pump, overloading of the motor and operation of pumps at lower efficiency leading to high power consumption. Careful analysis of the system requirements have to be assessed while installing new pumps.

Capacity regulation by opting for variable speed drives (VSD) in place of throttling control: Quite often, there is variable flow requirement from pumps due to changes or flexibilities in output, varying water requirements, etc. In such cases normally plant personnel throttle the discharge valve of the pumps to meet the water requirement. Under such circumstances, a more accurate and energy saving method is to install variable speed drive for the pumps during variable requirements, the speed of the pump can be varied as per the requirements. As per affinity laws, speed is proportional to the cube of power, hence any reduction in speed will result in reduction in power consumption by one third thereby resulting in power savings.

Avoiding recirculation/bypass by opting for two way valves and VSD: In pump systems sometimes a bypass valve is provided, which returns a portion of the water back to the tank and the balance of the water required is only utilised for process requirements. This method of flow control is inefficient as some portion of water is recirculated consuming energy. Providing two way valves and installing variable speed drives will result in pumps operating specific to process requirements, thereby resulting in energy savings.

Adequate sizing of suction and discharge pipe: When pump manufacturers design a pump, the inlet and the outlet pipe dimensions connected to the casing are specified. During installation of the pump it so happens that the guidelines are not strictly followed which results in pressure losses at the inlet and outlet of the pump in excess of what would have been if the pipes were designed adequately. Hence it is advisable to size the suction and discharge pipes as specified by the manufacturer. This will help in operating the pump at its peak efficiency.

Parallel operation of pumps: Pumps are operated in parallel if different requirements exist throughout the day. This is so when capacity of pumps is large and also when variable speed drives don't suit the system requirements due to high static head. In such cases, parallel operation of pumps is the solution. However in most of the industries, parallel operation of pumps is carried out to meet the water requirements constantly round the clock throughout the year. Any parallel operation of pumps with three or four pumps in combination leads to efficiency loss by 15% to 20%. Installing higher capacity pumps will improve the operating efficiency and hence reduction in power consumption is achievable for the same application.

Trimming of impellers: Trimming of impellers is desired if already a higher capacity pump is installed and flow required is marginally less compared to the design parameters. In such cases trimming of impellers will result in required output which also results in energy savings. Here care has to be taken that trimming of impellers should be done upto 5% to 7% of the original diameter of the impeller. Trimming in excess of 7% will result in recirculation of water inside the casing thereby resulting in low output. In case the output of the pump required is 20% to 30% less than the pump installed, it is advisable to downsize the pump to required capacity. Another concern while selecting a new pump is that at times the required capacity of the pump impeller will not be available with the manufacturer. In such cases, the manufacturer trims the original impeller to suit the required capacity as specified by the buyer. This leads to inefficiency at the initial stage itself.

This issue has to be thoroughly discussed with the manufacturer while selecting new pumps. As far as possible it is better to opt for a correct pump by analysing the family of curves from different manufacturers and suit to the requirements. This will avoid trimming of impeller in excess of 7% of the original diameter.

Reduction of frictional losses by selection of low friction pipes: Sometimes it is often observed that even for a small duty application high friction loss pipes are installed. Pumps being continuously operated result in operation at normally high heads due to high frictional losses in the pipes. Selection of low friction pipes like PVC will result in operation of pumps at lower head and thereby resulting in power savings.

Reduction of frictional losses by adequate sizing of pipes: This is by far the most neglected area. For optimum pressure drop in the pipes the velocity in the pipes should be within 1.5 to 2 m/s. It has been noticed in several instances that the velocity in the pipes are in excess of 2 m/s. This results in increased pressure drop thereby resulting in operation of pumps at higher head. The design of the pipes has to be carried out in such a way that the velocity is well within acceptable limits.

Also care has to be taken for not over designing the pipes. The pressure drop will be minimal in this case; though the initial expenditure will be high. Hence an optimum solution has to be found out for minimising pressure drop in the pipe lines.

Downsizing of pumps with low capacity pumps: There are cases in industries where a higher capacity pump is installed for a low application duty. The flow is controlled either through variable speed drive, throttling control or trimming of impellers. All these three methods will result in energy savings, however the accurate and optimum solution is downsizing of the pumps as per flow requirements. Energy savings will be more in down-sizing compared to the other options of flow control.

Providing sufficient net positive suction head: When a pump is to be installed, the net positive suction head (NPSH) has to be estimated accurately. The NPSH available should always be more than NPSH required. If NPSH available is less than NPSH required, problems with respect to cavitation will occur which results in decreased output of the pumps.

Installation of overhead tanks: In industries, potable water requirements are met by installing a dedicated pump. Generally, these pumps are continuously operated irrespective of the requirements. This method of operation not only results in loss of energy but also water loss. Both energy and water losses can be minimised by installing an overhead tank. The pumps can then be operated by providing a level controller. Based on the level of the overhead tank required, programming can be done. This results in minimum hours of operation of the pumps thereby resulting in energy savings.

Matching of correct sized motor for the pumps: The motor coupled with the pump has to be sized as close to the brake horse power required to drive the pump for a given application. Installing high capacity motors for low capacity applications will result in operation of motors at low efficiency. Motor efficiency will be reduced if loading is less than 50%. The motors selected should be such that the loading is in the range of 80% to 100%. If motors are loaded more than 100%, the factor of safety as specified by the manufacturer has to be analysed clearly. If motors exceed the factor of safety, it leads to motor failures.

15.3 Parametric approach to energy conservation in pumping

Basically, energy-consumption is measured in kilowatt-hours, kWh. Obviously energy consumption can be reduced either by reducing kilowatts or by reducing hours or by reducing both.

15.3.1 Reducing hours

In the case of non-stop, continuous duty processes, e.g. power generation, there is no question of reducing hours. In case of pumping for water supply in a building, where total quantity to be pumped in a day is known, hours can be reduced by pumping at higher rate of flow. This can also be an option for reducing energy consumption, since efficiency of pump with higher rate of flow can often be better.

15.3.2 Reducing kW and overall efficiency

Reducing kilowatts means reducing power drawn. In pumping, power from the supply is drawn, first by the driver. Driver supplies it to the pump after its own efficiency, h_{motor}. Power supplied by the driver to the pump is the power input to the pump. The pump imparts the power to the liquid after its own efficiency h_{pump}. Actually, if the driver is not directly coupled to the pump, i.e. if the pump receives power from the driver through a transmission gear, there will be the efficiency of the transmission gear, h_{trans} which also needs to be looked into. Even when the pump is directly coupled, the transmission efficiency can be poor, if the alignment is not good or if components of the coupling have suffered wear and tear. Thus the right objective for energy conservation in pumping ought to be to improve overall efficiency $h_{overall}$.

15.3.3 Components of overall efficiency

To go into a finer detail, one would notice that supply of power to the driver itself would be happening across an unwarranted length of the cable. Or there can be an unwarranted length of cable between the driver and its controller. And the quality of power supply would also influence the energy-consumption. If one would think that this is stretching the logic too far, think of the bore well pumps. These pumps have a submersible motor and the supply cable has to reach to the motor terminals all the way down. So, there is the efficiency of power supply and control.

15.3.4 Bench-mark information and standards on pumps

When it comes to assessing, what the level of energy consumption should be in a given pumping system, one needs to have information on 'Bench-mark' values. Unfortunately, bench-mark values of energy consumption are not available directly in any standard or in any reference book on pumps. What is available is information on efficiencies of pumps.

Standards on motors also specify efficiencies of motors. In recent years there have been initiatives around the world to enhance efficiencies of motors. International specification of induction motor in USA are IEEE 112-B and Germany IEC 34-2 and Indian specification are IS: 13529. In India also there is Indian Standard IS-12516 for energy efficient motors.

15.4   10 Tips to save energy on pumping systems

1. Select the most efficient pump type for the application: A Finnish research study shows that the average pump efficiency is below 40% and that 10% of pumps are 10% efficient or less. Oversizing often comes in the design phase, since the practice for adding multiple safety factors is quite common. This means that both pressure and flow parameters for the pump design may be 25% more than the actual system operation. The specifying engineer may need to work closely with the pump manufacturer or distributor to optimally select the pump, in addition to its size, speed, power requirements, and type of drive, as well as the mechanical seal and ancillary equipment.

2. Right-size the pump: Right-sizing the pump represents a significant economic opportunity to reduce energy consumption. This is important because centrifugal pumps can consume up to 60% of motor energy in a facility, and also have the highest process equipment maintenance cost. When engineers add too much of a safety factor during the design phase, the pump can be oversized, resulting in higher energy and maintenance costs.

3. Trim the impeller: The impeller should not be trimmed any smaller than the minimum diameter shown on the manufacturer's pump curve. This is typically about 75% of a pump's maximum impeller diameter. Pump curves and affinity rules (which are valid for a maximum of approximately 5% change in diameter) can both provide information on impeller trim changes and the affected performance. In practice, impeller trimming is typically used to avoid throttling losses associated with control valves.

4. Minimise system pressure drop: A key way to reduce pressure drop is through pipe-sizing optimisation. Hydraulic friction loss creates a reduction in pressure from one end of a straight pipe to another. Factors such as the flow rate, pipe size (diameter), overall pipe length, pipe characteristics (surface roughness, material, etc.), and properties of the fluid being pumped all influence the system pressure drop.

5. Implement proper control valves: Control valves are typically used to control flow and/or pressure. They can help to reduce energy losses over non-controlled systems such as irrigation systems with a fixed-speed pump and multiple locations with different distances and elevations. The main functions of control valves are throttling flow or for bypassing flow. Throttling reduces the flow but increases the pressure. You can minimise excess pressure by bypassing excess flow back to the reservoir or another location.

6. Implement variable speed drives (VSDs): Drivers are used for either fixed-speed or variable-speed operation. For many applications, you can save energy by implementing variable speed drives. With a variable speed drive, the rotational speed of the pump is adjusted to achieve the desired head and flow necessary for the process application. A VSD can often be added to an existing pump motor system to slow the pump down to meet the actual requirements verses the theoretical requirements that were calculated at the start of the project. Once installed, the VSD can accommodate changing system demands, including many potential future expansion plans. This method often results in the highest energy efficiency with lowest life cycle costs.

7. Maintain pumping systems effectively: Effective pump maintenance allows facilities to keep their pumps operating efficiently. Regular maintenance may reveal deteriorations in efficiency and capacity, which can occur long before a pump fails. Wear ring and rotor erosion, for example, can be costly problems that reduce efficiency by 10% or more. Most maintenance activities can be classified as either preventive or predictive. Preventive maintenance addresses routine system needs such as lubrication, periodic adjustments, and removal of contaminants. Predictive maintenance focuses on tests and inspections that detect deteriorating conditions. Sometimes called 'condition assessment' or 'condition monitoring,' it has become easier to conduct with modern testing methods and equipment. This can help minimise unplanned equipment outages, which can be very costly.

8. Use higher efficiency/proper pump seals: Sealing systems impact efficiency, and mechanical friction losses are only the beginning. Leaks from static and dynamic seals waste fluid and can contaminate the environment. Leaks between the pump suction to the pump discharge reduce pump volumetric efficiency. Dynamic seals consume energy from the mechanical friction between the static and moving parts. Potential sealing system savings can exceed the energy savings obtained from

switching to variable frequency drives, trimming impellers, or resizing pumps in many applications.

9. Use multiple pumps: When multiple pumps operate as part of a parallel pumping system, there are opportunities for significant energy savings. A multiple pump parallel system works best when each pump is run individually, not concurrently, most or all of the time. Running multiple pumps simultaneously is appropriate as dictated by the flow requirements specific to the application and duty cycle.

10. Eliminate unnecessary uses: One of the most simple, but often over looked, measures to save energy is to eliminate unnecessary use. Pumping system efficiency measures include shutting down unnecessary pumps and using pressure switches to control the number of pumps in service when flow-rate requirements vary. Each pump system is different and there are many opportunities to save energy.

Energy conservation in iron and steel industry

16.1 Introduction

The iron and steel industry is one of the oldest and largest manufacturing operations in India. There are three basic types of steel manufacturers: integrated mills, mini-mills, and specialty-steel mills.

Integrated steel mills use iron ore and coal as the raw materials for manufacturing various steel products. Mini-mills produce steel from scrap, typically using electric arc furnaces. Speciality-steel mills are similar in operation to mini-mills; they manufacture stainless, tool and high-alloy steels. Few people outside the steel industry are aware of the enormous volumes of water and air required to produce iron and steel from raw materials. About 3.5 tonnes of air must be supplied to the Blast furnace to reduce the ore to metal, and this amount of air produces about 5 tonnes of Blast furnace gas. So, it is apparent that the tonnage of air and gas far exceeds the solid charge to the furnace and the products tapped from the furnace.

16.2 Iron and steel making processes

Three process steps—coking, iron-making and steel-making—are fundamental to the iron and steel industry.

16.2.1 Coking

Coking is the process of converting coal to pure carbon or coke. The coking process yielding coke as well as by-products, such as coal tar, ammonium sulphate and light oils. A typical coke plant consists of two major areas; the coke oven batteries and the by-product recovery plant.

16.2.2 Iron-making

In iron-making, coke is combined with iron ore and limestone, and supplied to the top of a Blast furnace (Fig. 16.1). Air is heated to about 1800°F and blown countercurrently from the bottom of the furnace through circumferential openings called tuyeres. In the furnace, coke reacts with iron ore, releasing iron and generating CO and CO_2 gas. Iron sinks to the furnace hearth. Impurities in the charge combine with the limestone, forming slag that floats to the top of the iron. The combustible exhaust gases from the furnace have considerable heating value and are used to pre-heat the incoming air and to generate steam.

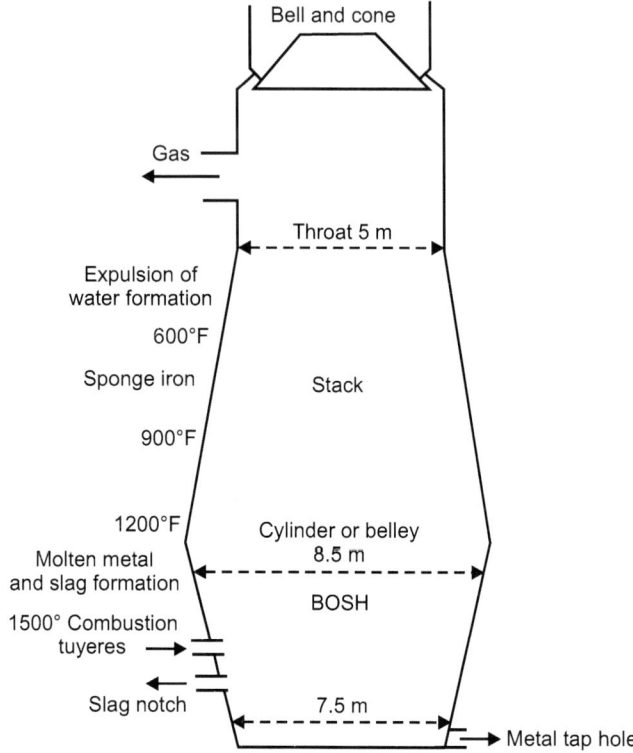

Figure 16.1: Blast furnace.

Blast furnaces produce iron, commonly known as pig iron, with a carbon content of about 4%.

16.2.3 Steel making

Steel is an alloy of iron, containing less than 1% carbon. Steel is made by three different methods: the basic oxygen furnace (BOF), the open hearth furnace (OHF), and the electric arc furnace (EAF) methods. These furnaces reduce impurities, especially the carbon content by oxidation, and add various alloying elements, depending on the desired grade steel. Because BOF and EAF are currently the most widely-used techniques, only these have been described briefly here. In the BOF method, (Fig. 16.2), hot molten pig iron from the Blast furnace, scrap steel, mill scale, and slag conditioning materials (such as limestone, burned lime, dolomite and fluorspar) are charged to the furnace. Oxygen is injected into the furnace through a lance positioned a few inches over the charge. Oxygen reacts to oxidise the charge, which releases large amounts of heat that melts the charge. During the melting process, the

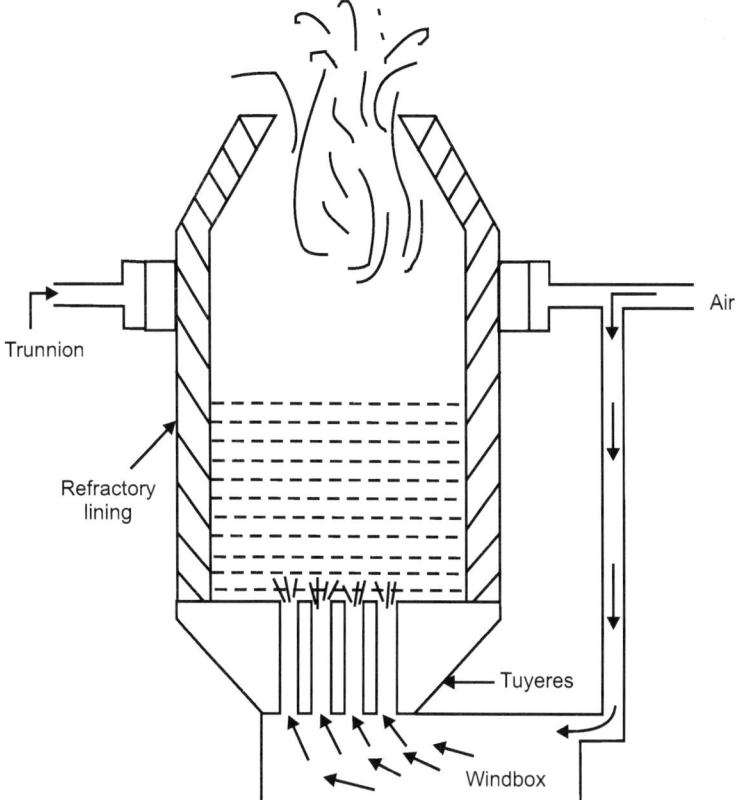

Figure 16.2: Bassemer converter.

fluxing materials, called slag, collect and float on top of the molten metal. The slag is decanted through a spout while the molten steel is either transferred to a teeming station or to a continuous casting station.

EAFs are used for manufacturing common grades of low-carbon steel as well as special alloy and tool steels. A chief advantage of the EAF method is that it is not dependent on hot molten pig iron from Blast furnaces and can typically operate, if necessary, with scrap steel as the sole iron source in the initial charge to the furnace. EAFs also need the same or similar flux conditioning materials as are required by BOFs. Slag and molten steel separation is similar to that in a BOF.

16.2.4   Open hearth furnaces

The open hearth furnace produces steel (below 1% carbon) from a charge of steel scrap, iron (about 4% carbon) and slagging materials, such as lime and fluorspar. These furnaces produce about 100 to 300 tonnes per heat of 8–12 hour duration (Fig. 16.3).

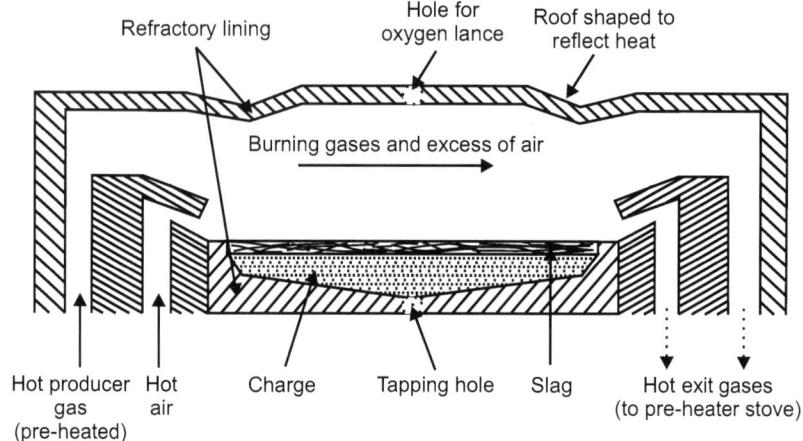

Figure 16.3: The open hearth furnace.

As in the Blast furnace, open hearth furnaces use large quantities of water for cooling, generally ranging from about 750 gpm for small furnaces to as much as 1500 gpm for larger units.

Also, as in the Blast furnace, the gases from the open hearth furnaces contain a high concentration of dust, which may be removed by washing, precipitation equipment, or both. This dust is removed from the wash water by clarifiers and may be combined with the recovered dust from the Blast furnaces for sintering if the zinc content is sufficiently low. The dust loading varies appreciably during the heat from a range of about 0.2 to 0.6 grain/standard cubic foot of gas (scf) during charging to about 2 to 3 grains/scf several hours after start of oxygen lancing.

The dust particles from open hearth furnaces are much finer than those from the Blast furnace, and chemical coagulation is needed to separate them from the waste-water. Oxygen lanced furnaces discharge a higher dust load than the older, conventionally-fired furnaces.

Whereas the wash water from the Blast furnace is alkaline, that from open hearth scrubbers is quite acidic, so special materials of construction are required in closing open hearth washer systems. However, because the pH of the circulated water is in the range of 2.5–3.0, the system can be completely closed without fear of deposits, the only water loss being that associated with the collected sludge of filter cake.

It may be necessary to select the scrap charge for the open hearth furnace carefully if the dust is recovered for sintering, since a high zinc content can be damaging to the refractory lining of the Blast furnace when the sinter is recharged to the furnace. The zinc, which comes from galvanised sheet or white metal in the scrap, may pass through the refractory and sublime on the metal shell, gradually expanding and loosening the refractory.

A third source of waste-water from the open hearth is the blow-down from the waste heat boilers, which recover heat from the exit gas and produce by-product steam. Water from the continuous blowdown is generally of such small volume that it does not significantly affect the composition of the collected waste-water from the open hearth area, where this includes the furnace cooling water. However, mass blowdown from the boilers may show up as periodic high pH or high conductivity in the combined waste-water. If there is a separate water treatment plant to supply make-up for the waste heat boilers, wastes from regeneration of the water treatment system may also constitute one of the pollution loads from the Blast furnace area.

Basic oxygen furnaces

The basic oxygen furnaces (BOF) shop is a facility for conversion of iron into steel at much higher rates than those attainable in the open hearth. In the conversion of iron to steel in the basic oxygen furnace, the molten metal charge is reacted with oxygen, introduced through a water-cooled lance, which burns off the impurities in a period of about 20 to 25 min. Total heat requires about 50 min compared with the 8–12 hours required for steel-making in the open hearth furnace. Most BOF units produce in the range of 100–300 tonnes of steel per heat. Because of the high heat release, gases leaving the furnace hood during lancing are very hot. The hood may be cooled with circulating water, or heat may be recovered from this gas through a boiler mounted directly above the hood. Because the heat is released at an uneven rate and because little heat is released during the charging and tapping period, the flow of steam from the generating unit is irregular and must be modulated by discharge through an accumulator. This then releases steam at a relatively steady rate into the plant steam system at a pressure below that of the waste heat boiler.

Because of the very high temperature service, the lance cooling water is specially treated to prevent corrosion or scale formation; this circuit may be completely closed and heat extracted from the lance cooling water through a secondary cooling water circuit using a surface type heat exchanger.

The gas in such furnaces is cleaned in equipment similar to that used in the open hearth, although there are variations in methods of wash-in and removing accumulated dust from the washers and precipitators. The dust loading is usually appreciably higher in the BOF discharge (about 50 to 60 lb/tonne of steel and in the range of 0.5–5.0 grains/scf in the gas entering the spark box) than in either the Blast furnace gas (about 30 lb/tonne) or the open hearth (about 20 lb/tonne). There is appreciable variation in temperature and pH of the wash water during the heat, as different ingredients are added to the charge during the oxygen blow.

Other important iron and steel operations

Other common and important operations in an iron and steel plant are sintering, vacuum degassing, casting, rolling and finishing.

Sintering: Sintering is the cementing of Blast furnace fines, mill scale and flue dust that cannot be directly charged to a Blast furnace. In this process, the fines are mixed with fine coal in a travelling gate, which makes them usable as a direct charge to the Blast furnace.

In this process, molten steel is subjected to a vacuum to remove gaseous impurities, such as hydrogen, nitrogen and oxygen.

Casting: At the teeming station, molten steel is poured into molds to form ingots. After the ingots are cooled and the molds are stripped, the ingots are transferred to a primary hot rolling mill where the ingots are reheated in soaking pit furnaces and rolled into billets, blooms, or slabs. The continuous casting operation eliminates the hot rolling operation by casting the molten metal directly into thin slabs or other desired shapes.

Hot-forming: In the secondary hot rolling mill, slabs or blooms are reduced to billets, shapes, or strips.

Cold-rolling: In cold rolling, unheated hot-rolled products are passed through rolls to produce steel with smooth dense surfaces and enhanced mechanical properties. Cold-rolled steel products are not as thick as those produced by hot rolling.

Surface preparation: The rolled-steel products are pickled in acid to prepare the surfaces for subsequent application of corrosion-protection coatings. Molten salt baths are used to descale stainless-steel products.

16.3 Energy conservation technologies

16.3.1 Arc furnace

Heat energy from steel melting in Arc furnace

Electric power is the major source of heat energy used for arc furnaces. In the steel industry, the arc furnace is mainly used to melt steel material by means of the arc and electric resistance heating and remove undesirable components such as phosphorus, sulphur, hydrogen and oxygen from the material through different chemical reactions including decarburisation, dephosphorisation, desulphurisation and deoxidation in order to impart it with required physical and mechanical characteristics while adjusting the contents of major components such as carbon so that steel with good properties can be obtained.

To achieve these objectives, it is essential to perform the whole process as quickly as possible because the above reactions may proceed reversibly as the material stays under the melting conditions in an arc furnace for a lengthy period of time.

Major methods currently used to accelerate the melting process and to save electric power required for the process include the use of an oil burner for auxiliary melting of the material in the furnace, use of a lance pipe to stop the supply of oxygen, blowing of oxygen into the metal bath and, in some cases, use of a heavy weight to compress bulky feed material in the furnace.

Energy-saving measures through the operation of Arc furnaces

In order to save energy one should:

1. Reduce operating hours.
2. Raise the resistance welding time in order to eliminate wasteful power consumption. Consequently, the use of excessively bulky materials should be avoided as much as possible, and they should be pressed and massed together.
3. Effectively use oxygen blow to quickly raise the temperature to over 1600°C. The use of essentials, such as a poker, is desirable.
4. As a charge will come two or three times, it is necessary, in order to secure operation speed, to have close contact between the crane operator and other related operators so that there is no waiting for the crane.
5. Install high-powered transformers and carry out rapid dissolution.

16.3.2   Arc furnace melting process

1. Feed material and feeding process: The type of feed material used in the arc furnace melting process depends on the product to be produced. Machine chips, pressed steel scraps, light steel scraps and steel casting scraps are generally used for producing bars and sections, while steel casting scraps, light steel scraps, machine chip scraps and steel casting scraps are employed for producing steel castings. In the former case, feed materials are bulky, and cannot be fed to the furnace in one step, generally requiring three or more steps instead. In any case, it is essential to prevent nonferrous metals, including copper and aluminum, and nonmetallic substances, including rust and oil, from getting into the feed material. Bulky feed materials are normally used for the production of general steel products, as stated above. In general, the canopy is opened and a charging basket is employed to feed them in several steps.

In the first step, machine chips are laid on the floor of the furnace, followed by the feeding of limestone, light steel scraps, returned scraps, pressed steel plate scraps, light steel scraps and machine chips in this order. Secondly, pressed steel scraps, light steel scraps and machine chips are fed onto the metal bath at the bottom. In the third process, returned scraps, steel casting scraps, pressed steel plate scraps, light steel scraps and machine chips, carried in a charging basket, are fed in this order from the top of the furnace onto the metal bath.

In feeding one charge, bulky materials should be at the bottom while lighter ones should be at the top. This permits the efficient use of electric power while preventing damage to the rod electrodes.

A bypass may be provided between the dust collector and the arc furnace, with a charging basket placed there for pre-heating in waste gas. This can reduce the power consumption by 20–50 kWh/T.

2. Operation of Arc furnace: After charging the arc furnace with materials, electric power is supplied and then melting operations are performed as described in the following example, where a furnace with a capacity of eight tons is used.

   (a) Melting period: The melting period accounts for more than 50% of the total power consumption used in the entire arc furnace melting process. The operations, therefore, require skilled workers.

   (b) Oil burner: For the saving of electric power, an oil burner is used to accelerate the melting of the fed material. The burner is fixed at a cold spot in the furnace so as to avoid the burning of the electrodes.

   (c) Compression of fed material under weight: When bulky material becomes slightly red hot in the furnace, the canopy is removed and the material is compressed under an appropriate weight suspended from a crane, with power supply stopped for saving electric energy. Power supply is resumed immediately after completing the compression. The duration of compression under the weight should generally be in the range of 15–25 min after the start of power supply depending on the total weight of the material. This operation should be carried out quickly and therefore should be performed in close coordination with the crane.

3. Removal of slag after meltdown: Steel scraps fed in the furnace often contain many undesirable components including soil, stones, scraped bricks and concrete debris, leaving large amounts of slag after the meltdown and reducing the fluidity. Slag should be removed as early as possible.

4. Oxidising period: Samples are taken from the molten metal bath and subjected to analysis of the contents of carbon, silicon, manganese and

sulphur to allow composition adjustment immediately before the start of the oxidising period.

The oxidising period is important in accelerating major processes including dephosphorisation, desulphurisation, decarburisation and deoxidation. This requires a metal bath temperature of above 1600°C. To achieve this, the voltage is decreased to increase the current. Oxygen blowing through the lance pipe is performed during this period. The metal bath temperature should generally be above 1620°C at the end of the oxidising period. Slag in the surface of molten metal bath should be removed completely at the end of the oxidising period.

5. Reducing period: During the reducing period, scouring is performed in the presence of basic slag to remove oxygen in the bath, whose content is increased during the oxidising period. At the same time, desulphurisation is carried out while adjusting the composition and temperature of the bath. The deoxidation process consists of diffused deoxidation, which uses reducing slag, and forced deoxidation. Steel is taken out when the reducing slag has become stable after the raise of the bath temperature showing the completion of the scouring.

6. Operations after removal of steel and reduction in duration: To reduce the time period until the start of the next melting process after the removal of steel is important in improving the steel removal efficiency and decreasing the power consumption. Furnace repair materials, limestone, feed steel materials, etc., should always be kept ready to permit quick repair and charging of the furnace. Coordinated operations with the crane, etc. should also be performed quickly and systematically. Adequate training of workers is essential to ensure these.

16.3.3   Reheating furnace

After billets are roughly rolled at a blooming mill or made by continuous casting, a reheating furnace reheats them at a given temperature according to its purpose before they are sent to the hot rolling process to make finished products. Reheating furnaces can be divided into batch-type furnaces and continuous furnaces.

Batch-type furnaces are mainly used as auxiliary equipment to reheat something of a special form. For mass production, continuous furnaces are used in general. The types of continuous furnaces include pusher-type furnace, walking hearth-type furnace, and walking beam-type furnace. In the past, pusher-type furnaces were used most widely, and walking hearth-type furnaces were used for special treatments. With increased heating capacity, walking beam-type furnaces are getting adopted widely.

Energy conservation technologies

The basic ideas for energy conservation on reheating furnaces are the rationalisation of combustion, the rationalisation of heating and cooling, the prevention of heat loss by radiation and transmission, and the recovery of waste heat.

Figure 16.4 shows a characteristic diagram of energy conservation for reheating furnace and Fig. 16.5 shows reduction point for fuel consumption rate.

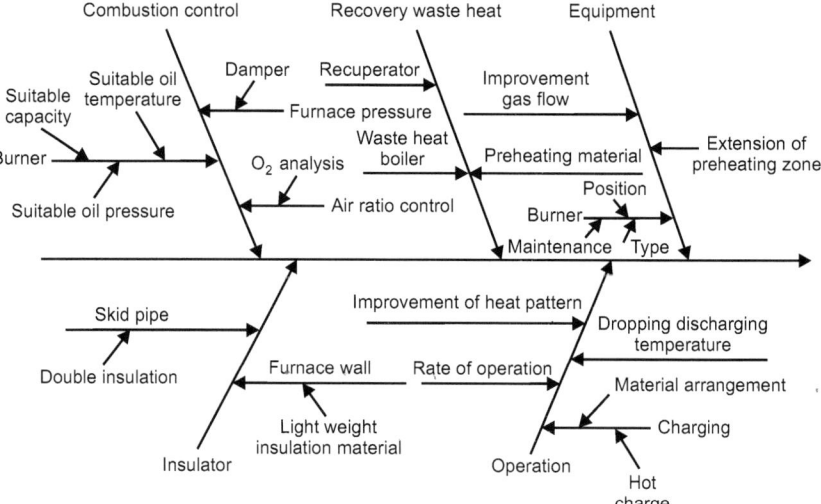

Figure 16.4: Energy conservation for reheating furnace.

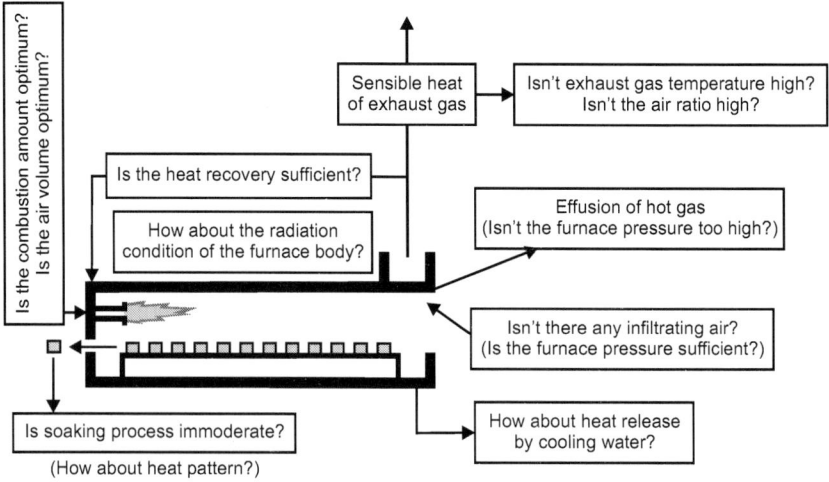

Figure 16.5: Reduction point for fuel consumption rate.

Heat balance

Heat balance is an effective means of promoting energy conservation. It allows to numerically grasp the present situation of heat loss and efficiency in heating. Then, based on these data, one can find out how to improve operating standards and facilities. Thus, implementation of heat balance is a precondition of promoting energy conservation.

Japanese Industrial Standard (JIS) GO702 'Method of heat balance for continuous furnaces for steel' was issued for the purpose of grasping the heat loss and efficiency of reheating furnace sufficiently.

Prevention of heat loss by radiation and transmission

The heat loss from a reheating furnace is largely divided into:

1. Radiation loss through openings and surface of the furnace body.
2. Cooling loss through water cooled skid pipes.
3. Heat accumulation loss to internal insulation and members composing the furnace body. Here, the heat accumulation loss can be ignored, if operation is continued for a certain period of time without much change in temperature as is the case with a continuous steel reheating furnace.

Prevention of radiation heat loss from surface of furnace: The quantity of heat release from surface of furnace body is the sum of natural convection and thermal radiation. This quantity can be calculated from surface temperatures of furnace. The temperatures on furnace surface should be measured at as many points as possible, and their average should be used. If the number of measuring points is too small, the error becomes large.

The quantity (Q) of heat release from a reheating furnace installed in the building of a factory is calculated with the following formula:

$$Q = a \times (t_1 \quad t_2)^{5/4} + 4.88\varepsilon \left\{ \left(\frac{t_1 + 273}{100} \right)^4 \quad \left(\frac{t_2 \quad 273}{100} \right)^4 \right\}$$

where,

    a : Factor regarding to the direction of the surface of natural convection ceiling = 2.8, side walls = 2.2, hearth = 1.5.

    t_1 : Temperature of external wall surface of the furnace (°C).

    t_2 : Temperature of air around the furnace (°C).

    E : Emissivity of external wall surface of the furnace.

Reinforcement of insulation: Improvement on radiation heat loss from surface of a furnace body can be achieved by reinforcing its insulation method. There are two ways to do this; one is to cover the internal wall surface with ceramic fibre, and the other is to cover the external wall surface with ceramic fibre or rock wool.

Prevention of heat loss through openings: Heat loss through openings consists of the heat loss by direct radiation from openings and the heat loss caused by combustion gas that leaks through openings.

1. Heat loss by radiation from openings: If a furnace body has an opening on it, the heat in the furnace escapes to the outside as radiant heat. The quantity of heat loss in this way depends on the thickness of furnace wall and the shape of the opening.

2. Heat loss caused by combustion gas that leaks through openings: Since the furnace pressure of a reheating furnace is slightly higher than outside air pressure during its operation, it is not avoidable that the combustion gas inside blows off through openings and heat is lost with that. It would cause more damage if outside air intruded into the furnace, making temperature distribution uneven and oxidising billets. As already mentioned, this heat loss is about 1% of the total quantity of heat generated in the furnace; if furnace pressure is controlled properly.

3. Prevention of cooling water loss: Cooling heat loss through water cooled skid pipes in a continuous reheating furnace amounts to as much as 10–15% of fuel consumption, In particular, a walking beam-type furnace structurally has about 1.5 times as much water cooled area as a pusher-type furnace does. Accordingly, the problem of cooling water loss is more significant.

To overcome this problem, the double insulation method for skid pipes has been widely applied to newly built as well as existing furnaces. This method uses high insulation ceramic fibre for internal surface and covers it with castable. Figure 16.6 shows an example of the double insulation and Fig. 16.7 shows a comparison between cooling loss values (calculated values) of single insulation and double insulation.

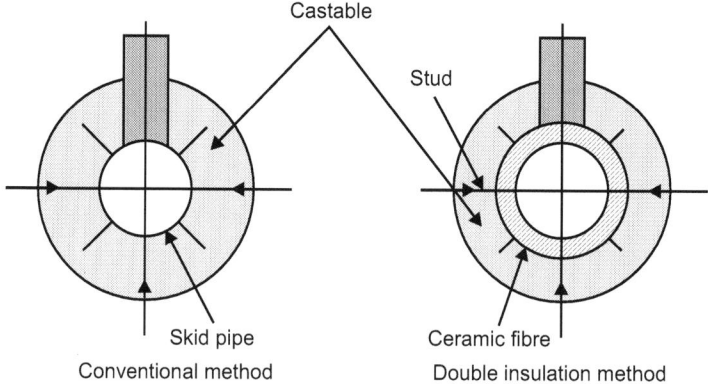

Figure 16.6: Double insulation method for skid.

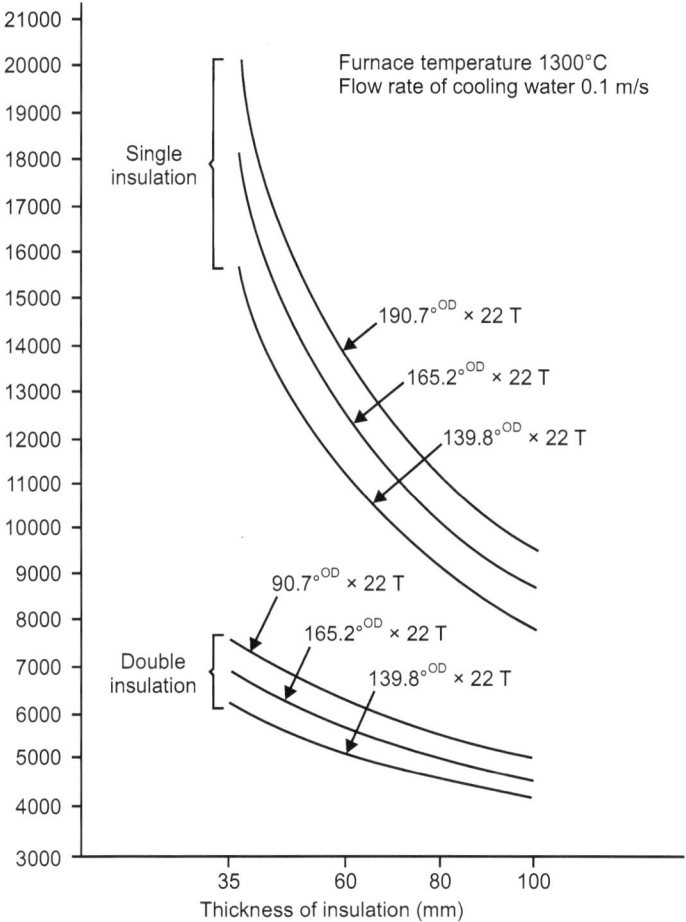

Figure 16.7: Comparison of water cooling loss (calculated values).

16.3.4 Waste heat recovery

The quantity of heat taken away from a reheating furnace by high temperature exhaust gas is very large. If this can be reduced, it has a large effect on fuel saving. There are two ways to reduce the quantity of heat taken away. One is to reduce the volume of exhaust gas, and the other is to reduce the temperature of exhaust gas. The former is the rationalisation of air ratio. The latter is achievable by recovering waste heat from exhaust gas.

Methods for recovering heat from exhaust gas include: (i) to pre-heat air for combustion by a recuperator, (ii) to generate steam or hot water by a waste heat boiler, (iii) to pre-heat materials by exhaust gas and (iv) cascade utilisation as a heat resource for others.

Table 16.1 shows standard waste heat recovery rates of industrial furnaces in Japan.

Table 16.1: Standard waste heat recovery rate of industrial furnace in Japan.

Exhaust gas temperature (°C)	Classification of capacity	Standard waste heat recovery rate (A)	Reference Exhaust gas temperature (°C)	Pre-heated air temperature (°C)
500	A . B	20	200	130
600	A . B	20	290	155
700	A	30	300	260
	B	25	330	220
	C	20	370	180
800	A	30	370	300
	B	25	410	250
	C	20	450	205
900	A	35	400	385
	B	25	490	285
	C	20	530	230
1000	A	40	420	490
	B	30	520	37s
	C	25	570	315
Over	A	40		
1000	B	30		
	C	25		

Note:
1. 'Exhaust gas temperature' means the temperature of exhaust gas discharged from the furnace chamber at the outlet of furnace.
2. The classification of the capacity of industrial furnace is as follows:
   A. Industrial furnace whose rated capacity is more than 20 MM kcal/hr.
   B. Industrial furnace whose rated capacity is from 5 MM kcal to not more than 20 MM kcal/hr.
   C. Industrial furnace whose rated capacity is from 1 MM kcal to not more than 5 MM kcal/hr.

Remark:
1. The values of standard waste heat recovery rate listed in the above Table are determined concerning the ratio of a recovered quantity of heat to a quantity of sensible heat in an exhaust gas discharged from the furnace chamber when a combustion is carried out under a load in the neighbourhood of a rating.
2. The values of standard waste heat recovery rate listed in the above Table shall be a standard for the continuous operating furnaces built on and after January 1, 1980:
3. The values of standard waste heat recovery rate listed the above Table shall not be a standard for the waste heat recovery rate of the undermentioned industrial furnaces:
   (i) Those whose rated capacity is not more than 1 MM kcal/hr.
   (ii) Those whose annual operating time does not exceed 1000 hours.

4. The values of exhaust gas temperature and pre-heated air temperature listed as references are values obtained by calculating the temperature of exhaust gas when the waste heat of standard waste heat recovery rate has been recovered and the temperature of pre-heated air when the air has been pre-heated by the aforementioned recovered waste heat, on the following conditions:

   (i) Temperature drop due to released heat loss, etc. from the furnace outlet to the heat exchanger for pre-heating air: 200°C.
   (ii) Fuel: liquid fuel.
   (iii) Atmospheric temperature: 20°C.
   (iv) Air ratio: 1.2

Now, we will describe pre-heating of the air for combustion, which is generally done for reheating furnaces.

Pre-heating the air for combustion by a recuperator

A recuperator is a device that recovers heat from exhaust gas exhausted from a reheating furnace. A metallic recuperator has heat transfer surface made of metal, and a ceramic recuperator has heat transfer surface made of ceramics. When the exhaust gas temperature is lower than 1000°C and air for combustion is pre-heated, a metallic recuperator is used in general.

By using pre-heated air for combustion, fuel can be saved. The fuel saving rate is given by the following formula:

$$S = \frac{P}{F+P-Q} \times 100(\%)$$

where,

S : Fuel saving rate

F : Low calorific value of fuel (kcal/kg fuel)

P : Quantity of heat brought in by pre-heated air (kcal/kg fuel)

Q : Quantity of heat taken away by exhaust gas (kcal/kg fuel)

By this formula, fuel saving rates for heavy oil and natural gas were calculated for various temperatures of exhaust gas and pre-heated air. The results are shown in Fig. 16.8 and Fig. 16.9.

When installing a recuperator in a continuous steel reheating furnace, it is important to choose a pre-heated air temperature that will balance the fuel saving effect and the invested cost for the equipment.

Moreover, the following points should be checked:

1. Draft of exhaust gas: When exhaust gas goes through a recuperator, its draft resistance usually causes a pressure loss of 5–10 mm H_2O. Thus, the draft of steak should be checked.

2. Air blower for combustion air: While the air for combustion goes through a recuperator, usually 100–200 mm H_2O of pressure is lost. Thus, the discharge pressure of air blower should be checked, and the necessary pressure should be provided by burners.

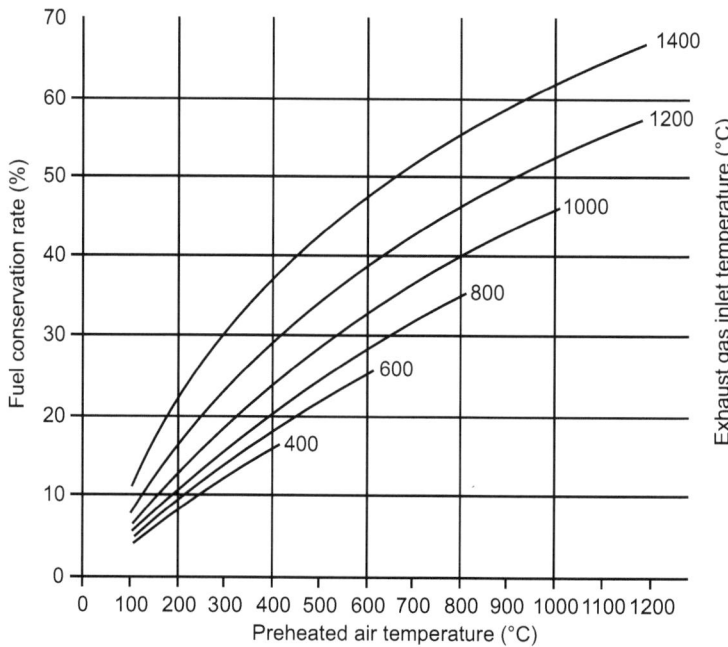

Figure 16.8: Fuel conservation rate when oil is used.

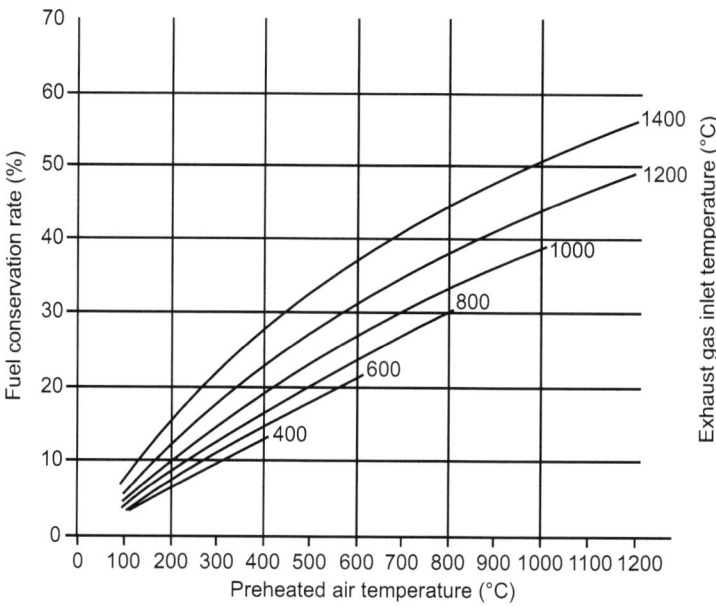

Figure 16.9: Fuel conservation rate when natural gas is used.

Energy conservation in furnaces

17.1 Introduction

A furnace is a piece of equipment used to melt metal for casting or to heat materials in order to change their shape (rolling, forging, etc.) or their properties (heat treatment). In other words furnace is an encloser in which energy in a non-thermal form is converted to heat, especially such as an encloser in which heat is generated by the combustion of a suitable fuel.

17.2 Types and classification of furnaces

Furnaces can be broadly classified into three categories based on mode of heat transfer, mode of charging, and mode of heat recovery. The furnace classification is given in Fig. 17.1.

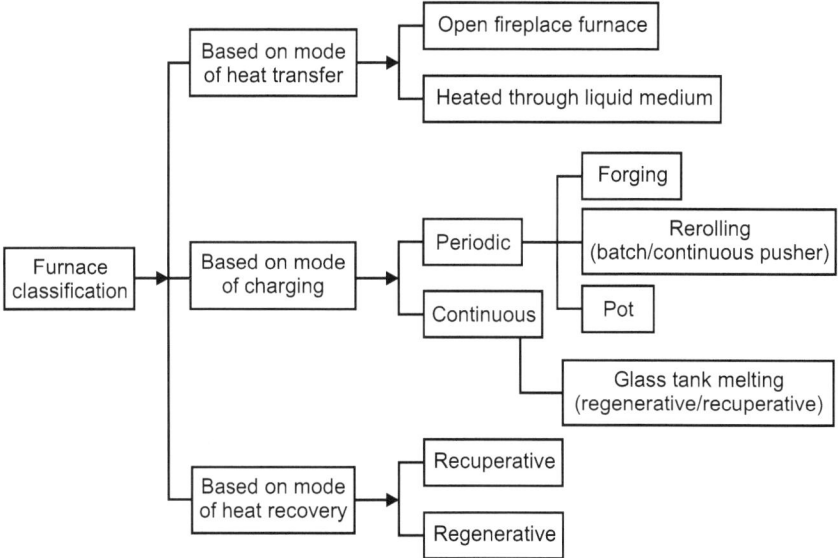

Figure 17.1: Furnace classification.

A furnace should be designed so that in a given time, as much of the material as possible can be heated to as uniform a temperature as possible with the least possible fuel and labour.

An industrial furnace or direct fired heater is an equipment used to provide heat for a process or can serve as reactor which provides heats of reaction. Furnace designs vary as to its function, heating duty, type of fuel and method of introducing combustion air. However, most process furnaces have some common features. Schematic diagram of an industrial process furnace is given in Fig. 17.2.

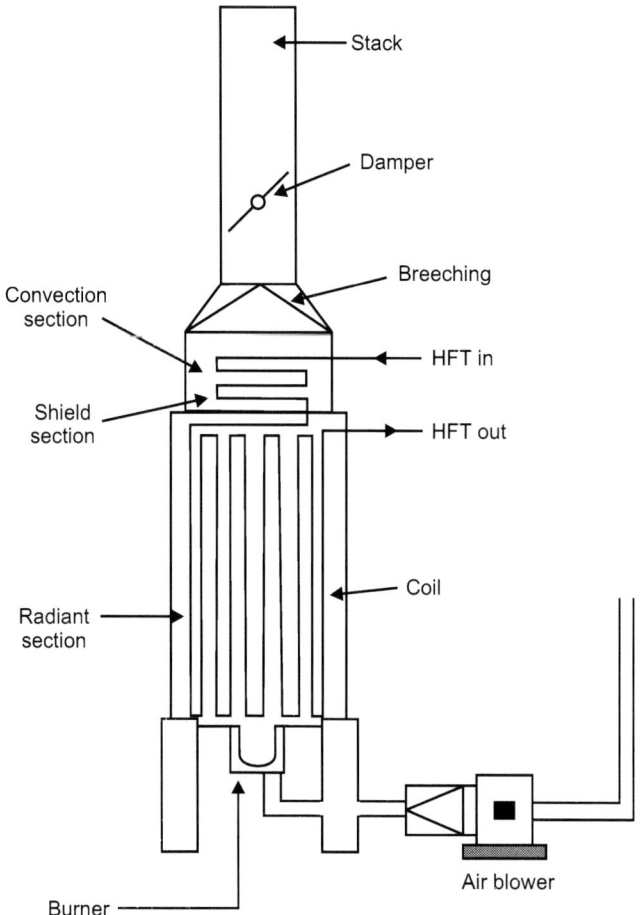

Figure 17.2: Schematic diagram of an industrial process furnace.

Fuel flows into the burner and is burnt with air provided from an air blower. There can be more than one burner in a particular furnace which can be arranged in cells which heat a particular set of tubes. Burners can also be floor mounted, wall mounted or roof mounted depending on design. The flames heat up the tubes, which in turn heat the fluid inside in the first part of the furnace known as the radiant section or firebox. In this chamber where combustion takes place, the heat is transferred mainly by radiation to tubes around the fire in the chamber. The heating fluid passes through the tubes and is thus heated to the desired temperature. The gases from the combustion are known as flue gas. After the flue gas leaves the firebox, most furnace designs include a convection section where more heat is recovered before venting to the atmosphere through the flue gas stack.

Industries commonly use their furnaces to heat a secondary fluid with special additives like anti-rust and high heat transfer efficiency. This heated fluid is then circulated round the whole plant to heat exchangers to be used wherever heat is needed instead of directly heating the product line as the product or material may be volatile or prone to cracking at the furnace temperature.

17.3 Furnace energy supply

The products of flue gases directly contact the stock, so the type of fuel chosen is of importance. For example, some materials will not tolerate sulphur in the fuel. In addition, use of solid fuels will generate particulate matter, which may affect the stock placed inside the furnace. Hence, the majority of furnaces use liquid fuel, gaseous fuel, or electricity as energy input. Melting furnaces for steel and cast iron use electricity in induction and arc furnaces. Non-ferrous melting furnaces utilise oil as fuel.

17.3.1 Oil-fired furnaces

Furnace oil is the major fuel used in reheating and heat treatment furnaces. Low Diesel Oil (LDO) is used in furnaces where the presence of sulphur is undesirable. Some furnaces operate with efficiencies as low as 7% as against up to 90% achieved in other combustion equipment such as boilers.

Factors reducing efficiency include the high temperatures at which the furnaces have to operate, huge thermal masses, high exit flue gas temperatures, standing losses during soaking, partial load operations, losses due to lack of temperature controls, opening losses, delays in stock removal even after heating has been achieved, and others. For example, a furnace heating the stock to 1200°C will have its exhaust gases leaving at least at 1200°C if no heat recovery is installed, resulting in a huge heat loss through the stack.

17.4 Typical furnace systems

17.4.1 Forging furnaces

Forging furnaces use an open fireplace system, and most of the heat is transmitted by radiation. It is used for preheating billets and ingots to attain forging temperature. Thus, furnace temperature is maintained at 1200–1250°C. The typical loading in a forging furnace is 5–6 tons, with the furnace operating for 16 to 18 hours daily. The total operating cycle can be divided into: (i) heat-up time (ii) soaking time and (iii) forging time.

Specific fuel consumption depends upon the type of material and number of such 'reheats' required. Basic diagram of furnace operation is shown in Fig. 17.3.

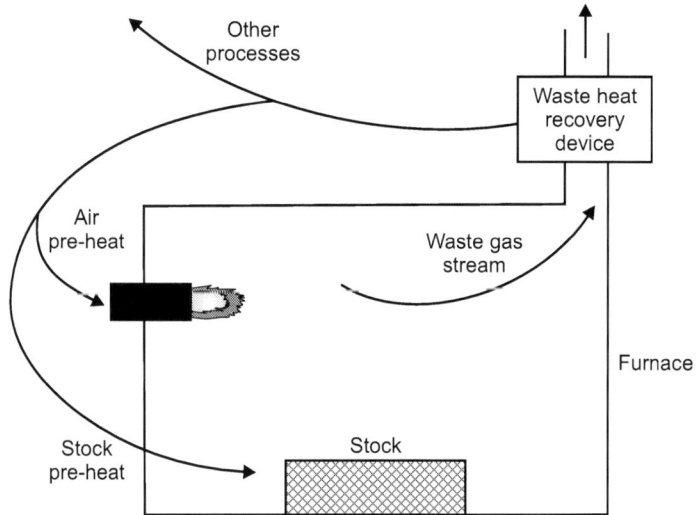

Figure 17.3: Basic diagram of furnace operation.

17.4.2   Rerolling mill furnaces

Batch type furnaces

Batch type furnaces are used for heating scrap, small ingots, and billets weighing 2–20 kg. For batch type rerolling, charging and discharging of the 'material' is done manually, and the final product is in the form of rods, strips, etc. The operating temperature is 1200°C. The total cycle time can be divided into heat-up time and rerolling time.

Continuous pusher type furnaces

The process flow and operating cycles of a continuous pusher type of furnace is the same as that of a batch furnace. The operating temperature is 1250°C. The material or stock recovers a part of the heat in flue gases as it moves down the length of the furnace. Heat absorption by the material in the furnace is slow, steady, and uniform throughout the cross-section compared with a batch type furnace.

17.5 Performance evaluation of a typical furnace

A furnace heat balance will have the components shown in Fig. 17.4. It can be seen that some of these losses do not exist in equipment like boilers, which is the reason why boilers are often much more efficient than furnaces.

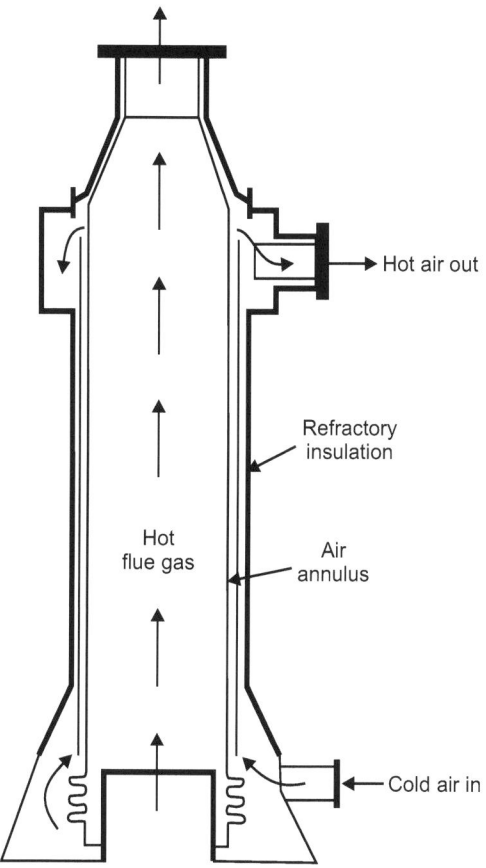

Figure 17.4: Furnace efficiency.

Normal furnace losses include the following:

1. Heat storage in the furnace structure.
2. Losses from the furnace outside walls or structure.
3. Heat transported out of the furnace by the load conveyors, fixtures, trays.
4. Radiation losses from openings, hot exposed parts, etc.
5. Heat carried by cold air infiltration into the furnace.
6. Heat carried by the excess air used in the burners.

17.5.1 Furnace efficiency

The efficiency of a furnace is the ratio of useful output to heat input. The furnace efficiency can be determined by both a direct and an indirect method.

Direct method testing

The efficiency of the furnace can be computed by measuring the amount of fuel consumed per unit weight of material produced from the furnace.

$$\text{Thermal efficiency of the furnance} = \frac{\text{Heat in the stock}}{\text{Heat in the fuel consumed}}$$

The quantity of heat to be imparted (Q) to the stock can be found from the formula:

$$Q = m \times C_p \times (t_2 - t_1)$$

where,

Q = Quantity of heat in kCal

m = Weight of the material in kg

C_p = Mean specific heat, in kCal/kg°C

t_2 = Final temperature desired, in °C

t_1 = Initial temperature of the charge before it enters the furnace, in °C

Indirect method testing

Similar to the method of evaluating boiler efficiency by indirect methods, furnace efficiency can also be calculated by an indirect method. Furnace efficiency is calculated after subtracting sensible heat loss in flue gas, loss due to moisture in flue gas, heat loss due to openings of the furnace, heat loss through the furnace skin, and other unaccounted losses from the input to the furnace.

The parameters that must be taken into account in order to calculate furnace efficiency using the indirect method include hourly furnace oil consumption, material output, excess air quantity, temperature of flue gas, temperature of the furnace at various zones, skin temperature and hot combustion air temperature. Efficiency is determined by subtracting all the heat losses from 100.

Measurement parameters

The following measurements should be made to calculate the energy balance in oil-fired reheating furnaces (e.g. heating furnaces):

Instruments like infrared thermometer, fuel consumption monitor, surface thermocouple and other measuring devices are required to measure the above parameters.

Reference manuals should be referred to for data like specific heat, humidity, etc.

1. Weight of stock/number of billets heated.
2. Temperature of furnace walls, roof, etc.
3. Flue gas temperature.
4. Flue gas analysis.
5. Fuel oil consumption.

17.6   General fuel economy measures in furnaces

17.6.1   Combustion with minimum excess air

To obtain complete combustion of fuel with the minimum amount of air, it is necessary to control air infiltration, maintain the pressure of combustion air, and monitor fuel quality and excess air. Higher excess air will reduce flame temperature, furnace temperature, and heating rate.

On the other hand, if the excess air is reduced, then unburnt components in flue gases will increase and will be carried away in the flue gases through the stack. Therefore the optimisation of combustion air is the most attractive and economical measure for energy conservation. The impact of this measure is higher when the temperature of the furnace is high.

The amount of heat lost in the flue gases depends upon amount of excess air. In the case of a furnace carrying away flue gases at 900°C, percentage of heat lost is shown in Table 17.1.

Table 17.1: Heat loss in flue gas based on excess air level.

Excess air	% of total heat in the fuel carried away by waste gases (flue gas temperature 900°C)
25	48
50	55
75	63
100	71

17.6.2   Correct heat distribution

For proper heat distribution, when using oil burners, the following measures should be taken:

1. Prevent flame impingement by aligning the burner properly to avoid touching the material. The flame should not touch any solid object and should propagate clear of any solid object. Any obstruction will disturb the atomised fuel particles, thus affecting combustion and creating black

smoke. If flame impinges on the stock, there will be an increase in scale losses, which are often hidden in nature but precious in value.

2. If the flames impinge on refractories, the incomplete combustion products can settle and react with the refractory constituents at high flame temperatures.

3. The flames of different burners in the furnace should stay clear of each other. If they intersect, inefficient combustion will occur. It is desirable to stagger the burners on the opposite sides.

4. The burner flame has a tendency to travel freely in the combustion space just above the material. In small reheating furnaces, the axis of the burner is never placed parallel to the hearth but always at an upward angle. Flames should not hit the roof.

5. Designs with few large-capacity burners produce a long flame, which may be difficult to contain within the furnace walls. More burners of less capacity give a better distribution of heat in the furnace and also increase furnace life.

6. For small reheating furnaces, it is desirable to have a long flame with golden yellow colour while firing furnace oil for uniform heating. The flame should not be so long that it enters the chimney and comes out through the top or through the doors. This will happen if excess oil is fired. In such cases, a major portion of additional fuel is carried away from the furnace.

17.7   Furnace waste heat recovery

Since furnaces operate in high temperature domains; the stack temperatures are very high, and waste heat recovery is a major energy efficiency opportunity in furnaces. The recovered waste heat is conventionally used for either pre-heating of combustion air or to heat the material itself.

17.7.1   Some common waste heat recovery systems

Simple double pipe type heat exchangers

Simple double pipe type heat exchangers (Fig. 17.5) generally take the form of concentric cylinders, in which the combustion air passes through the annulus and the exhaust gases from the furnace pass through the center. Such recuperators are very cheap to make, are suitable for use with dirty gases, have a negligible resistance to flow, and can replace the flue or chimney if space is limited.

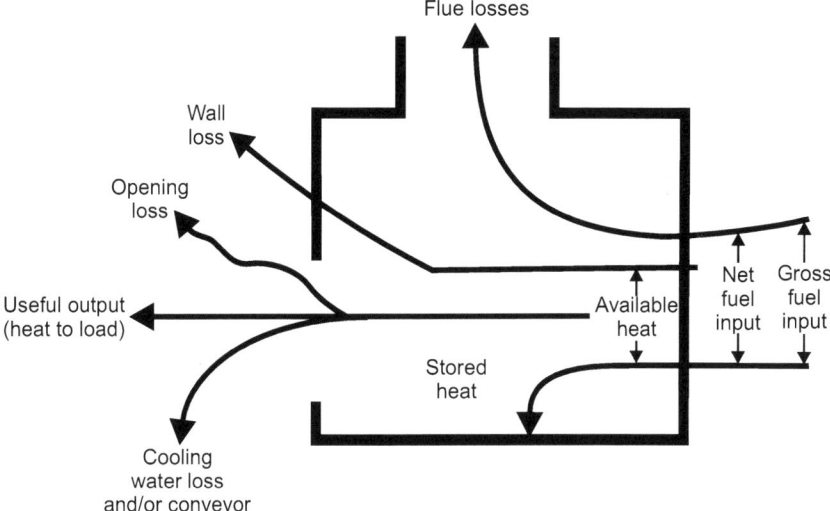

Figure 17.5: Heat exchanger.

Convection recuperators

Convection recuperators consist essentially of bundles of drawn or cast tubes. Internal and/or external fins can be added to assist with heat transfer. The combustion air normally passes through the tubes and the exhaust gases outside the tubes, but there are some applications where this is reversed. For example, with dirty gases, it is easier to keep the tubes clean if the air flows on the outside. Design variations include 'U' tube and double pass systems. Convection recuperators are more suitable for exhaust gas temperatures of less than about 900°C. Beyond 900°C, ceramic recuperators can be used which can withstand higher temperatures.

Regenerative burners

In regenerative burners, the heat of the gases is stored before they exit and incoming combustion air is preheated from this stored heat, switching between two chambers alternately. The cycle time is accurately controlled for switchover. Options include regenerative burners and regenerative air preheaters for heat recovery.

17.7.2   Ceramic fibre for reducing thermal mass

Thermal mass in a furnace contributes significantly to storage heat loss, especially in those batch furnaces with huge refractory brickwork. Every time the furnace goes through a heating cycle, the whole refractory mass needs to be heated all over. Introduction of ceramic fibre to replace conventional fire brick lining helps to reduce thermal mass to almost 15% of the original, helping to lower fuel consumption and batch time reduction as well.

17.7.3 Ceramic coatings to improve emissivity

Heat transfer in a furnace depends mainly on radiation, which in turn depends on emissivity. At higher temperatures, the emissivity decreases and is on the order of 0.3. High emissivity refractory coating, if applied on the internal surface of the furnace, increases the emissivity to 0.8, thus contributing to increased heat transfer.

The application of high-emissivity coatings in furnace chambers promotes rapid and efficient transfer of heat, uniform heating, and extended life of refractories and metallic components such as radiant tubes and heating elements. For intermittent furnaces or where rapid heating is required, use of such coatings is found to reduce fuel consumption up to 10%. Other benefits are temperature uniformity and increased refractory life.

17.7.4 Waste heat reduction and recovery from furnaces

Thermal efficiency of process heating equipment, such as furnaces, ovens, melters, heaters, and kilns is the ratio of heat delivered to a material and heat supplied to the heating equipment. For most heating equipment, a large amount of the heat supplied is wasted in the form of exhaust or flue gases. These losses depend on various factors associated with the design and operation of the heating equipment.

17.7.5 Heat losses from fuel-fired heating equipment

Waste-gas heat losses are unavoidable in the operation of all fuel-fired furnaces, kilns, boilers, ovens, and dryers. Air and fuel are mixed and burned to generate heat, and a portion of this heat is transferred to the heating device and its load. When the energy transfer reaches its practical limit, the spent combustion gases are removed (exhausted) from the furnace via a flue or stack to make room for a fresh charge of combustion gases. At this point, the exhaust flue gases still hold considerable thermal energy, often more than what was left behind in the process. In many fuel-fired heating systems, this waste heat is the greatest source of heat loss in the process, often greater than all the other losses combined.

Reducing these losses should be a high priority for anyone interested in improving the energy efficiency of furnaces and other process heating equipment. The first step in reducing waste heat in flue gases requires close attention and proper measures to reduce all heat losses associated with the furnace. Any reduction in furnace heat losses will be multiplied by the overall available heat factor. This could result in much higher energy savings. The multiplier effect and available heat factor are explained in greater detail in the following sections.

These furnace losses include:

1. Heat storage in the furnace structure.
2. Losses from the furnace outside walls or structure.
3. Heat transported out of the furnace by the load conveyors, fixtures, trays, etc.
4. Radiation losses from openings, hot exposed parts, etc.
5. Heat carried by the cold air infiltration into the furnace.
6. Heat carried by the excess air used in the burners.

All of these losses can be estimated by using the Process Heating Assessment and Survey Tool (PHAST) software tool.

Reducing waste heat losses brings additional benefits, among them:

1. Lower energy component of product costs.
2. Improved furnace productivity.
3. Lower emissions of carbon monoxide (CO), nitrogen oxides (NO_x) and unburned hydrocarbons (UHCs).
4. May contribute to more consistent product quality and better equipment reliability.

17.7.6   Waste-gas losses

To answer this, the flow of heat in a furnace, boiler, or oven must be understood. The purpose of a heating process is to introduce a certain amount of thermal energy into a product, raising it to a certain temperature to prepare it for additional processing, change its properties, or some other purpose. To carry this out, the product is heated in a furnace or oven.

First, the metal structure and insulation of the furnace must be heated so their interior surfaces are about the same temperature as the product they contain. This stored heat is held in the structure until the furnace shuts down, then it leaks out into the surrounding area. The more frequently the furnace is cycled from cold to hot and back to cold again, the more frequently this stored heat must be replaced.

In addition, because the furnace cannot run production until it has reached the proper operating temperature, the process of storing heat in it causes lost production time. Fuel is consumed with no useful output.

Wall losses: Additional heat losses take place while the furnace is in production. Wall or transmission losses are caused by the conduction of heat

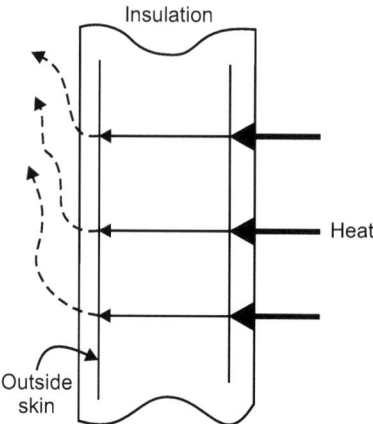

Figure 17.6: Wall loss.

through the walls, roof and floor of the heating device, as shown in Fig. 17.6. Once that heat reaches the outer skin of the furnace and radiates to the surrounding area or is carried away by air currents, it must be replaced by an equal amount taken from the combustion gases. This process continues as long as the furnace is at an elevated temperature.

Material handling losses: Many furnaces use equipment to convey the work into and out of the heating chamber, and this can also lead to heat losses. Conveyor belts or product hangers that enter the heating chamber cold and leave it at higher temperatures drain energy from the combustion gases. In car bottom furnaces, the hot car structure gives off heat to the room each time it rolls out of the furnace to load or remove work. This lost energy must be replaced when the car is returned to the furnace.

Cooling media losses: Water or air cooling protects rolls, bearings, and doors in hot furnace environments, but at the cost of lost energy. These components and their cooling media (water, air, etc.) become the conduit for additional heat losses from the furnace. Maintaining an adequate flow of cooling media is essential, but it might be possible to insulate the furnace and load from some of these losses.

Radiation (opening) losses: Furnaces and ovens operating at temperatures above 1000°F might have significant radiation losses, as shown in Fig. 17.7. Hot surfaces radiate energy to nearby colder high-temperature furnace can attest to the huge amount of thermal energy beamed into the room.

Anywhere or anytime there is an opening in the furnace enclosure, heat is lost by radiation, often at a rapid rate. These openings include the furnace flues and stacks themselves, as well as doors left partially open to accommodate oversized work in the furnace.

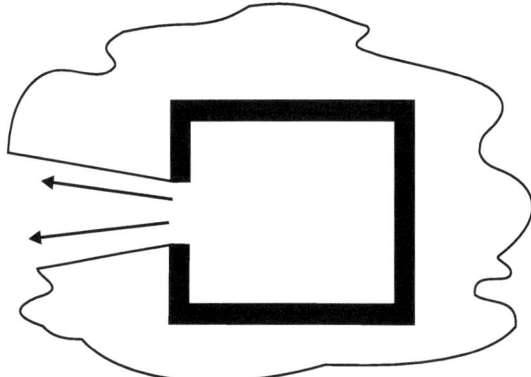

Figure 17.7: Radiation loss from heated to colder surface.

Waste-gas losses: All the losses mentioned above–heat storage, wall transmission, conveyor and radiation–compete with the workload for the energy released by the burning fuel-air mixture. However, these losses could be dwarfed by the most significant source of all, which is waste-gas loss.

Waste-gas loss, also known as flue gas or stack loss, is made up of the heat that cannot be removed from the combustion gases inside the furnace. The reason is heat flows from the higher temperature source to the lower temperature heat receiver.

In effect, the heat stream has hit bottom. If, for example, a furnace heats products to 1500°F, the combustion gases cannot be cooled below this temperature without using design or equipment that can recover heat from the combustion gases. Once the combustion products reach the same temperature as the furnace and load, they cannot give up any more energy to the load or furnace, so they have to be discarded. At 1500°F temperature, the combustion products still contain about half the thermal energy put into them, so the waste-gas loss is close to 50%. The other 50%, which remains in the furnace, is called available heat. The load receives heat that is available after storage in furnace walls, and losses from furnace walls, load conveyors, cooling media and radiation have occurred.

This makes it obvious that the temperature of a process, or more correctly, of its exhaust gases, is a major factor in its energy efficiency. The higher the temperature, the lower the efficiency. Another factor that has a powerful effect is the fuel-air ratio of the burner system.

Fuel-air ratios: For every fuel, there is a stoichiometric, amount of air required to burn it. One cubic foot of natural gas, for example, requires about 10 cubic feet of combustion air. Stoichiometric, or on-ratio combustion will produce the highest flame temperatures and thermal efficiencies.

However, combustion systems can be operated at other ratios. Sometimes, this is done deliberately to obtain certain operating benefits, but often, it happens simply because the burner system is out of adjustment. The ratio, as shown in Fig. 17.8, can go either rich (excess fuel or insufficient air) or lean (excess air). Either way, it wastes fuel. Because there is not enough air for complete combustion, operating the burners at rich combustion conditions wastes fuel by allowing it to be discarded with some of its energy unused. It also generates large amounts of carbon monoxide (CO) and unburned hydrocarbons (UHCs).

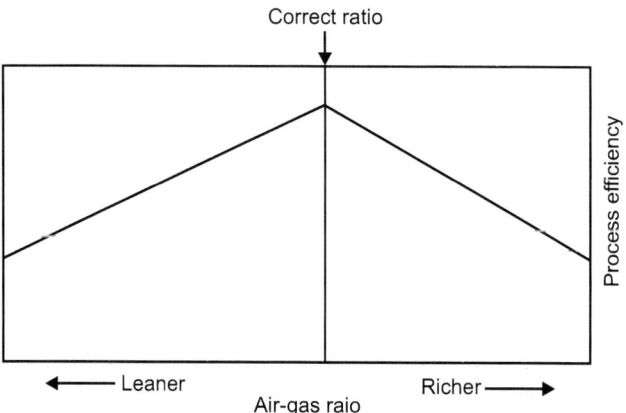

Figure 17.8: Effect of off-ratio operation on furnace efficiency.

At first glance, operating lean might seem to be a better proposition because all the fuel is consumed. Indeed, a lean operation produces no flammable, toxic by-products of rich combustion, but it does waste energy. Excess air has two effects on the combustion process. First, it lowers the flame temperature by diluting the combustion gases, in much the same way cold water added to hot produces warm water. This lowers the temperature differential between the hot combustion gases and the furnace and load, which makes heat transfer less efficient. More damaging, however, is the increased volume of gases that are exhausted from the process.

The products of stoichiometric combustion and the excess air at the same temperature. The excess air becomes one more competitor for the energy demand in the process. Because this is part of the combustion process, excess air goes to the head of the line, taking its share of the heat before the furnace and its contents. The results can be dramatic. In a process operating at 2000°F, available heat at stoichiometric ratio is about 45% (55% goes out the stack). Allowing just 20% excess air into the process (roughly a 12-to-1 ratio for natural gas) reduces the available heat to 38%. Now, 62% of the total heat input goes out the stack, the difference being carried away by that relatively small amount of excess air. To maintain the same temperatures and production rates in the furnace, 18% more fuel must be burned.

Air infiltration: Excess air does not necessarily enter the furnace as part of the combustion air supply. It can also infiltrate from the surrounding room if there is a negative pressure in the furnace. Because of the draft effect of hot furnace stacks, negative pressures are fairly common, and cold air slips past leaky door seals and other openings in the furnace. Figure 17.9 illustrates air infiltration from outside the furnace.

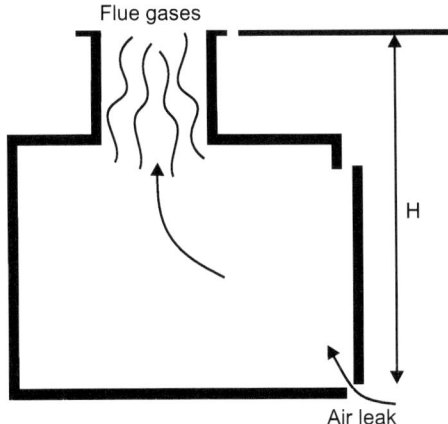

Figure 17.9: Air infiltration from furnace opening.

Once in the furnace, air absorbs precious heat from the combustion system and carries it out the stack, lowering the furnace efficiency. A furnace pressure control system may be an effective way to deal with this.

17.7.7 Furnace scheduling and loading

A commonly overlooked factor in energy efficiency is scheduling and loading of the furnace. 'Loading' refers to the amount of material processed through the furnace or oven in a given period of time. It can have a significant effect on the furnace's energy consumption when measured as energy used per unit of production, for example, in British thermal units per pound (Btu/lb).

Certain furnace losses (wall, storage, conveyor and radiation) are essentially constant regardless of production volume; therefore, at reduced throughputs, each unit of production has to carry a higher burden of these fixed losses. Flue gas losses, on the other hand, are variable and tend to increase gradually with production volume. If the furnace is pushed past its design rating, flue gas losses increase more rapidly, because the furnace must be operated at a higher temperature than normal to keep up with production. Total energy consumption per unit of production will follow the curve in Fig. 17.10, which shows the lowest at 100% of furnace capacity and progressively higher the farther throughputs deviate from 100%. Furnace efficiency varies inversely with the total energy consumption. The lesson here is that furnace operating schedules and load sizes should be selected to keep the furnace operating as near to 100% capacity as possible. Idle and partially loaded furnaces are less efficient.

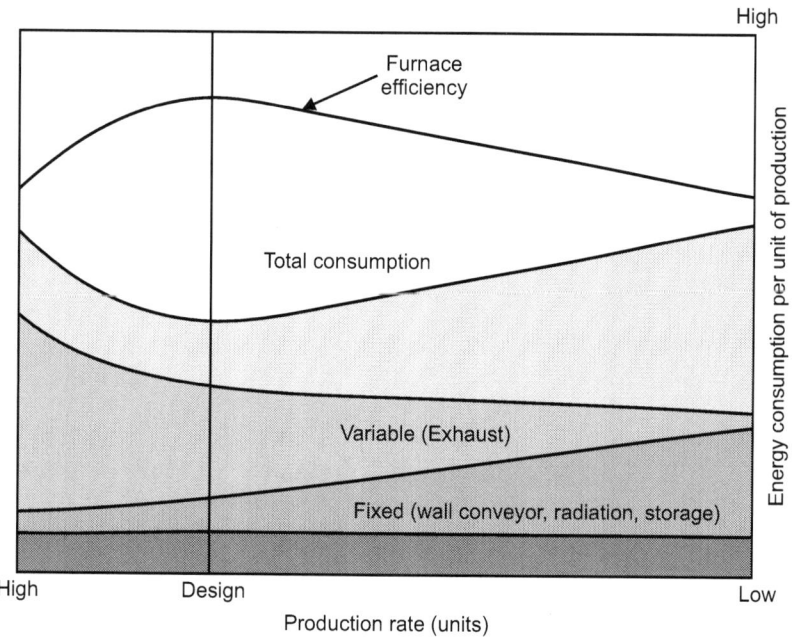

Figure 17.10: Impact of production rate on energy consumption per unit of production.

Minimise exhaust gas temperature and mass or volume of exhaust gases: The highest priority is to minimise exhaust gas temperature and mass or volume of exhaust gases.

1. The furnace exhaust gas temperature depends on many factors associated with the furnace operation and heat losses discussed above. It can be measured directly or can be assumed to 100°–200°F above the control temperature for the furnace zone where the flue gases are exhausted.

2. The exhaust mass flow depends on the combustion air flow, fuel flow and the air leakage into the furnace. Measurement of fuel flow together with the percentage of oxygen (or carbon dioxide [CO_2]) in the flue gases can be used to estimate mass or volume of exhaust gases.

3. The flue-gas specific heat (Cp) for most gaseous fuel-fired furnaces can be assumed to be 0.25 Btu/lb per °F or 0.02 Btu/(standard cubic foot per °F) for a reasonably accurate estimate of flue gas heat losses.

Minimise exhaust gas temperatures: Excessive exhaust gas temperatures can be the result of poor heat transfer in the furnace. If the combustion gases are unable to transfer the maximum possible heat to the furnace and its contents, they will leave the furnace at higher temperatures than necessary. Optimising heat transfer within the furnace requires different methods for different situations. Heat transfer will provide greater insight into how transfer takes place and what can be done to improve it.

Overloading a furnace can also lead to excessive stack temperatures. To get the proper rate of heat transfer, combustion gases must be in the heating chamber for the right amount of time. The natural tendency of an overloaded furnace is to run colder than optimal, unless the temperature is set artificially high. This causes the burners to operate at higher than normal firing rates, which increase combustion gas volumes. The higher gas flow rates and shorter time in the furnace cause poor heat transfer, resulting in higher temperature for the flue gases. Increased volumes of higher temperature flue gases lead to sharply increased heat losses. Overly ambitious production goals might be met, but at the cost of excessive fuel consumption.

Minimise exhaust gas volumes: Avoiding overloading and optimising heat transfer are two ways to lower waste gas flows, but there are others.

The most potent way is to closely control fuel-air ratios. Operating the furnace near the optimum fuel-air ratio for the process also controls fuel consumption. The best part is that it can usually be done with the existing control equipment. All that is required is a little maintenance attention.

Some reduction in exhaust volumes will be the indirect result of efficiencies applied elsewhere. As mentioned above, flue gas losses are a fixed percentage of the total heat input to the furnace. As shown in Fig. 17.11, any reduction in heat storage, wall, conveyor or radiation losses will be multiplied by the available heat factor. For example, on a furnace operating at 50% available heat (50% exhaust gas loss), lowering wall losses by 100,000 Btu per hour (Btu/hr) will permit a firing rate reduction of 200,000 Btu/hr. That is 100,000 Btu/hr for the wall loss and 100,000 Btu/hr for the accompanying exhaust gas loss.

Use of oxygen enriched combustion air: Ambient air contains approximately 21% oxygen with nitrogen and other inert gases as balance. The total volume of exhaust gases could be reduced by increasing the oxygen content of combustion air, either by mixing in ambient air or by using 100% oxygen. Reducing exhaust gases would result in substantial fuel savings. The exact amount of energy savings depends on the percentage of oxygen in combustion

Flue gas losses multiply wall, conveyor and radiation losses

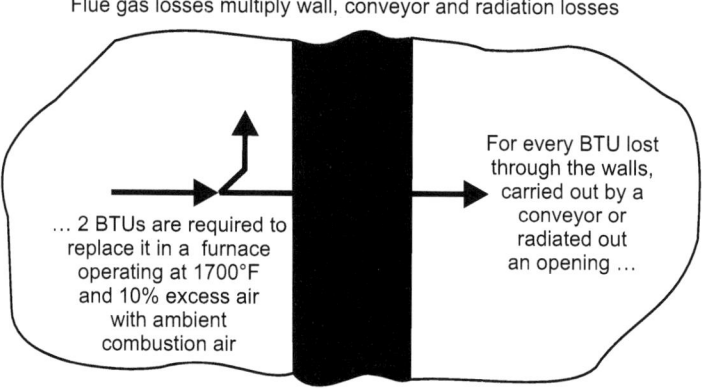

... 2 BTUs are required to replace it in a furnace operating at 1700°F and 10% excess air with ambient combustion air

For every BTU lost through the walls, carried out by a conveyor or radiated out an opening ...

Figure 17.11: Multiplying effect of available heat on furnace losses.

air and the flue gas temperature. Higher values of oxygen and flue gas temperature offer higher fuel savings. Obviously, the fuel savings would have to be compared to the cost of oxygen to estimate actual economic benefits.

17.7.8 Waste heat recovery

Reducing exhaust losses should always be the first step in a well-planned energy conservation programme. Once that goal has been met, consider the next level–waste heat recovery. Waste heat recovery elevates furnace efficiency to higher levels, because it extracts energy from the exhaust gases and recycles it to the process. Significant efficiency improvements can be made even on furnaces that operate with properly tuned ratio and temperature controls. There are four widely used methods:

Direct heat recovery to the product

If exhaust gases leaving the high-temperature portion of the process can be brought into contact with a relatively cool incoming load, energy will be transferred to the load and preheats the load. This reduces the energy that finally escapes with the exhaust (Fig. 17.12). This is the most efficient use of waste heat in the exhaust.

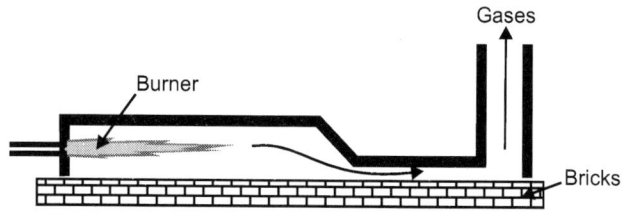

Figure 17.12: Direct preheating of incoming work.

Use of waste heat recovery to preheat combustion air is commonly used in medium- to high temperature furnaces. Use of preheated air for the burners reduces the amount of purchased fuel required to meet the process heat requirements. Figure 17.13 shows the effect of preheating combustion air on exhaust gas heat losses. Preheating of combustion air requires the use of a recuperator or a regenerator.

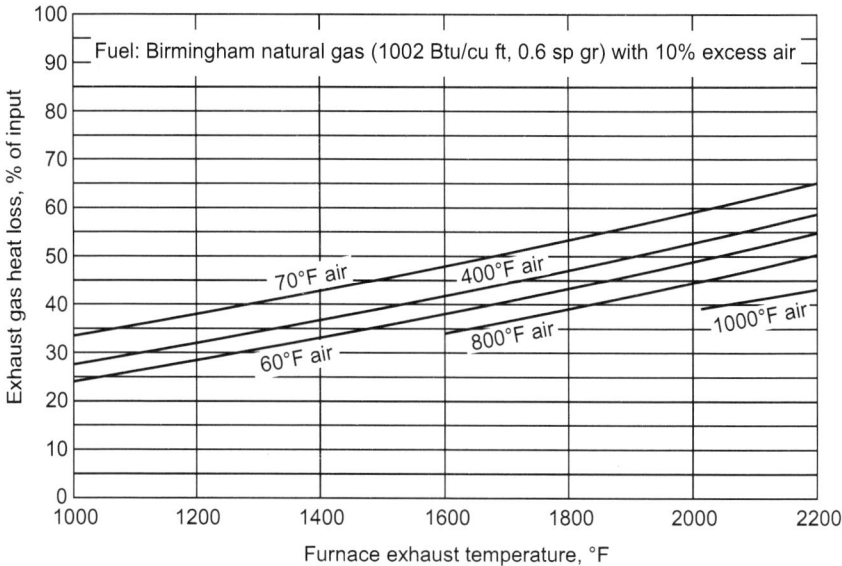

Figure 17.13: Exhaust gas losses with ambient and preheated air.

Recuperators

A recuperator (Fig. 17.14) is a gas-to-gas heat exchanger placed on the stack of the furnace. There are numerous designs, but all rely on tubes or plates to transfer heat from the outgoing exhaust gas to the incoming combustion air, while keeping the two streams from mixing. Recuperators are the most widely used heat recovery devices.

Regenerators

These are basically rechargeable storage batteries for heat. A regenerator (Fig. 17.15) is an insulated container filled with metal or ceramic shapes that can absorb and store relatively large amounts of thermal energy. During the operating cycle, process exhaust gases flow through the regenerator, heating the storage medium. After a while, the medium becomes fully heated (charged). The exhaust flow is shut off and cold combustion air enters the unit. As it passes through, the air extracts heat from the storage medium, increasing in

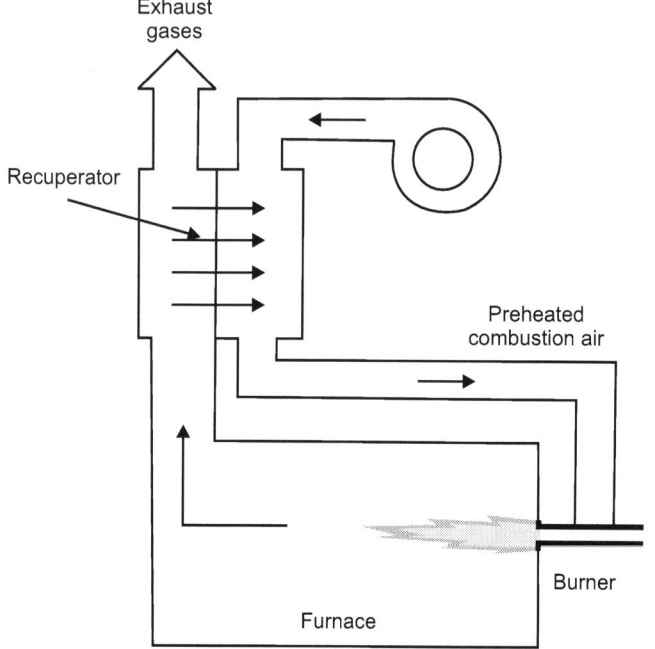

Figure 17.14: Recuperator system for preheating combustion air losses.

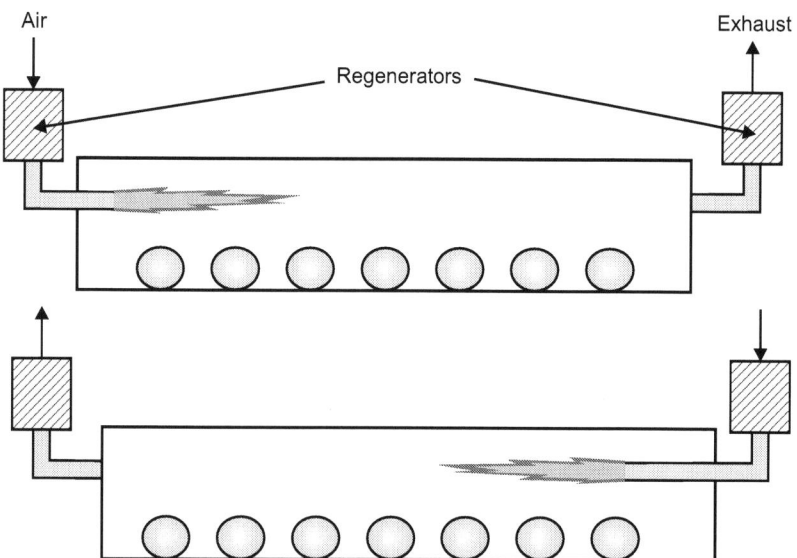

Figure 17.15: Regenerator system for storing thermal energy.

temperature before it enters the burners. Eventually, the heat stored in the medium is drawn down to the point where the regenerator requires recharging. At that point, the combustion air flow is shut off and the exhaust gases return to the unit. This cycle repeats as long as the process continues to operate.

For a continuous operation, at least two regenerators and their associated burners are required. One regenerator provides energy to the combustion air, while the other recharges. In this sense, it is much like using a cordless power tool; to use it continuously, one must have at least two batteries to swap out between the tool and the recharger.

An alternate design of regenerator uses a continuously rotating wheel containing metal or ceramic matrix. The flue gases and combustion air pass through different parts of the wheel during its rotation to receive heat from flue gases and release heat to the combustion air.

Use of waste heat boiler

Use of a waste heat boiler to recover part of the exhaust gas heat is an option for plants that need a source of steam or hot water. The waste heat boiler is similar to conventional boilers with one exception: it is heated by the exhaust gas stream from a process furnace instead of its own burner. Waste heat boilers may be the answer for plants seeking added steam capacity. Remember, however, that the boiler generates steam only when the process is running.

Not all processes are candidates for waste heat recovery. Exhaust volumes and temperatures may be too low to provide financial justification, but if the exhaust temperature is above 1000°F, waste heat recovery is worth investigating.

17.7.9 Energy reduction and recovery strategy

A comprehensive programme for reducing furnace energy consumption involves two types of activities. The first deals with achieving the best possible performance from the existing equipment. Equipment modifications, if required, are relatively modest. The second involves major equipment modifications and upgrades that can make substantial reductions in energy consumption. These techniques and their benefits are summarised in Table 17.2.

Thus, obtaining the maximum efficiency and productivity from industrial furnaces and ovens is a two-step process. First, get the equipment up to its peak performance by reducing heat losses, improving production scheduling and closely controlling gas-air ratios. Once the equipment has reached this level of performance, additional significant improvements may come from recapturing waste heat through direct load preheating, combustion air preheating or steam generation.

Table 17.2: Areas of potential waste heat reduction and recovery improvement.

Energy conservation technique	Heat transfer to load	Reduction of exhaust gas mass	Temperature uniformity	Productivity
Improving the performance of existing equipment				
Reducing heat storage		✓		✓
Reducing wall losses		✓		✓
Reducing material handling losses		✓		✓
Reducing cooling media losses	✓	✓	✓	✓
Reducing radiation losses	✓	✓	✓	✓
Optimising fuel-air ratio	✓	✓		✓
Reducing air infiltration	✓	✓	✓	✓
Improving scheduling and loading		✓		✓
Modifying and upgrading equipment				
Waste heat recovery				
Air preheating	✓	✓		✓
Load preheating		✓	✓	✓
To external processes*				
Oxygen-enhanced combustion	✓	✓		✓
Improving heat transfer with				
advanced burners and controls	✓	✓	✓	✓

* Process is not directly affected, but energy reduction can be achieved at the plant level.

17.8   Tips for improving energy efficiency in furnaces

17.8.1   General

1. Establish a management information system on loading, efficiency, and specific fuel consumption.
2. Prevent infiltration of air, using doors or air curtains.
3. Monitor $O_2/CO_2/CO$ ratios and control excess air level.
4. Improve burner design, combustion control, and instrumentation.
5. Ensure that the furnace combustion chamber is under slight positive pressure.
6. Use ceramic fibre linings in the case of batch operations.
7. Match the load to the furnace capacity.
8. Retrofit with heat recovery devices.
9. Investigate cycle times and avoid extended hours of runtime and excess heating.

10. Provide temperature controllers.
11. Ensure that the flame does not touch the stock.

17.8.2   Insulation

1. Repair damaged insulation.
2. Use an infrared gun to check for hot wall areas during hot weather.
3. Ensure that all insulated surfaces are clad with aluminum lining.
4. Insulate all flanges, valves, and couplings.

17.8.3   Waste heat recovery

1. Recover maximum heat from flue gases.
2. Ensure upkeep of heating surfaces by regular cleaning.

Energy conservation and nanotechnology

18.1   Introduction

Nanotechnology provide, the potential to enhance energy efficiency across all branches of industry and to economically leverage renewable energy production through new technological solutions and optimised production technologies. Nanotechnology innovations could impact each part of the value-added chain in the energy sector which are discussed below.

18.2   Energy sources

Nanotechnology provide, essential improvement potentials for the development of both conventional energy sources (fossil and nuclear fuels) and renewable energy sources like geothermal energy, sun, wind, water, tides or biomass. Nano-coated, wear resistant drill probes, for example, allow the optimisation of lifespan and efficiency of systems for the development of oil and natural gas deposits or geothermal energy and thus the saving of costs. Further examples are high-duty nanomaterials for lighter and more rugged rotor blades of wind and tidepower plants as well as wear and corrosion protection layers for mechanically stressed components (bearings, gear boxes, etc.).

Nanotechnologies will play a decisive role in particular in the intensified use of solar energy through photovoltaic systems. In case of conventional crystalline silicon solar cells, for instance, increases in efficiency are achievable by antireflection layers for higher light yield.

First and foremost, however, it will be the further development of alternative cell types, such as thin-layer solar cells (among others of silicon or other material systems like copper/irradium/selenium), dye solar cells or polymer solar cells, which will predominantly profit from nanotechnologies. Polymer solar cells are said to have high potential especially regarding the supply of portable electronic devices, due to the reasonably-priced materials and production methods as well as the flexible design. Medium-term development targets are an efficiency of approximately 10% and a lifespan of several years. Here, for example, nanotechnologies could contribute to the optimisation of the layer design and the morphology of organic semi-conductor mixtures in component structures. In the long run, the utilisation of nanostructures, like quantum dots and wires, could allow for solar cell efficiencies of over 60%.

18.3   Energy conversion

The conversion of primary energy sources into electricity, heat and kinetic energy requires utmost efficiency. Efficiency increases, especially in fossil-fired gas and steam power plants, could help avoid considerable amounts of carbon dioxide emissions.

Higher power plant efficiencies, however, require higher operating temperatures and thus heat-resistant turbine materials. Improvements are possible, for example, through nano-scale heat and corrosion protection layers for turbine blades in power plants or aircraft engines to enhance the efficiency through increased operating temperatures or the application of lightweight construction materials (e.g. titanium aluminides).

Nano-optimised membranes can extend the scope of possibilities for separation and climate-neutral storage of carbon dioxide for power generation in coal-fired power plants, in order to render this important method of power generation environmentally friendlier in the long run. The energy yield from the conversion of chemical energy through fuel cells can be stepped up by nano-structured electrodes, catalysts and membranes, which results in economic application possibilities in automobiles, buildings and the operation of mobile electronics.

Thermoelectric energy conversion seems to be comparably promising. Nano-structured semiconductors with optimised boundary layer design contributes to increase in efficiency that could pave the way for a broad application in the utilisation of waste heat, for example in automobiles, or even of human body heat for portable electronics in textiles.

18.4   Energy distribution

Regarding the reduction of energy losses in current transmission, hope exists that the extraordinary electric conductivity of nanomaterials like carbon nanotubes can be utilised for application in electric cables and power lines. Furthermore, there are nanotechnological approaches for the optimisation of superconductive materials for lossless current conduction. In the long run, options are given for wireless energy transport, e.g. through laser, microwaves or electromagnetic resonance. Future power distribution will require power systems providing dynamic load and failure management, demand-driven energy supply with flexible price mechanisms as well as the possibility of feeding through a number of decentralised renewable energy sources. Nanotechnologies could contribute decisively to the realisation of this vision, inter alia, through nano-sensory devices and power-electronical components able to cope with the extremely complex control and monitoring of such grids.

18.5   Energy storage

The utilisation of nanotechnologies for the enhancement of electrical energy stores like batteries and super-capacitors turns out to be downright promising. Due to the high cell voltage and the outstanding energy and power density, the lithium-ion technology is regarded as the most promising variant of electrical energy storage. Nanotechnologies can improve capacity and safety of lithium-ion batteries decisively, as for example through new ceramic, heat-resistant and still flexible separators and high-performance electrode materials.

The company Evonik of U.S.A. pushes the commercialisation of such systems for the application in hybrid and electric vehicles as well as for stationary energy storage.

In the long run, even hydrogen seems to be a promising energy storage for environmentally-friendly energy supply. Apart from necessary nanostructure adjustments, the efficient storage of hydrogen is regarded as one of the critical factors of success on the way to a possible hydrogen management.

Current materials for chemical hydrogen storage do not meet the demands of the automotive industry, which requires a hydrogen-storage capacity of up to ten weight per cent. Various nanomaterials, inter alia based on nanoporous metal-organic compounds, provide development potentials, which seem to be economically realisable at least with regard to the operation of fuel cells in portable electronic devices. Another important field is thermal energy storage. The energy demand in buildings, for example, may be significantly reduced by using phase change materials such as latent heat stores. Interesting, from an economic point of view, are also adsorption stores based on nanoporous materials like zeolites, which could be applied as heat stores in district heating grids or in industry. The adsorption of water in zeolite allows the reversible storage and release of heat.

18.6   Energy usage

To achieve sustainable energy supply, and parallel to the optimised development of available energy sources, it is necessary to improve the efficiency of energy use and to avoid unnecessary energy consumption. This applies to all branches of industry and private households. Nanotechnologies provide a multitude of approaches to energy saving.

Examples are the reduction of fuel consumption in automobiles through lightweight construction materials on the basis of nanocomposites, the optimisation in fuel combustion through wear-resistant, lighter engine components and nanoparticular fuel additives or even nanoparticles for optimised tyres with low rolling resistance.

Considerable energy savings are realisable through tribological layers for mechanical components in plants and machines. Building technology also provides great potentials for energy savings, which could be tapped, for example, by nanoporous thermal insulation material suitably applicable in the energetic rehabilitation of old buildings.

In general, the control of light and heat flux by nanotechnological components, as for example switchable glasses, is a promising approach to reducing energy consumption in buildings.

18.7 Nanotechnology helps solve the world's energy problems

The aim is to explain how nanotechnology can help address present and future sustainable energy needs. The main fields of sustainable energy policies and research are: renewables, conventional energy, more energy efficiency in industrial production, and energy saving.

The relevant technologies and applications include: solar cells, hydrogen and fuel cells, batteries, improvement of light bulbs, fossil fuel, etc., with nanostructured materials and nanopowders, isolation materials, membranes and catalysts, etc.

18.7.1 Solar photovoltaics

Solar Photovoltaic (PV) electricity production is the most obvious technology where nanostructured materials and nanotechnology are contributing to technology development. Solar PV is already competitive in electricity production for homes or villages in remote areas without a connection to the electricity grid. Governments in the U.S., Europe and Japan are subsidising both technology development and installation of PV modules on roofs and integrated in new buildings for private homes, companies, or even churches (in Germany). The dominant technologies are at the moment mono or multicrystalline silicon.

The solar cells are produced by sawing 0.2–0.3 mm thin wafers from lumps of silicon. The problem is that this uses a lot of expensive material, about half of which gets wasted in the sawing process.

Thin film nanostructured alternatives which are currently on the market use an active layer of microns thickness, deposited on a cheap substrate such as glass. These alternatives include amorphous silicon, which is best known from its use in pocket calculators, but is also used in solar panels, on the market for about 15 years. Amorphous silicon is cheaper than crystalline silicon, because it uses 300x less active material. The efficiency is much lower, less than 10% compared to 15%.

Two other available thin film alternatives which entered the market in 2001 are Copper Indium diSelenide (CIS), and Cadmium Telluride CdTe. The market chances of the CdTe technology may be diminished because of environmental concerns. Cadmium is a toxic material. Metallic III-V high performance cells are mostly used in space applications, but also in concentrator cells. Concentrator cells consist of a relatively expensive efficient solar cell, and a device which funnels the incoming sunlight from a wider area to the cell. In the lab, efficiencies up to 40% have been measured. But real world manufacturing never achieves the same high efficiencies. One problem with thin film solar PV based on nanotechnology is that energy conversion is even less efficient than in crystalline silicon. According to a spokesperson from BP Solar, the main bottleneck in thin film PV manufacturing is that nobody can produce large enough areas of the thin films on an industrial scale.

Longer term alternatives include the organic Grätzel cell, first invented in 1991 by prof. Michael Grätzel. The principle is also the basis for other research on solid state variants. Prof. Joop Schoonman at the Technical University of Delft, Netherlands aims to replace the liquid electrolyte with a conducting polymer or inorganic material such as FeS, CuS, CuInS.

18.7.2 Grätzel cells

The organic Grätzel solar cell consists of a 10 μm thin layer of Titanium Dioxide TiO_2 particles, which are 20 nm in diameter. Organic dye molecules are adsorbed in the pores between the TiO_2 particles, surrounded by an electrolyte fluid. The cell is completed by two transparent electricity conducting electrodes, and a catalyst. The efficiency of Grätzel cells is much lower than of commercial crystalline silicon (around 7–8% in stead of around 15%). Therefore they are not competitive in the main market for Solar PV. The EU Nanomax project aims to improve this performance to 15%. Prof. Wim Sinke of ECN in the Netherlands coordinates it. Some start-up companies are already producing Grätzel cells for niche markets.

The company Greatcell www.greatcell.com (now part of Leclanché, a battery producing company in Switzerland) has developed the technology further and now offers its first product. This solar powered clock can work indoors without a battery. It can work indoor, because the organic dye sensitive solar cells can convert low light intensities in electricity.

In Australia, the Sustainable Energy Development Authority (SEDA) is investing US$368,000 in a project to integrate Grätzel organic dye solar cells into the walls of the CSIRO Energy Centre in Newcastle. The start up Sustainable Technologies International Ltd. (STI) will deliver the 200 m^2 PV panels. Nanotechnology is not really difficult, as this example shows: Even a child can make organic solar cells including nanostructured material!

The company Mansolar in the Netherlands manufactures and sells educational kits for school children to make their own organic solar cell, using blackcurrant juice or hibiscus tea as the dye. The company started in 2000 as a spin-off from the Energy Centre Netherlands (ECNs) in Petten.

18.7.3 Hydrogen

There is a lot of discussion at the moment about the Hydrogen Economy, where hydrogen will be the dominant fuel, converted into electricity in fuel cells, leaving only water as waste product. The hydrogen is not freely available in nature in large quantities, so it must be produced by conversion of other energy sources, including fossil fuels and renewables. Only renewables based hydrogen production can contribute to CO_2 emission reduction. Current renewable production methods of hydrogen include H_2 production from biomass, from water by electrolysis (where the electricity has been produced by wind, solar or hydro energy), and the Millennium cell alternative, Hydrogen on demand.

This company is based in Eatontown, New Jersey, U.S.A. since 1998, and has a patented process in which a catalysed reaction between water and sodium borohydrate produce hydrogen for applications in cars. The advantage is that the storage of the sodium borohydrate is inherently safe. It is a derivative of borax, which is a natural raw material with substantial natural reserves.

Hydrogen storage
Hydrogen can be stored in different kinds of materials, in gaseous, liquid or more recently in solid form. Gaseous hydrogen can either be transported through natural gas pipelines, mixed into the natural gas, or stored in gas tanks. Liquid hydrogen is stored in metal vessels at high pressures. In solid form, hydrogen is stored in metal hydrides. In the 1980s, the focus shifted to amorphous hydrides such as NiZr, and from 1990 the focus is on nanostructured hydrides including carbon nanotubes, nano-magnesium based hydrides, metal hydride-carbon nanocomposites, nanochemical hydrides and alanates. Many companies in U.S.A. can offer magnesium hydride and sodium sluminium hydride. At the Fraunhofer Institute for Solar Energy in Freiburg, Germany, researchers developed a hydrogen storage device and fuel cell system which is small enough to integrate in a portable digital camcorder.

18.7.4 Cleaner conventional energy

Nanotechnology can also contribute to the improvement of conventional energy sources including coal, oil, gas, and nuclear energy and electricity. The report covers both nanotechnology contributions to electricity production and to primary energy production. To start with electricity, the production from coal or natural gas can be made more efficient by using nanotechnology in turbine plants. In nuclear energy, nanotechnology can help improve the radiation resistance of the materials.

18.7.5 Batteries

Batteries are needed to supply electrical energy when you can't get it from the electricity grid. This includes mobile applications such as mobile phones, walkmans, but also home or even village power supply in remote areas and in backup systems in case the grid goes down. In the future, rechargeable batteries will be even more needed in combination with renewable electricity production such as by solar photovoltaics. The sun does not shine when you need the light the most: at night. Even though at the moment both rechargeable and non-rechargeable batteries are available on the market, the trend is towards rechargeables. There are basically two types of rechargeable batteries where nano-structured materials are applied and the focus of research. The first and most advanced is Lithium based, for example Li-ion batteries. These are dry batteries. The other type, wet batteries, uses basically the same materials as for hydrogen storage, and are based on metal hydrides, where hydrogen is the chemical energy carrier, or carbon nanotubes. The above mentioned Millenium cell system is also applied in batteries.

18.7.6 Transformation

There are many forms of primary energy, including fossil fuels such as oil and gas, biomass, nuclear energy, and renewables such as wind, sun and hydroenergy. These primary energy sources must be transformed into heat, electricity or mechanical power (movement, pressure, etc.). For some of these energy transformations, there is no efficient or cost effective solution. And for some of these needs for new energy transformation technologies, researchers are developing new nanostructured materials or nanocomponents. Fuel cells for transforming hydrogen or other gases (natural gas, methanol) into electricity is a well known example. But researchers are also working on less visible nanotechnologies such as catalysts and membranes for separating different types of gases. These can be used in fuel cells or other energy transforming technologies.

18.7.7 Greening industrial production

A lot of energy is applied in industrial production. This energy can be produced on site for instance by combined heat and power installations, or using the industrial waste as fuel. Industrial production can also contribute to energy saving by using less energy or materials for the same number of products, or by making the products such as cars lighter, hence more energy efficient in their use.

18.7.8   Energy saving

The most sustainable energy use is no energy use. Governments therefore also stimulate energy saving by consumers as well as industry. Some of these measures imply the use of new technologies, such as improved isolation materials. Nanostructured materials such as nanofoams may play a role here.

Thus, nanotechnology research in Europe can contribute to solving future needs for energy technologies, especially in new generations of solar photovoltaics, the hydrogen economy, more efficient conventional energy production and energy saving for industry as well as consumers. Considering the substantial budgets for research dedicated to nanoresearch including for energy applications, much of this potential is likely to be realised in the coming decades.

18.8   Application of nanotechnology to energy production

Here are some interesting ways that are being explored using nanotechnology to produce more efficient and cost-effective energy:

Generating steam from sunlight: Researchers have demonstrated that sunlight, concentrated on nanoparticles, can produce steam with high energy efficiency. The 'solar steam device' is intended to be used in areas of developing countries without electricity for applications such as purifying water or disinfecting dental instruments. Another research group is developing nano-particles intended to use sunlight to generate steam for use in running power plants.

Producing high efficiency light bulbs: A nano-engineered polymer matrix is used in one style of high efficiency light bulbs. The new bulbs have the advantage of being shatterproof and twice the efficiency of compact fluorescence light bulbs. Other researchers developing high efficiency LED's using arrays of nano-sized structures called plasmonic cavities. Another idea under development is to update incandescent light bulbs by surrounding the conventional filament with crystalline material that converts some of the waste infrared radiation into visible light.

Increasing the electricity generated by windmills: An epoxy containing carbon nanotubes is being used to make windmill blades. Stronger and lower weight blades are made possible by the use of nanotube-filled epoxy. The resulting longer blades increase the amount of electricity generated by each windmill.

Generating electricity from waste heat: Researchers have used sheets of nanotubes to build thermocells that generate electricity when the sides of the cell are at different temperatures. These nanotube sheets could be wrapped around hot pipes, such as the exhaust pipe of your car, to generate electricity from heat that is usually wasted.

Storing hydrogen for fuel cell powered cars: Researchers have prepared graphene layers to increase the binding energy of hydrogen to the graphene surface in a fuel tank, resulting in a higher amount of hydrogen storage and therefore a lighter weight fuel tank. Other researchers have demonstrated that sodium borohydride nanoparticles can effectively store hydrogen.

Clothing that generates electricity: Researchers have developed piezoelectric nanofibres that are flexible enough to be woven into clothing. The fibres can turn normal motion into electricity to power your cell phone and other mobile electronic devices.

Reducing friction to reduce the energy consumption: Researchers have developed lubricants using inorganic buckyballs that significantly reduced friction.

Reducing power loss in electric transmission wires: Researchers at Rice University are developing wires containing carbon nanotubes that would have significantly lower resistance than the wires currently used in the electric transmission grid. Richard Smalley envisioned the use of nano-technology to radically change the electricity distribution grid. Smalley's concept these upgraded transmission wires, which could transmit electricity thousands of miles with insignificant power losses, with local electricity storage capacity in the form of batteries in each building that could store power for 24 hours use.

Reducing the cost of solar cells: Companies have developed nanotech solar cells that can be manufactured at significantly lower cost than conventional solar cells.

Improving the performance of batteries: Companies are currently developing batteries using nanomaterials. One such battery will be as good as new after sitting on the shelf for decades. Another battery can be recharged significantly faster than conventional batteries.

Improving the efficiency and reducing the cost of fuel cells: Nanotechnology is being used to reduce the cost of catalysts used in fuel cells. These catalysts produce hydrogen ions from fuel such as methanol. Nanotechnology is also being used to improve the efficiency of membranes used in fuel cells to separate hydrogen ions from other gases, such as oxygen.

Making the production of fuels from raw materials more efficient: Nanotechnology can address the shortage of fossil fuels, such as diesel and gasoline, by making the production of fuels from low grade raw materials economical. Nanotechnology can also be used to increase the mileage of engines and make the production of fuels from normal raw materials more efficient.

Industrial waste heat recovery

19.1 Introduction

Process heating is a significant source of energy consumption in the industrial and manufacturing sectors, and it often results in a large amount of waste heat that is discharged into the atmosphere. Industrial waste heat refers to energy that is generated in industrial processes without being put to practical use. Waste heat losses arise both from equipment inefficiencies and from thermodynamic limitations on equipment and processes. Industrial process heat recovery effectively recycles this waste heat, which typically contains a substantial amount of thermal energy. The benefits of heat recovery include improving system efficiency, reducing fuel consumption, and reducing facility air emissions. While the type and cost-effectiveness of a heat recovery system are dependent on the process temperature and the facility's thermal requirements, many heat recovery techniques are available across low, medium, and high temperature ranges.

Process heating refers to the application of thermal energy to a product, raising it to a certain temperature to prepare it for additional processing, to change its properties, or for some other purpose. The energy required for process heating accounts for approximately 20% of total industrial energy use in the U.S., Europe and other development countries. As energy costs continue to rise, facilities are constantly in need of ways to improve the performance of their process heating systems and to reduce their energy consumption.

In many fuel-fired heating systems the exhaust gas that is emitted through a flue or stack is the single greatest heat loss. Process heat recovery saves energy by reusing this otherwise lost heat for a variety of thermal loads, such as pre-heated combustion air, boiler feedwater, and process loads, as well as for steam generation.

19.2 Industrial process heat recovery

Process heat recovery involves intercepting the waste streams before they leave the plant, extracting some of the heat they contain, and recycling that heat.

19.2.1 Applications

Heat recovery can be applied in a wide range of industries. For example, the pulp and paper industry can utilise heat recovery through several processes, from pre-heating milling water with steam to cooling effluent wastewater before sending it to waste treatment. The chemical industry can apply heat recovery to most processes, including chemical manufacturing, as well as to emissions control devices such as recuperative and regenerative thermal oxidisers. The petroleum industry can use heat recovery from production water and glycol regenerators, as well as using heat exchangers between wet and dry crude, in the natural gas cleaning process, and in waste treatment operations. The food and beverage industry can achieve savings through the installation of heat exchangers for food pasteurisers, blanch water heat recovery, boiler blow down heat recovery, heating feedstock in the distillation process, and recovery of waste heat from dryers and cookers. In both the commercial and industrial sectors, combined heat and power (CHP) systems can efficiently generate electrical power on-site as well as recovering waste heat to generate hot water or steam for process operations.

19.2.2 Benefits of waste heat recovery

The benefits of heat recovery are multiple: economic, resource (fuel) saving, and environmental. First, recovered heat can directly substitute for purchased energy, thereby reducing the facility's energy consumption and its associated costs; further, waste heat substitution can lower capacity requirements for energy generating equipment, thus reducing capital costs for new installation projects. Second, for a specific heating process, fuel efficiency can be improved through the use of heat recovery, thus reducing the cost of operation. For example, the use of exhaust gas from a fuelfired burner to pre-heat the combustion air can reduce heating energy use by as much as 30%. Third, due to improved equipment efficiency, smaller equipment capacity requirements, and reduced fuel consumption, heat recovery can produce environmental benefits through reductions in emissions of greenhouse gases and atmospheric pollutants.

Sources and quality of waste heat

Waste heat sources can be classified by temperature range, as shown in Table 19.1.

Table 19.1: Waste heat temperature categories.

Category	Temperature ($°F$)
High	1100 to 3000
Medium	400 to 1100
Low	80 to 400

About 92% of process heat energy used by industry is directly provided by fossil fuels. The waste heat generated from direct-fired processes falls in the high and medium temperature ranges. In the high temperature range, sources of waste heat include refining furnaces, steel heating furnaces, glass melting furnaces, and solid waste incinerators.

In the low-temperature range, sources of waste heat include process steam condensate, cooling water from refrigeration condensers, welding machines, boilers, and air compressors. In some applications low-temperature waste heat can be used for pre-heating through heat exchangers. For example, cooling water from a battery of spot welders can be used to pre-heat the ventilating air for winter space heating. In the medium temperature range, sources of waste heat include exhaust gases from steam boilers, gas turbines, reciprocating engines, water heating boiler furnaces, fuel cells, and drying and baking ovens. Potential heat recovery opportunities include, among others, low pressure steam generation and incoming product pre-heating.

High-temperature waste heat is the highest quality and most useful because it provides more heat recovery options and thus greater potential cost-effectiveness than lower temperature waste heat. It can be made available to do work through the utilisation of steam turbines or gas turbines to generate energy in a cogeneration plant. Table 19.2 lists examples of waste heat sources and their potential applications.

Table 19.2: Various waste heat sources and applications.

Sources	Temperature range	Application
Exhaust gas from refining furnaces, steel heating furnaces, glass melting furnaces, solid waste incinerators	High	Hazardous gas reduction Steam generation Water heating Water pre-heating Combustion air pre-heating Power generation
Exhaust from gas turbines, reciprocating engines, incinerators, furnaces	Medium	Pre-heating incoming product Steam generation
Steam boiler blown down		Water heating Water pre-heating Combustion air pre-heating
Exhaust gas from fuel burner Reciprocating engine jacket cooling Waste stream from condensers, boilers, and air compressors	Low	Absorption cooling Dehumidification Feedwater pre-heating Space heating Evaporation

19.3   Heat recovery methods

Various methods for recovering waste heat are given below:

1. High-temperature heat recovery through recuperators and regenerators.
2. Load pre-heating.
3. Combustion air pre-heating.
4. Steam generation in waste heat boilers.
5. Feedwater pre-heating.
6. Heat recovery in condensing boilers.
7. Heat recovery from boiler blow down.
8. Cascade heating.
9. Heat recovery using absorption chillers.
10. Heat recovery using desiccant dehumidifiers.

19.3.1   High temperature heat recovery through recuperators and regenerators

Recuperators and regenerators are two methods of recovering heat from high-temperature processes, such as incineration or thermal oxidation. Recuperators are essentially gas-to-gas heat exchangers, where as the gas coming into the process is pre-heated by the high-temperature gas going out of the process. In regenerators, refractory materials are utilised to absorb heat from the high-temperature gas and release it back to the process, thus reducing the combustion energy. Regenerators typically operate in alternate cycles between two chambers.

Recuperators

Recuperators are gas-to-gas heat exchangers that are installed in the stack of the furnace. There are numerous designs, but all rely on tubes or plates to transfer heat from the outgoing exhaust gas to the incoming combustion air, while keeping the two streams from mixing.

A simple configuration and low-cost recuperator is a metallic radiation recuperator, as shown in Fig. 19.1. The inner tube carries the hot exhaust gas while the external annulus carries the combustion air from the atmosphere to the air inlets of the furnace burners. The assembly is often designed to replace the exhaust stack.

In a recuperator, heat exchange takes place between the flue gases and the air through metallic or ceramic walls. Duct or tubes carry the air for combustion to be pre-heated, the other side contains the waste heat stream.

The simplest configuration for a recuperator is the metallic radiation recuperator, which consists of two concentric lengths of metal tubing. The

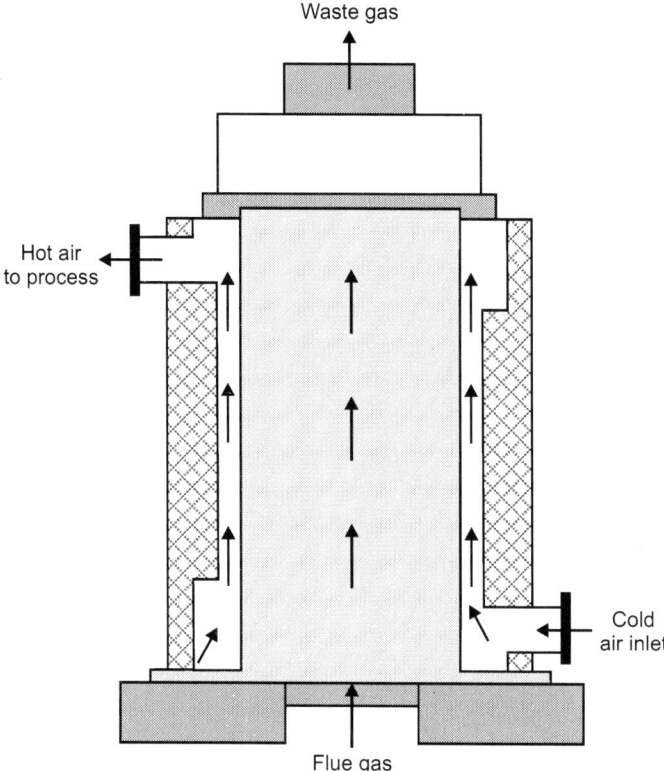

Figure 19.1: Metallic radiation recuperator.

inner tube carries the hot exhaust gases while the external annulus carries the combustion air from the atmosphere to the air inlets of the furnace burners. The hot gases are cooled by the incoming combustion air which now carries additional energy into the combustion chamber. This is energy which does not have to be supplied by the fuel; consequently, less fuel is burned for a given furnace loading. The saving in fuel also means a decrease in combustion air and therefore stack losses are decreased not only by lowering the stack gas temperatures but also by discharging smaller quantities of exhaust gas. The radiation recuperator gets its name from the fact that a substantial portion of the heat transfer from the hot gases to the surface of the inner tube takes place by radiative heat transfer.

The cold air in the annuals, however, is almost transparent to infrared radiation so that only convection heat transfer takes place to the incoming air. As already shown in Fig. 19.1, the two gas flows are usually parallel, although the configuration would be simpler and the heat transfer more efficient if the flows were opposed in direction (or counterflow). The reason for the use of parallel flow is that recuperators frequently serve the additional function of cooling the duct carrying away the exhaust gases and consequently extending its service life.

A second common configuration for recuperators is called the tube type or convective recuperator. The hot gases are carried through a number of parallel small diameter tubes, while the incoming air to be heated enters a shell surrounding the tubes and passes over the hot tubes one or more times in a direction normal to their axes.

If the tubes are baffled to allow the gas to pass over them twice, the heat exchanger is termed a two-pass recuperator; if two baffles are used, a three-pass recuperator, etc. Although baffling increases both the cost of the exchanger and the pressure drop in the combustion air path, it increases the effectiveness of heat exchange. Shell and tube type recuperators are generally more compact and have a higher effectiveness than radiation recuperators, because of the larger heat transfer area made possible through the use of multiple tubes and multiple passes of the gases.

Radiation/convective hybrid recuperator: For maximum effectiveness of heat transfer, combinations of radiation and convective designs are used, with the high-temperature radiation recuperator being first followed by convection type. These are more expensive than simple metallic radiation recuperators, but are less bulky.

19.3.2 Ceramic recuperator

The principal limitation on the heat recovery of metal recuperators is the reduced life of the liner at inlet temperatures exceeding 1100°C. In order to overcome the temperature limitations of metal recuperators, ceramic tube recuperators have been developed whose materials allow operation on the gas side to 1550°C and on the pre-heated air side to 815°C on a more or less practical basis. Early ceramic recuperators were built of tile and joined with furnace cement, and thermal cycling caused cracking of joints and rapid deterioration of the tubes. Later developments introduced various kinds of short silicon carbide tubes which can be joined by flexible seals located in the air headers.

Earlier designs had experienced leakage rates from 8 to 60%. The new designs are reported to last two years with air pre-heat temperatures as high as 700°C, with much lower leakage rates.

Regenerators

Regenerators are essentially rechargeable storage batteries for heat that utilise an insulated container filled with metal or ceramic shapes capable of absorbing and storing large amounts of thermal energy.

The regeneration which is preferable for large capacities has been very widely used in glass and steel melting furnaces. Important relations exist between the size of the regenerator, time between reversals, thickness of brick, conductivity of brick and heat storage ratio of the brick.

In a regenerator, the time between the reversals is an important aspect. Long periods would mean higher thermal storage and hence higher cost. Also long periods of reversal result in lower average temperature of pre-heat and consequently reduce fuel economy.

Accumulation of dust and slagging on the surfaces reduce efficiency of the heat transfer as the furnace becomes old. Heat losses from the walls of the regenerator and air in leaks during the gas period and out-leaks during air period also reduces the heat transfer. Figure 19.2 shows a plate-type regenerator. A plate-type regenerator is constructed of alternate channels that separate adjacent flows of heated and heating gases by a thin wall of conducting metal. Although their use eliminates cross-contamination, they are bulkier, heavier, and more expensive than recuperators.

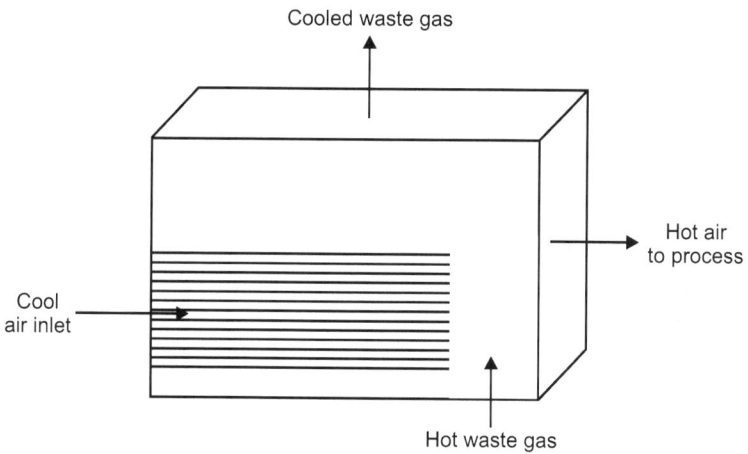

Figure 19.2: Plate-type passive gas-to-gas regenerator.

For the process to operate without interruption, at least two regenerators are required; one provides energy to the combustion air while the other is recharging. Regenerators can operate at temperatures beyond the range of recuperators and at higher efficiency ratings. They are resistant to corrosion and fouling, but because of their back-and-forth switching to maintain continuous operation, they require more complex, more expensive flow control systems than recuperators. These passive air pre-heaters are used in low- and medium-temperature applications.

19.3.3   Load pre-heating

In general, there are direct and indirect heat recovery methods. Direct heat recovery implies directly pre-heating incoming product using the process waste heat. If the high-temperature exhaust fluid can be brought into contact with a relatively cool incoming fluid, energy will be transferred to pre-heat the low-temperature fluid and reduce the energy that finally escapes with the exhaust. Direct heat recovery to the product has the highest potential efficiency because it does not require any 'carrier' to return the energy to the product. It does, however, require a furnace or oven configuration that permits routing the stream of exhaust counter-flow to incoming product or materials.

19.3.4   Combustion air pre-heating

Two methods of combustion air pre-heating use the technologies described above under high-temperature heat recovery through recuperators and regenerators. A large amount of energy is required to heat combustion air from atmospheric temperature to combustion temperature. Pre-heating results in the burners needing less fuel to heat the incoming air to combustion temperature. The most common means of transferring flue gas energy to combustion air is to use a recuperator placed in the exhaust stack or ductwork. This strategy can recover a sizable percentage of the exhaust heat that would otherwise be lost to the atmosphere.

Regenerators can be used for applications where cross-contamination cannot be tolerated. During part of the operating cycle, process exhaust gas flows through the regenerator, heating the storage medium. Once the medium becomes fully charged, the exhaust flow is shut off and cold combustion air enters the unit. As it passes through, the supply air extracts heat from the storage medium and rises in temperature before entering the burners. The operation cycles between a charge mode (top) and a discharge mode (bottom). Whether air pre-heating will be cost-effective is usually determined by the process temperature:

1. Processes at temperatures above 1,600°F are generally good candidates.
2. Processes operating in the range of 1,000°F to 1,600°F may still produce cost-effective savings, but must be evaluated case by case.
3. Processes operating below 1,000°F are typically not worth the cost of installing and maintaining the regenerator system. However, low-temperature processes should still be evaluated for heat recovery potential. If the exhaust gas flow rate is high enough, energy savings may still be achievable.

19.3.5   Heat wheels

A heat wheel is finding increasing applications in low to medium temperature waste heat recovery systems.

It is a sizable porous disk, fabricated with material having a fairly high heat capacity, which rotates between two side-by-side ducts: one is a cold gas duct, the other a hot gas duct. The axis of the disk is located parallel to, and on the partition between, the two ducts. As the disk slowly rotates, sensible heat (moisture that contains latent heat) is transferred to the disk by the hot air and, as the disk rotates, from the disk to the cold air. The overall efficiency of sensible heat transfer for this kind of regenerator can be as high as 85%. Heat wheels have been built as large as 21 metres in diameter with air capacities up to 1130 m^3/min. A variation of the heat wheel is the rotary regenerator where the matrix is in a cylinder rotating across the waste gas and air streams. The heat or energy recovery wheel is a rotary gas heat regenerator, which can transfer heat from exhaust to incoming gases.

Its main area of application is where heat exchange between large masses of air having small temperature differences is required. Heating and ventilation systems and recovery of heat from dryer exhaust air are typical applications.

19.3.6   Heat pipe

A heat pipe can transfer up to 100 times more thermal energy than copper, the best known conductor. In other words, heat pipe is a thermal energy absorbing and transferring system and have no moving parts and hence require minimum maintenance.

The heat pipe comprises of three elements–a sealed container, a capillary wick structure and a working fluid. The capillary wick structure is integrally fabricated into the interior surface of the container tube and sealed under vacuum. Thermal energy applied to the external surface of the heat pipe is in equilibrium with its own vapour as the container tube is sealed under vacuum. Thermal energy applied to the external surface of the heat pipe causes the working fluid near the surface to evaporate instantaneously. Vapour thus formed absorbs the latent heat of vapourisation and this part of the heat pipe becomes an evaporator region. The vapour then travels to the other end of the pipe where the thermal energy is removed causing the vapour to condense into liquid again, thereby giving up the latent heat of the condensation. This part of the heat pipe works as the condenser region. The condensed liquid then flows back to the evaporated region. Heat pipe is shown in Fig. 19.3.

Performance and advantage: The heat pipe exchanger (HPHE) is a lightweight compact heat recovery system. It virtually does not need

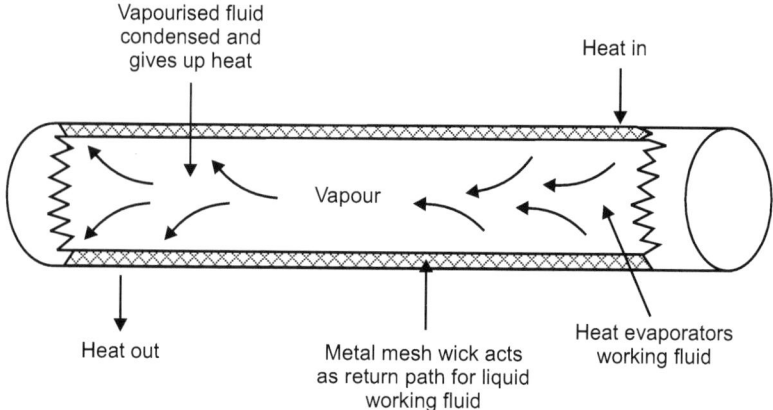

Vapourised fluid
condensed and
gives up heat

Heat in

Vapour

Heat out

Metal mesh wick acts
as return path for liquid
working fluid

Heat evaporators
working fluid

Figure 19.3: Heat pipe.

mechanical maintenance, as there are no moving parts to wear out. It does not need input power for its operation and is free from cooling water and lubrication systems. It also lowers the fan horsepower requirement and increases the overall thermal efficiency of the system. The heat pipe heat recovery systems are capable of operating at 315°C with 60%–80% heat recovery capability.

Typical application: The heat pipes are used in following industrial applications:

1. Process to space heating: The heat pipe heat exchanger transfers the thermal energy from process exhaust for building heating. The pre-heated air can be blended if required. The requirement of additional heating equipment to deliver heated make up air is drastically reduced or eliminated.

2. Process to process: The heat pipe heat exchangers recover waste thermal energy from the process exhaust and transfer this energy to the incoming process air. The incoming air thus become warm and can be used for the same process/other processes and reduces process energy consumption.

3. HVAC applications:

    (a) Cooling: Heat pipe heat exchangers precools the building make up air in summer and thus reduces the total tons of refrigeration, apart from the operational saving of the cooling system. Thermal energy supplied is recovered from the cool exhaust and transferred to the hot supply make up air.

    (b) Heating: The above process is reversed during winter to pre-heat the make up air. The other applications in industries are:

        (i) Pre-heating of boiler combustion air.

(ii) Recovery of waste heat from furnaces.

(iii) Reheating of fresh air for hot air driers.

(iv) Recovery of waste heat from catalytic deodorising equipment.

(v) Reuse of furnace waste heat as heat source for other oven.

(vi) Cooling of closed rooms with outside air.

(vii) Pre-heating of boiler feed water with waste heat recovery from flue gases in the heat pipe economisers.

(viii) Drying, curing and baking ovens.

(ix) Waste steam reclamation.

(x) Brick kilns (secondary recovery).

(xi) Reverberatory furnaces (secondary recovery).

(xii) Heating, ventilating and air-conditioning systems.

19.3.7   Thermal oxidising/combustion emission control

Many chemical facilities need to treat hazardous waste gas from their process lines. A thermal oxiser is used to decompose hazardous gases before releasing them to the atmosphere. In a thermal oxiser, the hazardous gas is passed through an oxidising burner (oxidiser) at a controlled and optimal temperature, typically above 1500°F, at which the volatile organic compounds (VOCs) are converted into safe gases such as water vapour and carbon dioxide. Because thermal oxidisers operate at such high temperatures, these systems have significant heat recovery savings potential.

The simplest technique for meeting regulatory VOC reduction requirements would be to heat the gas in an after burner to more than 1500°F and not recover any heat. Without heat recovery, however, the operation of an afterburner is cost prohibitive unless the gas stream is very rich in VOCs and has a low flow rate. Alternatively, thermal oxidiser efficiency can be optimised by utilising recuperative or regenerative types of heat recovery methods. A recuperative thermal oxidiser uses a gas-to-gas heat exchanger to recover some of the energy from the high-temperature exhaust gas. Regenerative thermal oxidisers are more energy efficient, reaching thermal efficiencies of up to 95%, while recuperative types typically achieve efficiencies of up to 80%. However, the capital cost for regenerative types is higher than for recuperative types. When examining the appropriate type of thermal oxidiser, some considerations are the gas stream volume, flow, temperature, and moisture content; the VOC concentrations; and the desired VOC destruction efficiency. Processes with higher gas flow rates and lower VOC concentrations, for example, are more suitably managed with a regenerative thermal oxidiser. An alternative type of recuperative or regenerative oxidiser is a catalytic oxidiser, in which a catalyst reduces the temperature required to destroy VOCs from around 1500°F to around 800°F. Because the catalyst accelerates VOC destruction and lowers the required operating temperature, a catalytic oxidiser can have a 20% to 30% gain in efficiency over thermal oxidisers.

19.3.8   Steam generation in waste heat boilers

While conventional boilers are fired by fossil fuel, waste heat boilers utilise an exhaust gas stream from external sources to heat the water instead of burning fuel in the burners. Figure 19.4 shows a waste heat boiler for steam generation. Waste heat boilers may be horizontal or vertical shell boilers, or water tube boilers, where hot exhaust gases are passed over parallel tubes containing water.

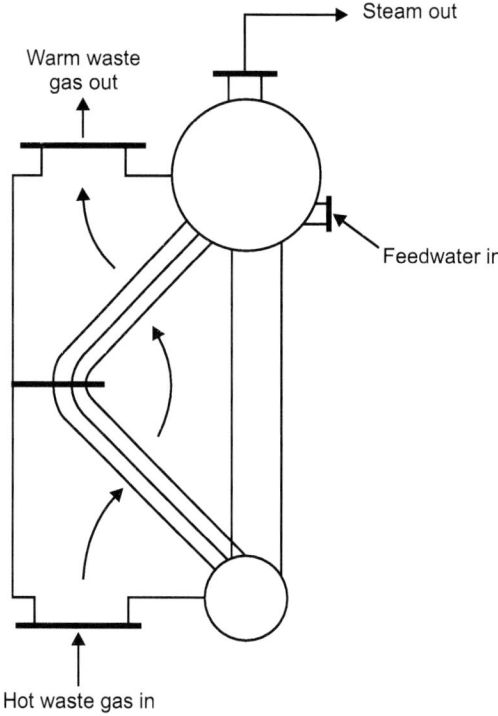

Figure 19.4: Waste heat steam boiler.

The water is vapourised and collected in a steam drum. These boilers can be designed to work with individual applications ranging from gas turbine exhaust to reciprocating engines, incinerators, and furnaces.

Waste heat boilers can be used with most furnace applications, as long as the exhaust gases contain sufficient usable heat to produce steam or hot water at the condition required. For steam generation, the exhaust gas should preferably be above 750°F. For water heating, the exhaust gas should be about 400°F or higher. When the heat source is in the low-temperature range, boilers become bulky. The use of finned tubes extends the heat transfer areas and allows a more compact size.

Waste heat boilers may be an option for facilities looking for additional steam capacity; however, these boilers only generate steam coincident with the process furnace operation. It should be noted too that the physical size of a waste heat boiler may be larger than that of a conventional boiler because the furnace exhaust gas temperature is lower than the flame temperature used in conventional systems. This may pose a disadvantage in retrofits where space is limited.

Boilers using exhaust gas from engines fired by heavy fuel oil must be carefully designed, because the exhaust gas may contain soot, which can form an insulation layer on the tubes and shells of the boiler. When this happens, heat transfer is impeded and the efficiency of the system can drop dramatically. Therefore, the gas exit temperature must be maintained at a predetermined level to prevent dew point from being reached and soot from accumulating inside the boiler.

The exhaust gas capacities of waste heat boilers can range from less than a thousand to almost a million cubic feet per minute. If the waste heat in the exhaust gas is insufficient to generate the required process steam in an application, it may be necessary to add auxiliary fuel burners to the waste heat boiler, or to add an afterburner. Because waste heat boilers do not use burners, they are less expensive to install and operate than a new combustion boiler. However, for an industrial facility to benefit from a waste heat boiler, the waste heat source must coincide with the steam demand that would otherwise be met with a combustion boiler.

19.3.9   Economiser

In case of boiler system, economiser can be provided to utilise the flue gas heat for pre-heating the boiler feed water. On the other hand, in an air pre-heater, the waste heat is used to heat combustion air. In both the cases, there is a corresponding reduction in the fuel requirements of the boiler. A economiser is shown in Fig. 19.5.

For every 22°C reduction in flue gas temperature by passing through an economiser or a pre-heater, there is 1% saving of fuel in the boiler. In other words, for every 60°C rise in feed water temperature through an economiser, or 20°C rise in combustion air temperature through an air pre-heater, there is 1% saving of fuel in the boiler.

Pre-heating boiler feedwater offers the following primary advantages:

1. Reduced fuel usage and increased boiler efficiency.

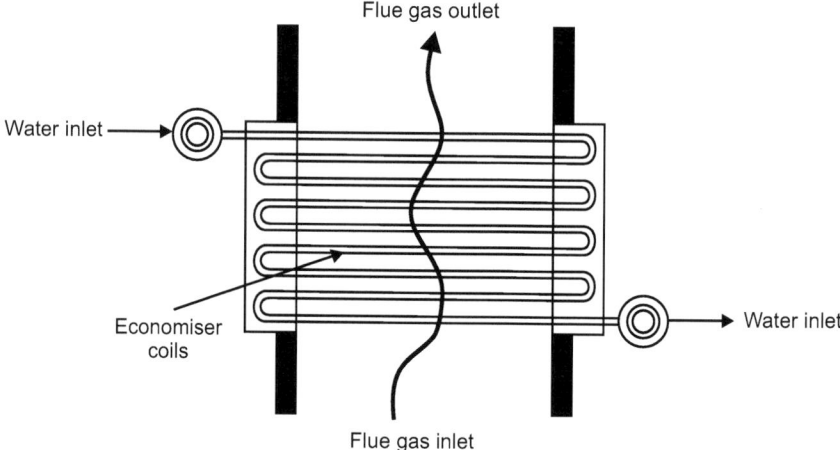

Figure 19.5: Economiser.

2. Reduced emissions due to less fuel use.
3. Quicker response to load changes.
4. Potentially increased steam production.
5. Potentially longer boiler life.

Economisers are available in two types of designs: water-tube and firetube. In a *fire-tube economiser*, flue gas flows inside the tubes heating the surrounding water. This type of economiser has a large water reservoir, which makes it extremely resistant to steaming and eliminates the need for expensive feedwater-proportioning systems. For boilers larger than 400 boiler horsepower or 13.4 MMBtu/hr (1 Boiler HP is about 33.5 MBtu/hr), a water-tube type economiser can be used. In a water-tube economiser like that, feedwater flows through a tube bundle that is heated by the surrounding flue gas. In many cases *water-tube economisers* can be fit directly into the exhaust stack, allowing cost-effective installation. When implementing a feedwater economiser, special consideration must be given to ensure that flue gas is not cooled beyond the low-temperature limit. The lowest temperature to which flue gasses can be cooled depends on the type of fuel being used: 250°F for natural gas, 300°F for coal and low sulphur content fuel oils, and 350°F for high sulphur fuel oils. Cooling below these limits can result in condensation and possible corrosion of the heat exchanger and the exhaust stack.

19.3.10 Waste heat boilers

Waste heat boilers are ordinarily water tube boilers in which the hot exhaust gases from gas turbines, incinerators, etc., pass over a number of parallel tubes containing water. The water is vapourised in the tubes and collected in a steam drum from which it is drawn off for use as heating or processing steam.

Because the exhaust gases are usually in the medium temperature range and inorder to conserve space, a more compact boiler can be produced if the water tubes are finned in order to increase the effective heat transfer area on the gas side. The pressure at which the steam is generated and the rate of steam production depends on the temperature of waste heat. The pressure of a pure vapour in the presence of its liquid is a function of the temperature of the liquid from which it is evaporated. The steam tables tabulate this relationship between saturation pressure and temperature. If the waste heat in the exhaust gases is insufficient for generating the required amount of process steam, auxiliary burners which burn fuel in the waste heat boiler or an after-burner in the exhaust gases flue are added. Waste heat boilers are built in capacities from 25 m^3 almost 30,000 m^3/min. of exhaust gas.

A condensing boiler is designed to resist corrosion and allow the flue gas to be cooled to its condensing temperature, releasing the latent heat contained in the water vapour. Approximately 10% of the energy content of natural gas is used in the latent heat of vapourisation. This latent heat content is not released unless the combustion gas is condensed. Under proper operating conditions, condensing boilers can be approximately 10% more efficient than efficient non-condensing boilers.

Inside the condensing boiler exhaust section, heat is transferred from the flue gas through an enlarged heat exchanger surface to pre-heat the boiler feedwater. In order to effectively extract the latent heat contained in the flue gas water vapour, the boiler inlet water temperature must be low enough to cool the flue gas to condensing temperature. Condensing boilers need to be operated at inlet water temperatures below 140°F.

19.3.11 Heat recovery from boiler blow down

Boiler blow down involves either periodic or continuous removal of water from a steam boiler in order to remove accumulated dissolved solids and/or sludge, which can have damaging effects on boiler efficiency and maintenance. However, boiler blow down wastes energy because the liquid blown down is at about the same temperature as the steam produced. Two methods are typically employed for recovering energy lost from boiler blow down, and both are often incorporated in one system. In the first method, the saturated liquid high-pressure blow down is discharged into a relatively low-pressure receiver, or flash vessel. In the receiver, a portion of the liquid flashes to steam, which can be used either in a low-pressure steam system or in the deaerator to pre-heat the boiler feedwater. Figure 19.6 shows a flash steam vessel recovering steam from condensate lines. By removing steam from the condensate system,

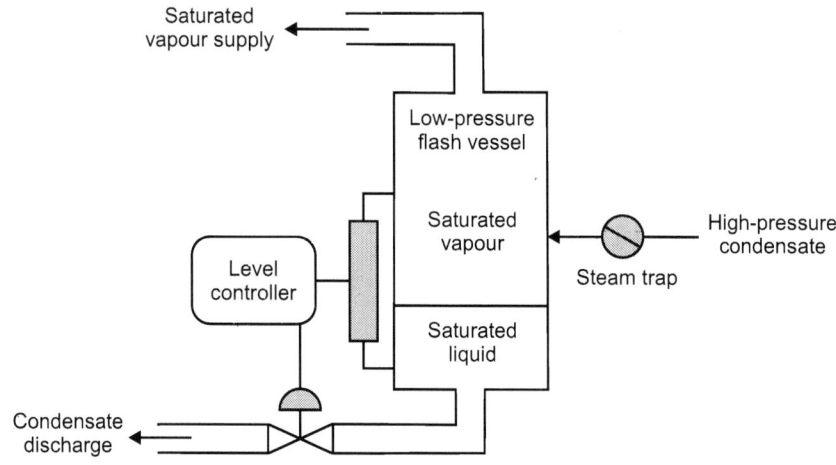

Figure 19.6: Flash vessel steam recovery.

flash steam vessels provide an efficient source of steam to low-pressure end uses. For example, 250°F condensate has a saturation pressure of about 15 psig. It can be used in low-pressure steam applications such as space heating and pre-heating.

The second method of boiler blow down heat recovery takes advantage of the significant temperature difference that exists between the saturated liquid from the flash vessel and the makeup water. The remaining liquid blow down is piped through a heat exchanger to pre-heat the makeup water before entering the deaerator. A combined flash steam and residual blow down heat recovery system, can recover up to 90% of heat energy that would otherwise be wasted.

Larger energy savings occur with high-pressure boilers.

19.3.12 Heat cascading

Heat cascading describes a broader application of recycling heat for external uses. Waste heat from a primary process may still contain enough energy to operate a secondary process, as long as its temperature is high enough to drive the energy to its intended destination. Cascading heat from preceding processes can reduce the amount of energy required in subsequent processes.

Some examples include: Water heating with waste heat boilers, drying or evaporating using exhaust gas from high-temperature furnaces, using multiple-effect evaporators in food processes, and using cooling tower water for space heating. The goal of cascading heat is to use a continuous flow of waste gas through process after process, serving many heat needs in the facility, until no usable heat is left before the gas finally exits.

In a heat cascading process, heat is transferred between sequentially smaller temperature differentials or steps, rather than a single large temperature differential—enabling efficient utilisation of thermal energy. In designing heat cascading, it is necessary that the heating load absorbing the waste heat be available during the periods of waste heat generation; otherwise, the waste heat may be useless, regardless of its quantity and quality. When source and load cannot be synchronised, either another heat load must be found, or an auxiliary heat source needs to be available to carry the load.

Tying two processes together using cascading heat requires more than just the correct temperatures and heat flows. To make the system operate effectively, the logistics must also be set up correctly. For example, if a chemical plant needs a constant supply of heated water for a specific process, and the water heater is totally dependent on the exhaust from an oven, then the oven must run continuously. If this is not feasible, an auxiliary burner can be installed on the water heater to carry the load when the primary process is not running. On the other hand, as long as the oven is operating there will be a supply of hot water, whether it is needed for the process or not. Another key consideration is the placement of equipment. The closer the proximity of primary and secondary processes, the better. Carrying exhaust gas through long runs of ductwork can create an expensive and difficult-to-maintain infrastructure, and the efficiency of energy recovery will be compromised by the heat losses between the two processes. This is of less concern if the primary energy source is liquid or hot oil because these heat transfer mediums can carry energy over greater distances.

19.3.13 Heat recovery using absorption chillers

Combined heat and power (CHP) plants are being utilised in many facilities. In particular they are becoming more common for facilities with large cooling loads and those with balanced simultaneous demands for electric power and heating. A CHP plant generates electrical power using an internal combustion engine, gas turbine, microturbine, or fuel cell. The waste heat from the power generator can be used for process heating and cooling through a waste heat recovery loop. Applications include space heating, absorption chillers, dehumidifiers, heat pumps, heat wheels, and other devices.

Absorption chillers use heat rather than mechanical energy to provide cooling. A thermal compressor consists of an absorber, generator, pump, and throttling device. In the evaporator, the refrigerant evaporates and extracts heat from the building. The refrigerant vapour then is absorbed by the absorbent. The combined fluids go to the generator, where heat is provided from a waste steam heat source to separate refrigerant from the absorbent. The refrigerant then goes to the condenser to be cooled back down to a liquid, while the absorbent is pumped back to the absorber. The cooled refrigerant is released through an expansion valve into the evaporator, and the cycle repeats. Low-pressure, steam-driven absorption chillers are available in capacities ranging from 100 to 1500 tons. Figure 19.7 illustrates an absorption chiller cycle.

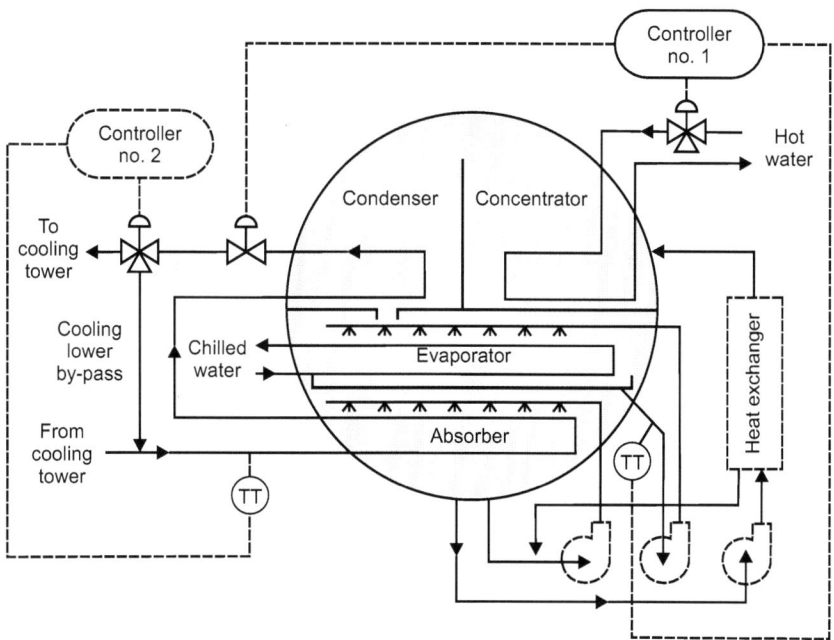

Figure 19.7: Absorption chiller cycle.

Absorption chillers generally have lower coefficients of performance (chiller load/energy input) than traditional chillers; however, they can substantially reduce operating costs because they are powered by using waste heat. Considering the energy efficiency from the source to the point of use, a waste heat absorption chiller can be comparable to a large watercooled electric chiller plant. Single-effect absorption chillers have a coefficient of performance of 0.7, double-effect absorption chillers are about 40% more efficient.

In an absorption chiller application in a CHP plant, the waste heat from the electrical generator is captured by a waste heat recovery boiler. The boiler provides steam for processes and also drives an absorption chiller that provides cooling to the facility. Considering the outputs of electricity, heating, and cooling, the fuel efficiency of CHP plants can be as high as 60%–80%, compared with the 30%–40% from conventional electrical generators.

19.3.14 Heat recovery using a desiccant dehumidifier

A desiccant dehumidifier uses a drying agent, or sorbent, to remove water from the air used to condition building space. Desiccants can run off the waste heat from distributed generation technologies, with system efficiency approaching 80% in CHP mode. The desiccant process involves exposing the desiccant material, such as silica gel, to a moisture-laden process air stream. Once the moisture is absorbed from this stream, another stream of regenerated air removes the moisture from the desiccant.

A solid desiccant dehumidifier is most commonly placed on the surface of a corrugated matrix in a wheel that rotates between the process and regeneration air streams. On the process side, the desiccant removes moisture from the air while releasing heat during the sorption process. As the wheel rotates onto the regeneration side, natural gas, waste heat, or solar energy can be used to regenerate the desiccant material.

Humidification applications are found in the chemical manufacturing industry where control of ambient temperature and moisture content are critical for product quality.

19.4   Cost considerations

In general, waste heat recovery methods can improve performance—i.e. increase the overall efficiency of a process heating system—by 5% to 30%. Table 19.3 provides a summary of the cost-saving potential and expected simple payback periods of waste heat recovery methods and applications described above.

Table 19.3 Summary of waste heat recovery methods.

Methods	Waste sources	Temp range	Applications	Savings potential	Simple payback
High-temperature heat recovery through recuperators/ regenerators	Exhaust gases from incineration or thermal oxidation processes	H,M	Incoming product pre-heating	20–40%	24–48 months
Load pre-heating	Exhaust gas from fuel-fired burner, after burner	H,M,L	Incoming product pre-heating	10–25%	6–24 months
Combustion air pre-heating	Exhaust gas from fuel-fired burner, after burner	H,M	Combustion air pre-heating	10–30%	

(Cont'd...)

Methods	Waste sources	Temp range	Applications	Savings potential	Simple payback
Waste heat boiler	Exhaust gas from gas turbines, reciprocating engines, incinerators, furnaces	H,M	Steam generation, Water heating	5–20%	6–24 months
Feedwater pre-heating	Exhaust gas from fuel-fired burner	H,M,L	Feedwater, make-up water pre-heating	2–20%	6–24 months
Heat recovery through boiler blow down	Steam boiler blow down	H,M,L	Steam generation, Feedwater pre heating	Up to 90%	6–12 months
Heat cascading	Various	H,M,L	Various	5–20%	
Absorption chiller	Waste steam from gas turbines	L	Absorption cooling		
Desiccant Dehumidifier	Waste steam from gas turbines	L	Air dehumidification and/or cooling		

Note: H (high), M (medium), L (low)

Generally, the payback period is a measure of the cost effectiveness of a project. The payback period is affected by the service life of the equipment installed. Heat exchangers generally have a service life of up to 20–25 years, although special applications or harsh environments can shorten that life. Waste heat boilers and turbines have a service life of about 30 years. A longer payback period is generally acceptable for projects having long-life equipment, but a payback period of three to five years is considered reasonable.

The cost of a heat exchanger varies with the temperature range to which it would apply: the higher the temperature range, the higher the cost, due to higher material cost and additional engineering requirements.

However, because a high-temperature source provides high-quality waste heat, the cost per unit of energy transferred can be less. Choosing appropriate heat exchange equipment is the key to high cost-effectiveness.

19.4.1 Heat exchangers

Shell and tube heat exchanger: When the medium containing waste heat is a liquid or a vapour which heats another liquid, then the shell and tube heat exchanger must be used since both paths must be sealed to contain the pressures of their respective fluids. The shell contains the tube bundle, and usually internal baffles, to direct the fluid in the shell over the tubes in multiple passes. The shell is inherently weaker than the tubes so that the higher-pressure fluid is circulated in the tubes while the lower pressure fluid flows through the shell. When a vapour contains the waste heat, it usually condenses, giving up its latent heat to the liquid being heated.

In this application, the vapour is almost invariably contained within the shell. If the reverse is attempted, the condensation of vapours within small diameter parallel tubes causes flow instabilities. Tube and shell heat exchangers are available in a wide range of standard sizes with many combinations of materials for the tubes and shells. A shell and tube heat exchanger is illustrated in Fig. 19.8.

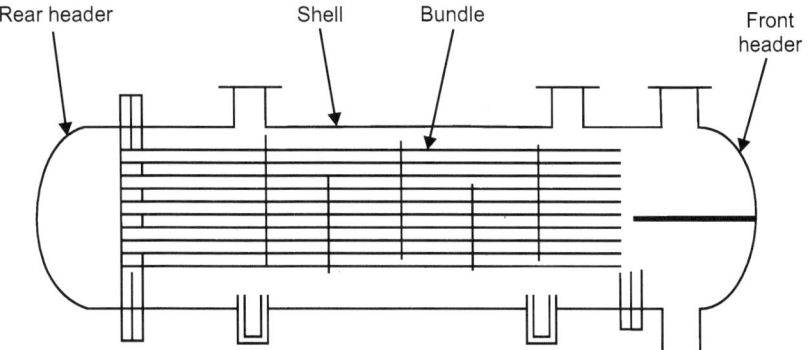

Figure 19.8: Shell and tube heat exchanger.

Typical applications of shell and tube heat exchangers include heating liquids with the heat contained by condensates from refrigeration and air-conditioning systems; condensate from process steam; coolants from furnace doors, grates, and pipe supports; coolants from engines, air compressors, bearings, and lubricants; and the condensates from distillation processes.

The cost of heat exchange surfaces is a major cost factor when the temperature differences are not large. One way of meeting this problem is the plate type heat exchanger, which consists of a series of separate parallel plates forming thin flow pass. Each plate is separated from the next by gaskets and the hot stream passes in parallel through alternative plates whilst the liquid to be heated passes in parallel between the hot plates. To improve heat transfer the plates are corrugated.

Hot liquid passing through a bottom port in the head is permitted to pass upwards between every second plate while cold liquid at the top of the head is permitted to pass downwards between the odd plates. When the directions of hot and cold fluids are opposite, the arrangement is described as counter current.

Typical industrial applications are:

1. Pasteurisation section in milk packaging plant.

2. Evaporation plants in food industry.

Run around coil exchanger: It is quite similar in principle to the heat pipe exchanger. The heat from hot fluid is transferred to the colder fluid via an intermediate fluid known as the heat transfer fluid. One coil of this closed loop is installed in the hot stream while the other is in the cold stream. Circulation of this fluid is maintained by means of a circulating pump. It is more useful when the hot land cold fluids are located far away from each other and are not easily accessible. Typical industrial applications are heat recovery from ventilation, air conditioning and low temperature heat recovery.

19.4.2   Heat pumps

In the various commercial options previously discussed, we find waste heat being transferred from a hot fluid to a fluid at a lower temperature. Heat must flow spontaneously 'downhill', that is from a system at high temperature to one at a lower temperature. When energy is repeatedly transferred or transformed, it becomes less and less available for use. Eventually that energy has such low intensity (resides in a medium at such low temperature) that it is no longer available at all to perform a useful function.

It has been taken as a general rule of thumb in industrial operations that fluids with temperatures less than 120°C (or, better, 150°C to provide a safe margin), as limit for waste heat recovery because of the risk of condensation of corrosive liquids. However, as fuel costs continue to rise, even such waste heat can be used economically for space heating and other low temperature applications. It is possible to reverse the direction of spontaneous energy flow by the use of a thermodynamic system known as a heat pump.

The majority of heat pumps work on the principle of the vapour compression cycle. In this cycle, the circulating substance is physically separated from the source (waste heat, with a temperature of T in) and user (heat to be used in the process, T out) streams, and is reused in a cyclical fashion, therefore called 'closed cycle'. In the heat pump, the following processes take place:

1. In the evaporator the heat is extracted from the heat source to boil the circulating substance.

2. The circulating substance is compressed by the compressor, raising its pressure and temperature. The low temperature vapour is compressed by a compressor, which requires external work. The work done on the vapour raises its pressure and temperature to a level where its energy becomes available for use.

3. The heat is delivered to the condenser.

4. The pressure of the circulating substance (working fluid) is reduced back to the evaporator condition in the throttling valve, where the cycle repeats.

The heat pump was developed as a space heating system where low temperature energy from the ambient air, water, or earth is raised to heating system temperatures by doing compression work with an electric motor-driven compressor.

The heat pumps have the ability to upgrade heat to a value more than twice that of the energy consumed by the device. The potential for application of heat pump is growing and number of industries have been benefited by recovering low grade waste heat by upgrading it and using it in the main process stream. Heat pump applications are most promising when both the heating and cooling capabilities can be used in combination. One such example of this is a plastics factory where chilled water from a heat is used to cool injection-molding machines whilst the heat output from the heat pump is used to provide factory or office heating. Other examples of heat pump installation include product drying, maintaining dry atmosphere for storage and drying compressed air.

19.4.3   Thermocompressor

In many cases, very low pressure steam are reused as water after condensation for lack of any better option of reuse. In many cases it becomes feasible to compress this low pressure steam by very high pressure steam and reuse it as a medium pressure steam. The major energy in steam, is in its latent heat value and thus thermocompressing would give a large improvement in waste heat recovery.

The thermocompressor is a simple equipment with a nozzle where HP steam is accelerated into a high velocity fluid. This entrains the LP steam by momentum transfer and then recompresses in a divergent venturi.

Thermocompressor is shown in Figure 19.9. It is typically used in evaporators where the boiling steam is recompressed and used as heating steam.

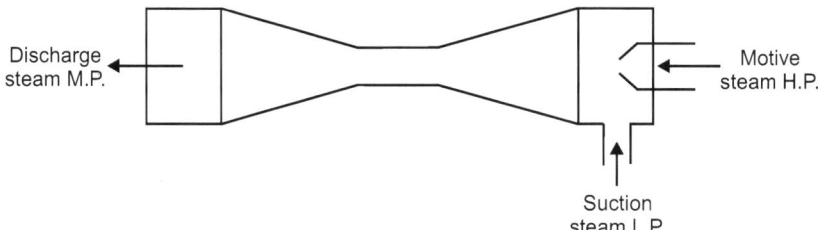

Figure 19.9: Thermocompressor.

Energy audit

20.1 Introduction

The 'energy audit' is the key to a systematic approach for decision-making in the area of energy management. It attempts to balance the total energy inputs with their use, and serves to identify all the energy streams in a facility. It quantifies energy usage according to its discrete functions. An industrial energy audit is an effective tool in defining and pursuing a comprehensive energy management programme within a business. Energy audit is defined as 'the verification, monitoring and analysis of use of energy including submission of technical reports containing recommendations for improving energy efficiency with cost benefit analysis and an action plan to reduce energy consumption.'

Need for energy audit: In any industry, the three top operating expenses are often found to be energy (both electrical and thermal), labour, and materials. In most assessments of the manageability of the cost or potential cost savings in each of the above components, energy would invariably emerge as a top ranker, and thus energy management function constitutes a strategic area for cost reduction. A well done energy audit will always help managers understand more about the ways energy and fuel are used in their industry, and help to identify areas where waste can occur and where scope for improvement exists. The energy audit would give a positive orientation to the energy cost reduction, preventive maintenance, and quality control programmes which are vital for production and utility activities. Such an audit programme will help to keep focus on variations that occur in the energy costs, availability, and reliability of supply of energy, help decide on the appropriate energy mix, identify energy conservation technologies, retrofit for energy conservation equipment, etc.

In general, the energy audit is the translation of conservation ideas and hopes into reality, by lending technically feasible solutions with economic and other organisational considerations within a specified time frame.

Objective of energy audit: The primary objective of the energy audit is to determine ways to reduce energy consumption per unit of product output or to lower operating costs. The energy audit provides a benchmark, or reference point, for managing and assessing energy use across the organisation and provides the basis for ensuring more effective use of energy.

20.2 Types of energy audits

The type of energy audit to be performed depends on:
1. Function and type of industry.
2. Depth to which a final audit is needed.
3. Potential and magnitude of cost reduction desired.

Thus energy audits can be classified into the following two types:
1. Preliminary audit.
2. Detailed audit.

20.2.1 Preliminary energy audit methodology

The preliminary energy audit uses existing or easily obtained data. It is a relatively quick exercise to:
1. Determine energy consumption in the organisation.
2. Estimate the scope for saving.
3. Identify the most likely (and easiest areas) for attention.
4. Identify immediate (especially no-cost/low-cost) improvements/ savings.
5. Set a reference point.
6. Identify areas for more detailed study/measurement.

20.2.2 Detailed energy audit methodology

A detailed energy audit provides a comprehensive energy project implementation plan for a facility, since it evaluates all major energy-using systems.

This type of audit offers the most accurate estimate of energy savings and cost. It considers the interactive effects of all projects, accounts for the energy use of all major equipment, and includes detailed energy cost saving calculations and project cost.

In a detailed audit, one of the key elements is the energy balance. This is based on an inventory of energy-using systems, assumptions of current operating conditions, and calculations of energy use.

This estimated use is then compared to utility bill charges.

Detailed energy auditing is carried out in three phases:
1. Phase I – Pre-audit.
2. Phase II – Audit.
3. Phase III – Post-audit.

20.3 Steps for conducting energy audit

Industry-to-industry, the methodology of energy audits needs to be flexible. A 10-step summary for conducting a detailed energy audit at the field level is listed in Table 20.1. The energy manager or energy auditor may follow these steps to start with and add/change as per their needs and the industry type.

Table 20.1: 10 Steps for a detailed energy audit.

Step	Action	Purpose
1	*Phase I – pre-audit* • Plan and organise • Walk-through audit • Informal interviews with energy manager, production/plant manager	• Resource planning; establish/organise energy audit team • Organise instrumentation and time frame • Macro data collection (suitable to type of industry) • Familiarisation of process/plant activities • First-hand observation and assessment of current level operation and practices
2	Conduct briefing/awareness session with all divisional heads and persons concerned (2–3 hrs)	• Building up cooperation • Issue questionnaire for each department • Orientation, awareness creation
3	*Phase II – audit* Primary data gathering, process flow diagram, and energy utility diagram	Historic data analysis; baseline data collection Prepare process flowchart(s) All service utilities system diagram (Example: Single line power distribution diagram, water, compressed air and steam distribution) Design, operating data and schedule of operation Annual energy bill and energy consumption pattern (refer to manuals, log sheets, equipment spec. sheets, interviews)
4	Conduct survey and monitoring	Measurements: • Motor survey, insulation, and lighting survey with portable instruments to collect more and accurate data • Confirm and compare actual operating data with design data
5	Conduct detailed trials/experiments for biggest energy users/equipment	Trials/experiments: • 24-hr power monitoring (MD, PF, kWh, etc.) • Load variation trends in pumps, fans, compressors, heaters, etc. • Boiler efficiency trials (4–8 hrs) • Furnace efficiency trials • Equipment performance experiments, etc.

(Contd...)

Step	Action	Purpose
6	Analysis of energy use	Energy and material balance and energy loss/waste analysis
7	Identification and development of energy conservation (ENCON) opportunities	Identification and consolidation of ENCON measures Conceive, develop, and refine ideas Review ideas suggested by unit personnel Review ideas suggested by preliminary energy audit Use brainstorming and value analysis techniques Contact vendors for new/efficient technology
8	Cost-benefit analysis	Assess technical feasibility, economic viability and prioritisation of ENCON options for implementation Select the most promising projects Prioritise by low-, medium-, long-term measures Documentation, report presentation to top management
9	Reporting and presenting to top management	Documentation, report presentation to top management
10	*Phase III – post-audit* Implementation and follow-up	Assist and implement ENCON measures and monitor performance Action plan, schedule for implementation Follow-up and periodic review

20.3.1 Phase I – pre-audit activities

A structured methodology to carry out the energy audit is necessary for efficient implementation. An initial study of the site should always be carried out, as the planning of the audit procedures is of key importance.

Initial site visit and preparation required for detailed auditing

An initial site visit may take one day and gives the energy auditor/manager an opportunity to meet the personnel concerned, to familiarise him or her with the site, and to assess the procedures necessary to carry out the energy audit.

During the initial site visit, the energy auditor/manager should carry out the following actions:

1. Discuss with the site's senior management the aims of the energy audit.

2. Discuss economic guidelines associated with the recommendations of the audit.
3. Analyse the major energy consumption data with relevant personnel.
4. Obtain site drawings where available – building layout, steam distribution, compressed air distribution, electricity distribution, etc.
5. Tour the site accompanied by engineering/production staff.

Main aims of this visit are

1. To finalise energy audit team.
2. To identify the main energy-consuming areas/plant items to be surveyed during the audit.
3. To identify any existing instrumentation or additional metering that may be required.
4. To decide whether any meters will have to be installed prior to the audit, e.g. kWh, steam, oil, or gas meters.
5. To identify the instrumentation required for carrying out the audit.
6. To plan the time frame.
7. To collect macro data on plant energy resources, major energy consuming centers.
8. To create awareness through meetings/programme.

20.3.2   Phase II – detailed energy audit activities

Depending on the nature and complexity of the site, a detailed energy audit can take from several weeks to several months to complete. Detailed studies to establish and investigate energy and material balances for specific plant departments or items of process equipment are carried out. Whenever possible, checks of plant operations are conducted over extended periods of time, at nights and at weekends as well as during normal daytime working hours, to ensure that nothing is overlooked. The audit report will include a description of energy inputs and product outputs by major departments or by major processing function, and will evaluate the efficiency of each step of the manufacturing process. Means of improving these efficiencies will be listed, and at least a preliminary assessment of the cost of the improvements will be made to indicate the expected payback on any capital investment needed. The audit report should conclude with specific recommendations for detailed engineering studies and feasibility analyses, which must then be performed to justify the implementation of those conservation measures that require additional capital investment.

Information to be collected during the detailed audit includes:

1. Energy consumption by type of energy, by department, by major items of process equipment, by end-use.
2. Material balance data (raw materials, intermediate and final products, recycled materials, use of scrap or waste products, production of by-products for reuse in other industries, etc.).
3. Energy cost and tariff data.
4. Process and material flow diagrams.
5. Generation and distribution of site services (e.g. compressed air, steam).
6. Sources of energy supply (e.g. electricity off the grid or self-generation).
7. Potential for fuel substitution, process modifications, and the use of cogeneration systems (combined heat and power generation).
8. Energy management procedures and energy awareness training programmes within the establishment.

Existing baseline information and reports are useful to understand consumption patterns, production cost, and productivity levels in terms of product per raw material inputs.

For this the audit team should collect the following baseline data:

1. Technology, processes used, and equipment details.
2. Capacity utilisation.
3. Amount and type of input materials used.
4. Water consumption.
5. Fuel consumption.
6. Electrical energy consumption.
7. Steam consumption.
8. Other inputs such as compressed air, cooling water, etc.
9. Quantity and type of wastes generated.
10. Percentage rejection/reprocessing.
11. Efficiencies/yield.

20.4   Data collection hints

It is important to plan additional data gathering carefully. Here are some basic tips to avoid wasting time and effort:

1. Measurement systems should be easy to use and provide information to the level of accuracy that is actually needed, not the accuracy that is technically possible.

2. Measurement equipment can be inexpensive (flow rates using a bucket and stopwatch).

3. The quality of the data must be such that correct conclusions are drawn (what the grade of product is in production is the production normal, etc.)

4. Define how frequent data collection should be to account for process variations.

5. Measurement exercises over abnormal workload periods (i.e. startup and shutdown).

6. Design values can be taken where measurements are difficult (i.e. cooling water through a heat exchanger).

20.4.1 Process flow diagram to identify waste streams and energy wastage

An overview of unit operations, important process steps, areas of material and energy use, and sources of waste generation should be gathered and should be represented in a flowchart as shown in Fig. 20.1. Existing drawings, records, and a shop floor walk-through will help in making this flowchart.

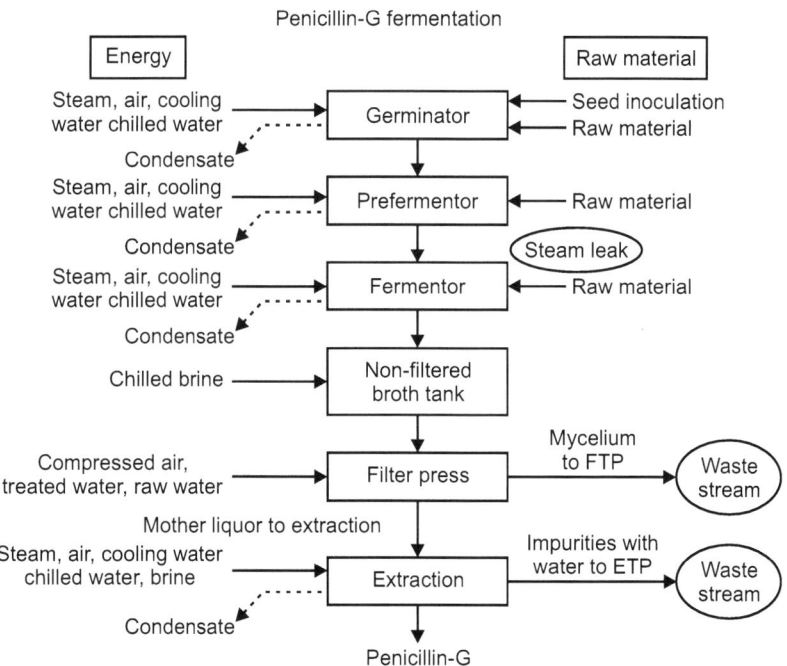

Figure 20.1: Process flow diagram for manufacturing penicillin-G.

Simultaneously the team should identify the various inputs and output streams at each process step.

Example: A flowchart of a production line for penicillin is given in Fig. 20.1. Note that a waste stream (Mycelium) and obvious energy losses such as condensate drainage and steam leakages are identified in this flowchart. The audit focus area depends on several issues such as consumption of input resources, energy efficiency potential, impact of specific process steps on the entire process, or intensity of waste generation/energy consumption. In the example process, the modularised operations such as germinator, prefermentor, fermentor, and extraction are the major conservation potential areas identified.

20.4.2 Identification of energy conservation opportunities

1. Fuel substitution: Identifying alternative fuels for efficient energy conversion.

2. Energy generation: Identifying efficiency opportunities in energy conversion equipment/utilities such as captive power generation, steam generation in boilers, thermic fluid heating, optimal loading of diesel generator sets, minimum excess air combustion with boilers/thermic fluid heating, optimising existing efficiencies, efficient energy conversion equipment, biomass gasifiers, cogeneration, high efficiency diesel generator sets, etc.

3. Energy distribution: Identifying efficiency opportunities networks such as transformers, cables, switchgears, and power factor improvement in electrical systems and chilled water, cooling water, hot water, compressed air, etc.

4. Energy usage by processes: This is one of the major opportunities for improvement and many of them are hidden. Process analysis is a useful tool for process integration measures that can greatly improve energy efficiency.

20.4.3 Technical and economic feasibility

Technical feasibility assessment should address the following issues:

1. Technology availability, space, skilled manpower, reliability, service, etc.

2. The impact of energy efficiency measures on safety, quality, production, or process.

3. Maintenance requirements and availability of spare parts and components.

Economic viability often becomes the key parameter for acceptance by top management. The economic analysis can be conducted by using a variety of methods. Examples include: payback method, internal rate of return method, net present value method, etc. For low investment, short-duration measures, which have attractive economic viability, the simplest of the methods–payback– is usually sufficient.

A sample worksheet for assessing economic feasibility is provided below:

20.4.4 Sample worksheet for economic feasibility

Energy efficiency measure:

1. Investment equipment, civil works, instrumentation, auxiliaries.
2. Annual operating cost, cost of capital maintenance, manpower, energy, depreciation.
3. Annual savings thermal energy, electrical energy, raw material, waste disposal.

Net savings/year = (Annual savings minus annual operating costs).

Payback period in months = (Investment/net savings/year)/12.

20.4.5 Classification of energy conservation measures

Based on the energy audit and analysis of the plant, a number of potential energy saving projects may be identified. These may be classified into three categories: (i) low cost-high return, (ii) medium cost-medium return and (iii) high cost-high return.

Normally the low cost-high return projects receive priority. Other projects have to be analysed, engineered, and budgeted for implementation in a phased manner. Projects relating to energy cascading and process changes almost always involve high costs coupled with high returns, and may require careful scrutiny before funds can be committed. These projects are generally complex and may require long lead times before they can be implemented.

To sum up, energy auditing procedures are different for different industries. Given below are the brief outline auditing procedures adopted for the following industries:

1. Electrical system network: This would include detailed study of all the transformer operations of various ratings/capacities, their operational pattern, loading, no load losses, power factor measurement on the main power distribution boards and scope for improvement if any. The study would also cover possible improvements in energy metering systems for better control and monitoring.
2. Study of motors and pumps loading: Study of motors (above 10 kW) in terms of measurement of voltage (V), Current (I), Power (kW) and power factor and thereby suggesting measures for energy saving like reduction in size of motors or installation of energy saving device in the existing motors. Study of pumps and their flow, thereby suggesting measures for

energy saving like reduction in size of motors and pumps or installation of energy saving device in the existing motors/optimisation of pumps.

3. Study of air conditioning plant: With respect to measurement of specific energy consumption, i.e. kW/TR of refrigeration, study of refrigerant compressors, chilling units, etc. Further, various measures would be suggested to improve its performance.

4. Cooling tower: This would include detailed study of the operational performance of the cooling towers through measurements of temperature differential, air/water flow rate, to enable evaluate specific performance parameters like approach, effectiveness, etc.

5. Performance evaluation of boilers: This includes detailed study of boiler efficiency, thermal insulation survey and flue gas analysis.

6. Performance evaluation of turbines: This includes detailed study of Turbine efficiency, Waste heat recovery.

7. Performance evaluation of air compressor: This includes detailed study of air compressor system for finding its performance and specific energy consumption.

8. Evaluation of condenser performance: This includes detailed study of condenser performance and opportunities for waste heat recovery.

9. Performance evaluation of burners/furnace: This includes detailed study on performance of furnace/burner, thermal insulation survey for finding its efficiency.

10. Windows/split air conditioners: Performance shall be evaluated as regards, their input power vis-a-vis TR capacity and performance will be compared to improve to the best in the category.

11. DG set: Study the operations of DG sets to evaluate their average cost of power generation, specific energy generation and subsequently identify areas wherein energy savings could be achieved after analysing the operational practices, etc., of the DG sets.

The entire recommendations would be backed up with techno-economic calculations including the estimated investments required for implementation of the suggested measures and simple payback period. Measurement would be made using appropriate instrumentation support for time lapse and continuous recording of the operational parameters.

Case studies

21.1   Honeywell farms dairy milk—(USA)

The Honeywell farms dairy milk processing plant is one of the largest in the New York city area. The plant processes milk which arrives daily by tank trucks from milk producing areas in New York. The milk is pasteurised and bottled at this 8361 m² (90,000 ft²) facility located in the Jamaica section of Queens. The plant started as a small operation over 50 years ago and has grown since then to become the largest independent milk processor in the long island area. By retrofitting an integrated cogeneration system, the honeywell farms dairy processing facility achieved energy and cost-saving goals. Innovative elements employed to reach these goals included: using cogeneration concepts for refrigeration prime movers; waste energy recovery using absorption refrigeration for subcooling; and exhaust heat recovery for steam generation.

Aim of the project: Honeywell farms dairy owners were concerned about rising electrical energy costs and sought ways to improve energy efficiency. They wished to augment an effective energy management programme with additional energy conservation efforts. Data collection and analysis was undertaken with the following goals:

1 Investigate the use of cogeneration technology for refrigeration applications.
2. Evaluate utilising waste heat recovery for subcooling.
3 Reduce the energy costs associated with the refrigeration plant while expanding processing capacity.

The principle: This project includes three elements which have proven successful in operation:

1. The use of cogeneration concepts for refrigeration prime movers.
2. Waste energy recovery using absorption refrigeration for subcooling to boost the coefficient of performance (COP) of the refrigeration plant.
3. Exhaust heat recovery for steam generation.

In a subcooling system, as used in this project, chilled water from an absorption chiller can be used to cool liquid refrigerant of the main refrigeration system far below its saturation temperature. An additional refrigeration effect equivalent to the enthalpy difference caused by the liquid temperature reduction is obtained by this process. This added refrigeration capacity is provided without additional work input to the compressor. Since the cooling capacity of the system is increased without a corresponding increase in energy consumption, the COP is improved. Allowing for a reasonable temperature approach as heat is transferred between the refrigerant and chilled water in the subcooling heat exchanger, energy savings of 8–10% can be expected if the thermal load remains constant. In this arrangement, the added capacity is equal to the capacity of the absorption machine.

The situation: A natural gas-engine-driven compressor was configured into an existing building by remodelling the equipment room and relocating some existing equipment. The system consists of a 450 kW (600 hp) turbocharged, twelve cylinder natural gas engine directly coupled to a screw type ammonia compressor. Heat from the engine jacket is used to produce chilled water in three 35 kW (10 tonnes) lithium bromide absorption refrigeration units. The chilled water is circulated through a heat exchanger to subcool liquid ammonia refrigerant after it leaves the refrigeration system condenser. The heat rejected from the absorption refrigeration units is rejected to an atmospheric cooling tower. Heat is furthermore recovered from the engine exhaust using a waste-heat steam generator rated at 147 kW (500,000 BTU/hr).

The steam is produced at a pressure of 4.5 atmospheres (65 psi) and is utilised for process steam (Fig. 21.1). The engine, compressor and heat recovery equipment combine to form a highly efficient cogeneration plant. Very high fuel efficiency is obtained and substantial savings are achieved when compared with the separate conventional steam generators and electric powered refrigeration systems used previously. This plant was installed parallel to the existing refrigeration plant. The existing plant consisted of seven electrically driven reciprocating ammonia compressors of 264 kW (75 tonnes) capacity each and two backup compressors of equivalent capacity. None of this equipment was removed from service.

The cogeneration plant's waste heat boiler was connected to the plant steam header in parallel with existing packaged, gas-fired boilers. When the engine is operating, the gas-fired boilers are shut down. A bypass valve was installed around the waste heat boiler to divert exhaust flow in the event that the waste heat boiler was unavailable. Normally, the absorption units are operated to remove heat from the engine jacket. In the event that the absorption units are not operated, engine cooling can be rejected to an engine radiator on the roof of the building. Thus, the plant has a maximum amount of efficiency and

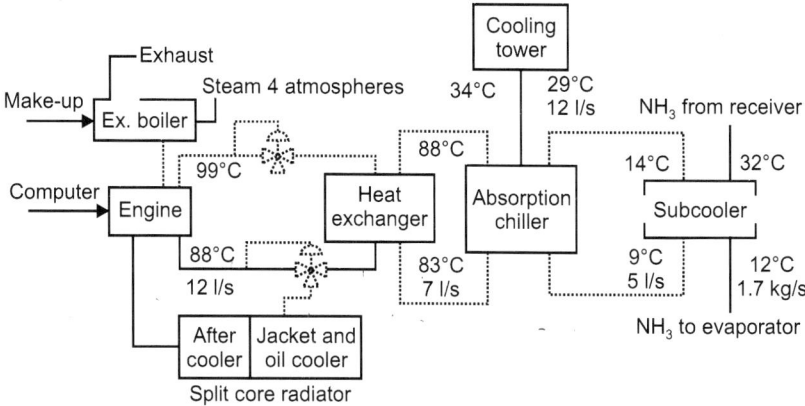

Figure 21.1: Schematic of the engine subcooling system.

flexibility. The engine/compressor system is now used for base load operation. When it is used in conjunction with the subcooling system the engine/compressor cogeneration system has been able to provide more than 80% of the required cooling capacity. This performance included downtime accumulated as the result of unexpected power failures, planned maintenance activities and operation performances.

21.1.1 Economics

The total installation costs for the engine-compressor based refrigeration plant with subcooling were USD 519,540. Compared with an all-electric plant the additional costs were USD 339,540.

Estimates of annual energy consumption and costs were developed using a model-based approach. The costs include all appropriate taxes, demand charge, fuel adjustments, a monetary credit for the waste heat recovered as steam from the engine exhaust, and maintenance costs.

These configurations include an all electric plant, an engine compressor based plant without subcooling and an engine-compressor based plant with subcooling. A summary of the energy consumption for the refrigeration plant is shown in Table 21.1, and an estimated simple payback based on energy savings is shown in Table 21.2.

Table 21.1: Annual energy consumption for the refrigeration plant.

Fuel source	All electric plant	Engine compressor based plant without subcooling	Engine compressor based plant with subcooling
Electricity (kWh)	1,853,462	655,633	178,298
Natural gas (MJ)	251,400	2,156,800	2,362,000

Table 21.2: Estimated simple payback based on energy savings.

| Option | Project costs (USD) | | Annual cost (USD) | | Payback |
	Total	Incremental cost	Total	Incremental savings	(years)
All electric plant	180,000	0	317,400	0	N/A
Subcooling engine-compressor	519,540	339,540	227,000	90,400	3.8
Non-subcooling engine-compressor	385,220	205,220	202,150	115,250	1.8
Subcooling increment	0	134,320	0	24,850	5.4

21.2 Hindustan petroleum corporation ltd.—Mumbai, India

Hindustan Petroleum Corporation Ltd. (HPCL) refineries use energy in various forms viz., fuel, steam, electricity, etc. in different processing units to convert crude oil to valuable petroleum products. Most of these refinery processes are energy intensive and use part of the finished products produced in the refinery to derive their energy requirement. Thus any reduction in consumption of energy directly results in higher availability of finished products, which in turn results in higher profit. Hence, energy conservation has direct impact on refinery's profits. HPCL Refineries have accorded highest priority to energy conservation and have dedicated Energy Conservation (ENCON) cell, consisting of managers and engineers to monitor ENCON measures on a daily basis. To optimise utilisation of energy with constant endeavor of the Corporation, several research and development initiatives have been undertaken to achieve the optimal level. The employees of the corporation across all ranks have been educated to improve the knowledge on the Encon improvement activities. Also upcoming projects have been tailored to be compatible with the corporation's energy efficiency standards.

21.2.1 Various ENCON measures at HPCL refineries

1. Energy conservation measures implemented in the past include:
   (a) Adoption of cogeneration principle for generation of steam/power, etc., which includes installation of FCCU CO - boiler.
   (b) Modernisation of fired heaters.
   (c) Maximisation of crude pre-heat by optimisation of heat exchanger train using pinch technology.
   (d) Effective use of waste heat.
   (e) Modernisation of instrumentation and advanced control strategies.

2. HPCL was one of the first oil companies in the country to initiate and implement full-fledged automation of its offsite facilities at both the refineries. The facilities comprise of the following:

   (a) Automation of tank gauging and inventory management system.
   (b) Advanced on-line blend control.
   (c) Continuous on-line monitoring of quality of critical products/streams.
   (d) Accurate measurement and monitoring of custody transfer operations.
   (e) Continuous monitoring of critical offsite transfer pumps.

      The offsite automation facilities have been designed to reduce loss, enhance quality, optimise allocation of various refinery streams to improve overall yield of distillation as well as product accounting. The new projects have usually been implemented in multistage processes taking updates from every step to avoid any untoward loss of efficiency in the overall performance. As a part of these projects, customised software packages have been widely used and customised for implementation in refinery projects. The software have helped to decide optimal crude mix based on prevailing prices and demand of various products and optimise plant performance and product slate to get maximum outputs within the stipulated capacity. The projects implemented in recent years for the maintenance of energy efficiency have been path breaking in the Indian perspective. HPCL has paved the way for efficient operational methods in the Indian context to be followed by organisations for years to come.

3. A detailed hydrocarbon loss study has been carried out by M/s British Petroleum for Mumbai refinery and the various recommendations have been implemented.

4. Both refineries have carried out energy conservation activities to improve ENCON performance as details given below:

   (a) Sonic soot blowers have been installed at boiler house for using the kinetic energy of sound waves to avoid soot deposition. This results in greater steam savings as compared to the conventional steam soot blower.
   (b) On-line oxygen analysers on furnaces/boilers have been provided to control excess air in the refineries. These instruments measure the oxygen content of flue gas continuously. The checking of oxygen content in flue gas helps in improving the furnace efficiencies.
   (c) Ultrasonic flare meters have been provided to measure the flow rates as well as molecular weights of different hydrocarbons going to the

flare. This helps to identify the source, type and quantity of hydro-carbon going to flare.

(d) Mass flow meters have been installed in various furnaces/heaters for monitoring individual furnace fuel consumption.

(e) CCTV for round-the-clock viewing/controlling of flare stack has been installed in both refineries.

(f) Electronic themoprobes have been installed for registering accurate tank temperature from time to time. The accurate measurement of temperature reduces the flaws in accounting due to faults in the reading of temperature which were earlier reflected in hydrocarbon loss.

(g) A CO-boiler is being installed as part of FCCU-I revamp under the second Expansion Project of Visakh refinery. The CO-boiler will recover heat from FCCU flue gas and the required amount of steam.

As a result of the above mentioned ENCON projects the Fuel and Loss for Visakh refinery has reduced to 5.7 wt% during 2008–09. Specific energy consumption has subsequently reduced to 85.78 MBTU/BBL/NRGF during the same period.

There has been significant improvement in Mumbai refinery in the same line and currently the Fuel and Loss for Mumbai refinery is restricted to 6.5 wt% and specific energy consumption is held at a steady 92.5 MBTU/BBL/NRGF. These figures represent efficiency levels compatible with the international standard.

21.3   Tata power company ltd.—Mumbai, India

21.3.1   Energy conservation techniques

Energy conservation in power utility

A power plant produces electrical energy and also consumes a substantial amount of energy in the form of auxiliary consumption required for various plant equipments and services

Management attitude: The top management of the company is committed to and has the right attitude towards energy conservation. This is passed on to the plant staff for ensuring success by quality management systems such as Tata business excellence model and ISO.

1. Consistent achiever of business results.
2. Vision statement to be an excellent and efficient organisation.
3. Focus on enhancing business and customer orientation.

Trombay generating station

1. Installed capacity-1350 MW.
2. ISO 9001:2000 certified.
3. Perception has changed to strike a balance between reliability and energy conservation.
4. Key drivers for change.
5. Tata business excellence model (TBEM).
   (a) Leadership.
   (b) Customer and market focus.
   (c) Process management.
   (d) Business results.

21.3.2 Energy conservation in power utility

Energy conservation in power utility is achieved mainly by:

1. Operating the equipments at maximum efficiency.
2. Reduction of auxiliary power consumption.
3. Energy conservation measures adopted at Tata power company limited have taken both these factors into consideration.

Factors affecting auxiliary consumption

1. Unit generation and load pattern.
2. Operation of plant auxiliaries.
3. Service auxiliary such as illumination, air conditioning, etc.
4. Unit startups/shutdowns.

Criteria used for adopting new energy saving methods

1. Reliability.
2. Sustainability.
3. Method and quality of verification.
4. Investment required.

21.3.4 Energy saving techniques

Energy saving techniques adopted are classified in the following categories:

1. Design and conceptual stage.
2. Techniques developed through O&M experience.
3. Energy auditing by external agencies.

Design and conceptual stage

1. Use of steam turbine driven boiler feed pumps (2 × 9 MW).
2. Pioneered concept of using variable speed drives for large motors by using LCI drives for 4 × 2 MW ID fan motor of 500 MW Unit.
3. Use of fluid coupling for motor driven boiler feed pump.

Techniques developed through O&M experience

1. In addition to techniques adopted at design stage various initiatives have been taken over the years to optimise the auxiliary power consumption these are:
   (a) Monitoring of auxiliary power consumption–continuously improving the levels and accuracy.
   (b) Formation of cross functional teams–to analyse and optimise process.
   (c) Modification in operation strategy to achieve optimisation.
   (d) Knowledge sharing and benchmarking with other power utilities.

21.3.5   Prioritisation

Ideas and suggestions captured from various initiatives are then prioritised into three categories:

1. Without investment.
2. With minor investment.
3. With major investment.

With minor investment

1. Installation of vacuum pumps in fly ash removal system in place of hydro ejector based ash water pumps.
2. Replacement of conventional central air conditioning system by vapour absorption system to utilise waste steam
3. Improving efficiency of electro chlorination plant by regular acid cleaning
4. Replacement of Electro Static Precipitators (ESP) controllers by improved version having option of energy efficient mode.
5. Use of energy efficient lighting system.

With major investment

1. Replacement of HP turbine module of 500 MW Unit.
2. Proposal of retrofitting of constant speed ID fan motor with variable speed drives.

21.3.6 Energy audit by external agency external audit

1. Detailed energy audit of trombay thermal power station was conducted by Confederation of Indian Industry (CII).
2. CII studied various main and sub processes (Table 21.3).
3. Good energy conservation practices at TPCL as per CII's viewpoint are.

Table 21.3: Confederation of Indian Industry (CII) proposals.

Some of the CII Proposals under consideration for implementation	Annual saving	Investment required period	Simple Payback
	Rs. Lakhs	Rs. Lakhs	Months
Install next lower size impeller for the FGD spray pumps	10.48	2.50	3
Install variable frequency drive for Hot-well make-up water pump in Unit # 7	0.65	1.00	19
Avoid daytime lighting and install transmission roofing sheets in LC oil pressure filter area	0.34	0.20	7
Replace conventional chokes with electronic chokes in fluorescent lamps	4.95	8.75	21

Encon at tata power

1. Installation of load commutator invertors (LCI) for ID fans.
2. Optimisation of auxiliary consumption during non-peak hours.
3. Good excess air control in boilers.
4. Optimum loading of motors.
5. Efficient lighting systems.
6. Optimum ÄT in condensing water system.
7. Excellent control systems.

Comparison of auxiliary power consumption of thermal power utilities:

1. India benchmark – 6.33%.
2. International benchmark – 3.99%.
3. TPCL – 4.33%.

TPCL's auxiliary power consumption has been acknowledged by CEA as one of the best in India.

Saving due to energy conservation measures is shown in Table 21.4.

Table 21.4: Savings due to energy conservation measures.

Optimisation of auxiliary power in	Savings in MU
ESP	0.241
FGD	0.236
FAA	0.302
Ash plant	2.510
GR fans	4.180
CW pumps	2.460
SGWCPs	1.070
Total	11.000

Total saving in – 11 MU
Saving – Rs 2.23 Crores
Reduction in Cost – 0.25 Paise/KWH

References

Adams Douglas, *Principles of Energy Conversion*, Academic Press, New York.

Amanda Bishop, *Energy Conservation*, by Cavendish Square Publishing.

Ballard, J.G., *Mechanical Engineers Handbook,* Interscience, New York, USA.

Bauer, E., *Energy Conservation Equipment*, Van Nostrand Reinhold, USA.

Beason Doug, *Industrial Heat Recovery,* Interscience, New York.

Brian, F.W., *Thermal Power Plants,* Affiliated East-West Press Pvt. Ltd.

Carstens, H. Harold, *Photovoltaic Solar Devices*, Reinhold, USA.

Charlie Wing, *Handbook of Energy Conservation*, by Taunton Press.

David, A.R., *Industrial Energy Conservation*, Pergamon Press

Daniel, S.D., *Direct Heating and Cooling for Energy Conservation*, Van Nostrand Reinhold, USA.

Fardo, W.S., *Pumps and Motors*, Cambridge University Press, Cambridge.

Gaddis William, *Heat Recovery Equipments*, Henser, USA.

Garrett Randall, *Waste Heat Recovery*, Van Nostrand Reinhold, USA.

Jacques Brian, *Energy Options: Challenge for the Future*, Noyes, UK.

James, M.R., *Energy Management*, Noyes, UK.

Joan, D. Vinge, *Industrial Pumps*, Reinhold, USA.

Katzenbach John, *Elements of Energy Conversion*, Reinhold, USA.

Morton, M.J, *Electrical Energy Conservation*, Van Nostrand, Reinhold, USA.

Ojikutu Bayo, *Energy Utilisation in Air Conditioning*, Academic Press, New York.

Palmer, R.A., *Industrial Energy Conservation,* Maclaren, London.

Patrick, R.D., *Energy Conservation in Industrial Sector,* University of Wisconsin Press, Madison.

Paul, W. O'Callaghan, *Handbook of Energy Conservation*, Pergamon Press,

Penn, W.S., *Industrial Energy Conservation,* Maclaren, London.

Richardson, E.R., *Energy Conservation in Textile Industries,* Humana Press Inc., New Jersey.

Revonna, M. Bieber, *The Science of Renewable Energy*, Maclaren, London.

Roger, W.R., *Energy Efficiency by Audit,* Academic Press, New York.

Saberhagen Fred, *Direct Energy Conversion*, Noyes, UK.

Saroyan William, *Energy Costs in Small Business and Industries*, Reinhold, USA.

Solmes, Leslie A., *Energy Efficiency*, Springer Netherlands.

Steve, D.T., *Energy Management Handbook*, by Fairmont Press, New York

Turner, W.C., *Boiler and Turbines,* Ellis Harwood, New York.

Vargic Martin, *Developments in Heat Transfer,* Elsevier, USA.

Wallace Bronwen, *Handbook of Fuel Cell Technology,* Maclaren, London.

Wright, P., *Handbook of Fuel Cell Technology,* Maclaren, London.

Washington, T. Booker, *Energy Conversion and Utilisation*, Mcgraw-Hill, New York.

Index